Lotka-Volterra-Approach to Cooperation and Competition in Dynamic Systems

Mathematical Research · Mathematische Forschung

Wissenschaftliche Beiträge
herausgegeben von der
Akademie der Wissenschaften der DDR
Institut für Mathematik

Band 23

Lotka-Volterra-Approach
to Cooperation and Competition
in Dynamic Systems

Lotka-Volterra-Approach to Cooperation and Competition in Dynamic Systems

Proceedings of the 5th Meeting of UNESCO's Working Group on System Theory held on the Wartburg, Eisenach (GDR) March 5-9, 1984

edited by
Werner Ebeling
and
Manfred Peschel

Akademie-Verlag · Berlin 1985

Herausgeber
Prof. Dr. Werner Ebeling
Humboldt-Universität zu Berlin

Prof. Dr. Manfred Peschel
Akademie der Wissenschaften der DDR
Forschungsbereich Mathematik/Kybernetik

Die Titel dieser Schriftenreihe werden vom Originalmanuskript der Autoren reproduziert

ISSN 0138 - 3019

Erschienen im Akademie-Verlag, DDR-1086 Berlin, Leipziger Straße 3-4

Lizenznummer: 202 · 100/526/84
Printed in the German Democratic Republic
Gesamtherstellung: VEB Kongreß- und Werbedruck, 9273 Oberlungwitz
Lektor: Dipl.-Phys. Gisela Lagowitz
LSV 1095
Bestellnummer 763 415 8 (2182/23)
3400

Preface

This book contains all invited papers of an interdisciplinary workshop of the UNESCO Working Group on Systems Analysis of the European and North American Region entitled 'Lotka-Volterra Approach To Cooperation and Competition'.
The meeting took place in the Wartburg-Hotel on the Wartburg near Eisenach GDR one year after the famous anniversary of the 500-th birthday of Martin Luther.

This workshop brought together some 20 mathematicians, physicists, and ecologists from different European countries with the common interest to study complex interaction phenomena and properties of dynamic systems from the point of view of the Lotka-Volterra approach in a broadened sense.

The meeting on the Wartburg continued a series of similar small conferences of the UNESCO working group on Systems Analysis which started in 1977 in Bucharest and was continued in Cambridge, U.K., 1981, Lyon, France, 1982, Flattnitz, Austria, 1983.

The title was chosen to focus the interest on cooperation-competition phenomena from different point of views. Our colleagues from 'Forschungsschwerpunkt Dynamische Systeme' University of Bremen gave us an excellent frame in form of a computer-graphics exposition 'Harmony in Chaos and Cosmos'.

A poster exhibition of 12 posters completed the impression about the activities in the field of Dynamic Systems in the GDR. Short contributions of the poster-papers are included in this monograph.

The meeting was sponsored by UNESCO, Paris and Academy of Sciences of the GDR. We are greatly indebted to the President of the Academy of Sciences of the GDR, Prof.Dr.sc.med. Dr. h.c. mult. W. Scheler, and to the Head of the Scientific Cooperation Bureau for the European and North American Region, Prof.Dr. J. Jaz, who made this workshop possible through their interest and financial support.

We thank the Akademie-Verlag and especially Mrs. Lagowitz for preparing very quickly the Proceedings for publication.

Berlin

Werner Ebeling

Manfred Peschel

List of Contributors

Contents

Part V: Poster Contributions

Late Paper:

Introductory Remarks

W. Ebeling Sektion Physik, Humboldt-Universität Berlin, 1040 Berlin, Invalidenstr. 42, GDR

M. Peschel Forschungsbereich Mathematik/Kybernetik, Akademie der Wissenschaften der DDR, 1199 Berlin-Adlershof, Rudower Chaussee 5

The idea for the 5-th Meeting of the UNESCO Working Group on Systems Analysis was to discuss theory and applications of the Lotka-Volterra Approach for nonlinear systems and the connection to the general theory of strange attractors.
In fact most of the papers are dealing more or less with this topic.
SVIRESHEV, LOGOFET analyse qualitative properties of a generalised 2-dimensional predator-prey model and study the qualitative behaviour in socalled trophic chains. They demonstrate the relationship to strange attractors and consider the notion of ecological stability.
AMANN, HOFBAUER continued their study of the permanence in ecological systems, especially in the replicator representation. A permanence criterion is derived and applied to interesting examples of 3-dimensional Lotka-Volterra systems.
COSTE has shown, that usually with increasing dimension of a generalised Lotka-Volterra System the probability of permanence rather quickly drops to zero. He studies this phenomenon for predator-prey models with Michaelis-Menten interactions.
In PESCHEL, MENDE, JANSEN, ALBRECHT based on the formerly developped theory of EXPONENTIAL CHAINS signal generator models are derived for chains with higher aggregated modules. The STRUCTURE DESIGN PRINCIPLE is extended for the Riccati-module as basic module. This leads to a special case of a Riccati network showing the extraordinary role of Cobb-Douglas production functions.
EBELING, FEISTEL study qualitative properties of continuous Lotka-Volterra models which can be interpreted as mechanisms of evolution. Darwinian selection appears now as a collapse process, mutation phenomena are observed spreading the distributions. Species move hill-climbing in a variant landscape depending itself on the species distribution.
With Evolution in an adaptive landscape deals also the paper of SIGMUND. By a modification of the notion of a gradient the replicator dynamics can be interpreted as gradient dynamics under certain conditions. A two populations interaction game is considered.

Under homogenity conditions MOLTCHANOV distinguishes between a multiparticle system and its corresponding macrodynamic system, which reflects higher aggregated dynamic behaviour. This is generalised very much, if some structural relations hold.
PEITGEN, PRÜFER, RICHTER performed detailed studies of Julia sets for solving polynomial equations with Newton's method. They used their approach for the explanation of singularity phenomena of ISING-POTT models of magnetism based on the work of DERRIDA, who discovered the connection between the YANG-LEE zeros and Julia sets.
BOTHE gives according to the present state-of-art an overview of the geometrical structure of strange attractors, elaborates the connection to CANTOR sets and showed; that for a deeper understanding of strange attractors it must be differed between wild and tame CANTOR sets.
The next 3 papers are concentrated on problems about the role of informatics in the theory of dynamic systems.
In JANSEN, KRIEGEL it is demonstrated, that parameter identification of a system with strange attractors is not possible with the usual approach of comparison of trajectories of object and model.
By a modification of the metrics and with application of the shadowing theorem the identification of strange attractors is possible.
NICOLIS, TSUDA show on the base of MILLERs work in cognitive psychology that brains are able to differentiate at the same time only a relatively small number of 'classes'. The potential of strange attractors for differentiation between 'classes' and storage of information is studied with the help of the Lyapunov exponent.
In CZAJKOWSKI for a stochastic system description the mean values as 'accompanying ensemble' are considered together with all probability distributions possessing the same 'accompanying ensemble'. Information optimizing distributions are determined out of this family. This is a method for determination of approximate solutions of autonomous stochastic system equations.
Starting with EULER - LAGRANGE formalism of mechanics the connection to driving forces of 1-st order equations is established and by LIOUVILLE formalism a Lotka-Volterra like description is found from the point of view of statistical mechanics. This concept is applied on the Lotka-Volterra transport-metabolism in a cell.
In HERZEL, EBELING, SCHIMANSKY-GEIER, SELKOV three-dimensional Lotka-Volterra systems are studied with random perturbations arising from a biochemical reactor. The corresponding probability distributions are determined by simulations. Phenomena of noise-induced oscillations can be obverved.
For a long time HADELER studied together with DIETZ models for parasitic deseases. The paper represents new knowledge for diseases from helminth infections. A semistochastic model is won which takes into account the age structure of the host population, the discrete

structure of the parasite population and the transmission by vectors. By fast integration of stiff reaction-diffusion systems very interesting evolution patterns are discovered by HUNDING for reasonable models of mitosis and cytokinesis which reflect very well findings from experiments being not explained by the usual theory of nuclear division. By wave-form approximation of the motion of a reaction-diffusion system SCHNEIDER could formulate existence conditions for wave trains. This can explain the pattern-formation in the famous BELOUSOV - ZHABOTINSKII reaction.

ALLEN, ENGELEN offer a reasonable global model for socio-economic problems based on micro-bilance-equations or on a network similar to a net of logistic modules. This approach is applied on forecasting problems for the population and the employment in the United States.

Most of the Poster contributions are dealing with special questions in connection with the Lotka-Volterra-approach.

GROTE considered the possibilities for bifurcations in Lotka-Volterra equations and in the corresponding Riccati-representation.

FUNKE analysed the stochastic robustness of Exponential Chains by using the ITO description.

STRAUBE transformed Lotka-Volterra equations into a fuzzy description to study the qualitative behaviour of the original system in a simplified way.

SCHULMEISTER, MEISKE found interesting examples for chaotic behaviour in a model of autocatalytic, second-order in product reaction with reversible deposition of the substrats.

VOIGT used the replicator form of the Lotka-Volterra equations to follow structuring processes under the influence of controls of the replication rates. In comparison with IIASA studies on the development of world population and energy production KRIEGEL, MENDE, GRAUER approximated the corresponding time-series by solutions of the hyperlogistic differential equation and got quite reasonable values for the future development of population and energy.

ALBRECHT, MENDE, PESCHEL analysed analytically and by simulation chains with logistic modules and hyperlogistic modules and found quick convergence and frequently robustness against random disturbances.

HEINRITZ, BLEY proposed a relevant model for fermentation processes. With this model the synchronization by substrate pulses or oscillations by low dilution rates could be successfully investigated.

In a finite Volterra world SCHILLING analysed the phenomena of competing species by reaction-diffusion equations, especially under the aspects of emigration and imigration.

ENGEL determined mean transition time and mean transition distance in a continuous model of biological evolution using an analogue to the SCHRÖDINGER equation.

MALCHOW considered dissipative structures in an ionic Lotka-Volterra type chemical reactor leading to oscillations in space and time, which could offer a new possibility to store energy or information.
The Pattern Formation problem was also studied by SCHIMANSKY-GEIER, MIKHAILOV in connection with the effects from fluctuations on nonequilibrium systems.

COMPLICATED DYNAMICS IN SIMPLE MODELS OF ECOLOGICAL SYSTEMS

Yu.M. Svirezhev[1] and D.O. Logofet[2]

INTRODUCTION

The world of mathematical ecology is principally a non-linear one, where the models, though having apparently simple structures, can exhibit a complicated dynamic behaviour. A discrete analogue of the logistic model gives a "chaotic" behaviour [1]; limit cycles occur in a model of a predator-prey system; under random environment conditions those cycles either fail or give rise to some new ones [2]. Evolution of adaptability results in emergence of catastrophes [3]. More complex models for a community of a trophic-chain type demonstrate an interesting property of "quantization" which means that, under continuous alteration of an external resource inflow or of the total community biomass (in the case of a materially-closed chain), the length of the trophic chain changes by leaps [3]. In closed chains, strange attractors can emerge of some intermediate type (between the Lorentz and Feigenbaum types) [4]. The conditions determining their emergence are so common that an idea willy-nilly arises that strange attractors are not "exotic" at all but rather widespread in ecological systems.

There is one more aspect, namely the problem of stability, in mathematical ecology. The widespread concept of Liapunov stability is too rigid and too limited for ecological systems. From an ecological point of view, the Lagrange concept of stability is more appropriate one though it is not provided with such a powerful and well-developed analytical apparatus as the Liapunov concept. What is to do?

1. A PREY-PREDATOR SYSTEM, THE CLASSICAL OBJECT OF MATHEMATICAL ECOLOGY

Let x(t) and y(t) be population sizes of a prey and a predator species, respectively. The model of the "prey-predator" system can be written then in the following form:

$$\frac{dx}{dt} = \alpha(x)x - V(x)y$$
$$\frac{dy}{dt} = kV(x)y - my, \qquad (1.1)$$

where $\alpha(x)$ is a Malthus function of the prey, V(x) is a trophic func-

[1] Laboratory of Mathematical Ecology, Computer Centre of the USSR Academy of Sciences, 40 Vavilov Street, Moscow, 117333, USSR
[2] Centre of International Projects, P.O.Box 438, Moscow, 107053, USSR

tion, k is an efficiency of converting the prey biomass, m is a rate of predator's natural mortality. When $V(x) = \lambda x$ we have the classical Volterra model, however a realistic V(x) has a form shown in Fig. 1. The function of type 1 is characteristic of a "silly" predator incapable of adaptation whereas the type 2 function is characteristic of a "clever" predator, which is capable of adaptation.

Let us set $\alpha(x)$ = const, i.e. there is no selfregulation in the prey population. It turns out that in this case, with a trophic function of type 1, there exist neither stable equilibria nor stable limit cycles in the system, i.e. a "silly" predator cannot control a prey population. However, if the predator is "clever", i.e. its trophic function belongs to the type 2, then either stable equilibria or a limit cycle, or even a whole system of limit cycles, can exist. Hence, a "clever" predator can control a prey population. Note, that a cycle arises here, under continuous alteration of the parameter $\beta = kV(\infty)/m$, according to a "rigid" type.

Let the trophic function now be of the Volterra type and let the Malthus function have the form shown in Fig. 2, i.e. the prey population is one of the Alle type. System (1.1) has the equilibrium with $x^* = m/k\lambda$, $y^* = \alpha(x^*)/\lambda$. It is clear that if $x^* < x^\circ$, then the equilibrium is unstable, but it is stable if $x^* > x^\circ$.

Let us set $\lambda_0 = mk/x_0$ and variate the parameter λ within a small vicinity of λ_0 in such a way that

$$\lambda^2 > \frac{m}{4k^2} \cdot \frac{[\alpha'(x^*)]^2}{\alpha(x^*)} \tag{1.2}$$

In this case we have a typical Hopf bifurcation and by a standard technique we obtain that

a) if $\lambda > \lambda_o$ and $\alpha'''(x_o)x_o < -2\,\alpha''(x_o)$, (1.3)

then a stable limit cycle arises in the vicinity of (x^*, y^*);

b) if $\lambda < \lambda_o$ and $\alpha'''(x_o)x_o > -2\,\alpha''(x_o)$, (1.4)

then an unstable limit cycle arises in the vicinity of (x^*, y^*).

Note that $x^* > x_o$ implies $\lambda < \lambda_o$. Since the trajectories of the system are bounded and $(x_k, 0)$ is a saddle point, the unstable cycle must be encircled by a stable one. The question remains open whether the 3rd and further cycles exist.

Let now the prey population be a logistic one (in the generalized sense), so that its Malthus function can be represented in the form of

$$\alpha(x) = \alpha_o[1 - (x/x_k)^p],\ p > 0, \tag{1.5}$$

where the parameter x_k corresponds to a "carrying capacity" of the environment for the prey population. Parametrizing the trophic function by the formula

$$V(x) = \frac{Ax^n}{k^n + x^n}, \tag{1.6}$$

one may say that a value of n greater than 1 conforms to a good adaptation of the predator population (a "clever" predator). System (1.1) with these expressions for $\alpha(x)$ and $V(x)$ was extensively studied in [3], but now we consider a special case, namely, a pure logistic model (p=1) and a trophic function with n = 2. It will be interesting to see an emergence of a "catastrophe" of the "cusp" type under evolution of the "carrying capacity", x_k, of the prey environment.

After introducing the new variables $X = x/K$ and $Y = Ay/\alpha_o K$ as well as the new parameter $\beta = kA/m$, we obtain from (1.1) the following system:

$$\frac{1}{\alpha_o}\frac{dx}{dt} = X\left[(1 - X/X_k) - \frac{XY}{1+X^2}\right],$$

$$\frac{dY}{dt} = mY\left[\frac{\beta X^2}{1 + X^2} - 1\right]. \qquad (1.7)$$

The equilibrium manifold equation that depends on a single parameter X_k, the "carrying capacity" of the environment, has the form

$$Y^* = \frac{1 + X^2}{X^*}\cdot\left(1 - \frac{X^*}{X_k}\right) \qquad (1.8)$$

where X^* is determined from the equation $\beta X^{*2} = 1 + X^{*2}$ resulting in $X^* = \sqrt{1/\beta - 1}$. The equilibrium manifold is shown in Fig. 3. This apparently typical example for a "cusp" - type catastrophe. We leave the reader to give an ecological interpretation for the situation depicted.

2. CLOSED TROPHIC CHAINS CONSERVING THE TOTAL AMOUNT OF SUBSTANCE IN THE SYSTEM

The structure that is shown in Fig. 4 and described by the following system of differential equations:

$$\frac{dN_0}{dt} = Q - V_0 N_1 + \sum_{i=1}^{n} c_i m_i N_i,$$

$$\dots$$

$$\frac{dN_i}{dt} = k_i V_{i-1} N_i - V_i N_{i+1} - m_i N_i, \qquad (2.1)$$

$$\dots$$

$$i = \overline{1,n}; \quad N_{n+1} \equiv 0, \quad N_i \in P^n ,$$

is called a <u>closed trophic chain</u>. Here, Q designates an external resource inflow to the system, $V_i = V_i(N_i)$ are trophic functions, c_i are closure coefficients, k_i is an efficiency of converting the biomass of the (i - 1)th level into that of the i-th one. We assume the dead organics decomposition to proceed sufficiently fast resulting in replenishment of the resource level N_0. In the case where $c_i = k_i = 1$ and $Q = 0$, the system (2.1) has the first integral

$$M = \sum_{i=0}^{n} N_i = \text{const}$$

representing the matter conservation law for the system. In [3] the system (2.1) is proved to have the numbers $0 < M_{cr} \leq M_1^* < M_2^* < \ldots$ such that for $M \in (0, M_{cr})$ the ecosystem contains no living components for $M \in (M_1^*, M_2^*)$ there exists a trophic chain of length 1, for $M \in (M_2^*, M_3^*)$ a chain of length 2 and so on. In other words, the total amount of substance in the system is "quantised", each "quantum" generating a trophic chain of a certain length. Let us set $V_i = \alpha_i N_i$, i.e. the trophic functions are those of the Volterra type. Investigation of stability for the equilibrium states $\overrightarrow{N_q^*} = \{N_0^*, N_1^*, \ldots, N_q^*, 0, \ldots, 0\}$ shows that the conditions of stability are equivalent to those of existence of the equilibria themselves for q = 1,2, and 3. But even for N_4^* the stability conditions can be broken although the inequality $M_4^* < M < M_5^*$ still takes place. For example, this situation can occur under a sharp increase in mortality of the top trophic levels (due to intensive harvesting or a specific pollution). Since the trajectories, however, do not leave a certain domain in the interior of the positive orthant [3] the situation testifies the existence of either a limit cycle or a strange attractor. From numerical simulation we have seen that in this case there can be fairly complex systems of limit cycles.

Limit cycles can also emerge even for smaller lengths of a trophic chain provided though the trophic functions are no longer of the Volterra type (but of type 1 or 2).

Let us consider an ecosystem where no more than three trophic levels can exist (n = 3). Let $V_0 = \alpha_0 N_0$ and $V_i = \frac{AN_i}{K + N_i}$, i = 1,2. For this case B.I. Yatsalo has shown that a cycle can appear as a result of a Hopf bifurcation; but if $\alpha_0 \gg 1$, then a stable limit cycle arises in the system.

The numerical experiments for $m_1 = 0.1$, $m_2 = m_3 = 0.2$, $A = 1$, $K = 5$, and variable parameters M and α_0 $(0.2 < \alpha_0 < 0.38)$ have shown that a strange attractor does appear in the system. Dynamics of the system evolves in the following way. At the passage through a point $M = M_1^o(\alpha_0)$ a stable limit cycle, $\gamma_1 = \gamma_1(M)$, emerges as a result of the Hopf bifurcation; at the passage through $M = M_2^o(\alpha_0)$ $(M_2^o > M_1^o)$ the two cycles emerge, a stable (γ_2) and an unstable one (γ_3), which merge all together via +1 when $M \longrightarrow M_{2+}^o$. When $M = M_3^o(\alpha_0)$ they merge and annihilate themselves. Hence, a doubling bifurcation takes place at the points $M_1, M_2, M_3, \ldots$, i.e. when $M = M_1$ the cycle γ_2 loses its stability, its multiplicator passes through -1 and in the vicinity of γ_2, a stable limit cycle of a doubled period arises, which, in turn, goes through a similar bifurcation at $M = M_2$ and so on. For $n \rightarrow \infty$ the sequ-

ence $M_n \rightarrow M_\infty(\alpha_0)$. Taking $\alpha_0 = 0.34$ we obtain the following values:
$M_1^0 = 9,447$; $M_2^0 = 30,55$; $M_3^0 = 36,251$;...

$M_1 = 33,04$; $M_2 = 34,835$; $M_3 = 35,41$; $M_4 = 35,541$;... M 35,58.

Using the values of M_3, M_4, and M_5 to estimate the Feigenbaum constant, we have

$$F_5 = \frac{M_4 - M_3}{M_5 - M_4} = 4.52,$$

which is rather close to theoretical value,

$$F = \lim_{n \to \infty} \frac{M_n - M_{n-1}}{M_{n+1} - M_n} \approx 4.66$$

From the graphs of the functions $M_\infty(\alpha_0)$ and $M_3^0(\alpha_0)$ we observe an interesting phenomenon (Fig. 5): in the hatched region there exists a so-called "preturbulence" (or "prestochasticity") regime where both a strange attractor arising as a result of an infinite chain of Feigenbaum doublings, and a stable limit cycle take place. The complete stochasticity appears when $M > M_3^0$, i.e. when the stable cycle vanishes. Although the prestochasticity was observed earlier in the classical Lorentz system, the strange attractor emerged there in a way other than the bifurcation doubling.

These considerations reveal the possibility for strange attractors of a new type to occur even in a simple ecological system.

A few words about ecological interpretation of these results. In an ecological pyramid the mortality rate normaly decreases when the number of a trophic level increases ($m_i \sim 1/T_i$ where T_i is an average life span of an individual of the i-th species). Under such normal hierarchy the system behaviour is "quiet". Suppose, however, that we commence an intensive harvesting of a species in one of the top levels. Then its mortality rate will be increased and the natural hierarchy of those rates will be disturbed. As a result of overexploitation the own structure of the ecosystem is usually destroyed, the length of the trophic chain becoming smaller. However, this destruction does not occur immediately but rather the system "quietness" is disturbed first when the harvesting load increases. Here, complex dynamical phenomena occur such as population fluctuations and probably a "chaos", though the structure of the system is still preserved. And it is the further increase in harvesting load that sharply changes the structure whereafter a "quietness" takes place again, but now for a new structure with a smaller number of trophic levels.

3. LAGRANGE STABILITY AND THE CONCEPT OF ECOLOGICAL STABILITY

One of the most common definitions of ecological stability for a biological community requires all of the species in the community to be preserved for sufficiently long time [3], i.e. a population size for any species should not tend to zero or take close-to-zero values. From the other hand, an infinite increase in population sizes of one or several species may be one the model artifacts indicating inadequacy of the model. It is so-called Lagrange stability that serves as a formal analogue of such an ecological stability in the framework of a dynamical model for an ecosystem.

Let a model of a biological community consisting of n populations have the form

$$\frac{dN_i}{dt} = F_i(N_1, \ldots, N_n), \quad i = \overline{1,n} \qquad (3.1)$$

The model community will be called <u>ecologically stable</u> if in the interior of the positive orthant P^n of the phase space there exists such a finite closed domain Ω_o^n that for any $N_o = \{N_1^o, \ldots, N_n^o\} \in \Omega_o^n$ there exists the finite closed domain $\Omega^n(\Omega_o^n) \subset \operatorname{Int} P^n$ such that a trajectory originating from N_o does not fall outside the limits of Ω^n. The domain Ω^n is naturally called an <u>ecological stability domain</u>.

We propose an effective algorithm to determine this domain or, at least, to obtain some minimal estimates for its size. The idea is as follows.

By the change of variables,

$$\xi_i = \ln(N_i/N_i^o), \; i = \overline{1, n} \qquad (3.2)$$

we transform system (3.1) into the form

$$\frac{d\xi_i}{dt} = \Phi_i(\xi_1, \ldots, \xi_n; N_1^o, \ldots, N_n^o) + B_i, \; i = \overline{1,n} \; ; \qquad (3.3)$$

where

$$\Phi_i = \varphi_i(\xi_1, \ldots, \xi_n; N_1^o, \ldots, N_n^o) - \varphi_i(0, \ldots 0; N_1^o, \ldots, N_n^o),$$

$$B_i = \varphi_i(0, \ldots, 0; N_1^o, \ldots, N_n^o).$$

A solution for (3.3) is defined in the whole phase space R_ξ^n (rather than in the positive orthant only) and depends upon $N_1^o, \ldots, N_n^o$ as parameters. If we pose now the problem of stability under the permanent perturbations B_i for the trivial solution of (3.3), $\xi_i^* = 0 (i = \overline{1,n})$, and determine the domain of that stability in the parameter space $\{N_1^o, \ldots, N_n^o\}$, then the problem of ecological stability for system (3.3) will be solved too. The proof follows from the definition of Liapunov stability and from the properties of transformation (3.2).

According to the Chetaev - Malkin theorem [5] the trivial solution $\xi_i^* = 0$, $i = \overline{1,n}$, is stable under permanent perturbations if it is asymptotically stable in the system $d\xi_i/dt = \Phi_i (i = \overline{1,n})$ and the terms B_i are sufficiently small. So, the problem of the Lagrange stability is reduced to that of the stability under permanent perturbation for the trivial solution. In spite of the fact that there are no effective methods to estimate the smallness of B_i in a general case, application of the well-developed techniques for Liapunov stability analysis in specific "ecological" examples turns out effective to give some reasonable estimates of an ecological stability domain. We consider as an example the equation

$$\frac{dN}{dt} = rN(1 - \frac{N}{K})(N - K_{min}), \quad 0 < K_{min} < K, \tag{3.4}$$

governing dynamics of a single population with a so-called "lower threshold" for its size. As one can see from Fig. 6, for $N_o < K_{min}$ the population size $N(t) \to 0$, i.e. for $N_o < K_{min}$ the population is ecologically unstable. From the other hand, as can be easily shown by a direct method, the equilibrium state $N^* = K$ is asymptotically stable in the domain of attraction $\Omega_a^o = \{ N_o : K_{min} < N_o < \infty$. Applying the algorithm described above to analyse ecological stability we obtain

$$\Phi(\xi, N_o) = \frac{rN_o}{K}(K + K_{min} - 2N_o)\xi + O(\xi), \tag{3.5}$$

$$B = r(1 - \frac{N_o}{K})(N_o - K_{min}).$$

The solution $\xi^* = 0$ of the equation $d\xi/dt = \Phi(\xi, N_o)$ is asymptotically stable for $N_o > (K + K_{min}/2)$, i.e. as an ecostability domain we have.

$$\Omega^o = \{ N_o : (K + K_{min})/2 < N_o < \infty \} \tag{3.6}$$

Naturally, the domain Ω^o is substantially less than the actual domain of attraction Ω_a^o, since the method itself to estimate Ω_o is based upon the sufficient conditions for ecostability.

Let us now consider the classical "prey-predator" system, viz. the equation (1.1) with $\alpha(x) = const$. By the same technique we obtain

$$\frac{d\xi}{dt} = \alpha - \omega(\xi, x_o)e^{\eta}, \tag{3,7}$$

$$\frac{d\eta}{dt} = k\omega(\xi, x_o)x_o e^{\xi} - m.$$

Here $\xi = \ln(x/x_o)$, $\eta = \ln(y/y_o)$, and $\omega(\xi, x_o) = \frac{V[x(\xi)]}{x_o e}$.

Rewriting (3.7) into the form (3.3) we obtain the system

$$\frac{d\xi}{dt} = y_o[\omega(0,x_o) - \omega(\xi, x_o)e^{\eta}] + B_{\xi},$$

$$\frac{d\eta}{dt} = kx_o[\omega(\xi, x_o)e^{\xi} - \omega(0,x_o)] + B_{\eta}. \qquad (3.8)$$

where $B_{\xi} = \alpha - \omega(0,x_o)y_o$ and $B_{\eta} = kx_o\,\omega(0,x_o) - m$. For $B_{\xi} = B_{\eta} = 0$ the solution $\xi^* = \eta^* = 0$ is asymptotically stable if $\omega'(0,x_o) > 0$, coming back to the original variables, $[V(x_o)/x_o]' > 0$. So,we may assert that the band $\{y_o > 0,\ 0 < x_o < x_m\}$ (see Fig. 1) is an ecological stability domain for the prey-predator system. Actually, the domain is wider, although the sufficiency of the conditions gives, in essence, a lower estimation for the actual sizes. The more thorough investigation of the Chetaev - Malkin theorem in the application to the ecological stability problem, together with the examples considered, reveals that the smallness restrictions imposed previously on the functions B_i, are unessential.

CONCLUSION

In the present article we have examined several examples of application of the dynamic system analysis methods to some models of biological population and community dynamics. Not claiming for a complete review those examples were intended rather for giving an idea of new directions in development of this field of applied mathematics. It is appropriate to cite here the words delivered by H. Poincare at the First International Congress of Mathematicians in Zurich, 1897: " It needs to forget completely of science history in order to refute the permanent and the most beneficial influence that an aspiration to a cognition of Nature exerts on the development of mathematics". That H. Poincare's idea anticipated the posterior development of mathematical ecology, the field of application of mathematical methods that now can be rightfully regarded as quite a formed and specific field of applied mathematics.

REFERENCES

1. May R.M. Biological populations obeying difference equations: stable points, stable cycles and chaos. - J. Theor. Biol., 1975, 51, N 2, pp. 511 - 524.
2. Sidorin A.P., Svirezhev Yu.M. The dynamics of predator-prey system under random disturbances. - Izvestia Akad. Nauk SSSR, biol. ser. 1980, 4, pp. 573 - 579 (in Russian).
3. Svirezhev Yu.M., Logofet D.O. Stability of Biological Communities. - Moscow, Mir Publ., 1983, 319 pp.
4. Svirezhev Yu.M. Modern problems of mathematical ecology. - Proc. of Int. Congress Math., 18 sect., Warsaw, 1983, 24 pp.
5. Malkin I.G. Teoria ustoichivosti dvizheniya (Motion Stability Theory) - Moscow, Nauka Publ., 1966, 530 pp. (in Russian).

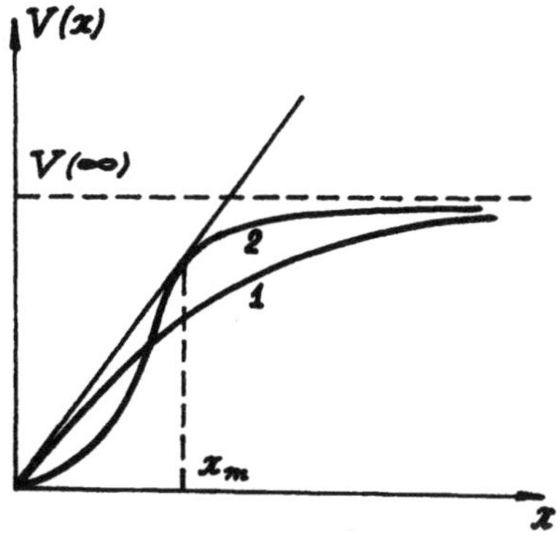

Fig 1. Trophic functions:
(1) a "silly" predator;
(2) a "clever" predator

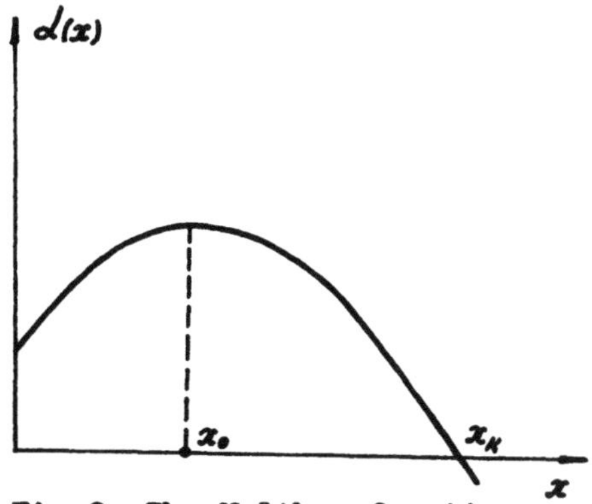

Fig.2. The Malthus function for a prey population (a population of the Alle type)

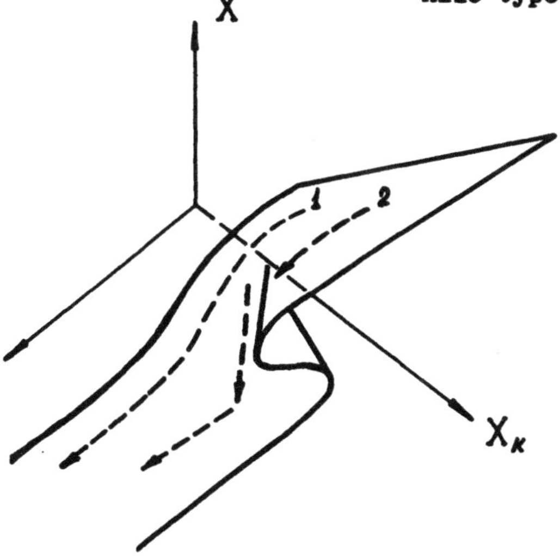

Fig. 3. Structure of the equilibrium manifold for the prey-predator system. If the "carrying capacity" of the prey environment, X_k, is small, then, under increase in the equilibrium size, Y^*, of the predator population, that of the prey X^*, decreases smoothly; but if X_k is sufficiently large (X_k 27), then, under increasing Y^*, the size X^* decreases by a jump (curve 2)

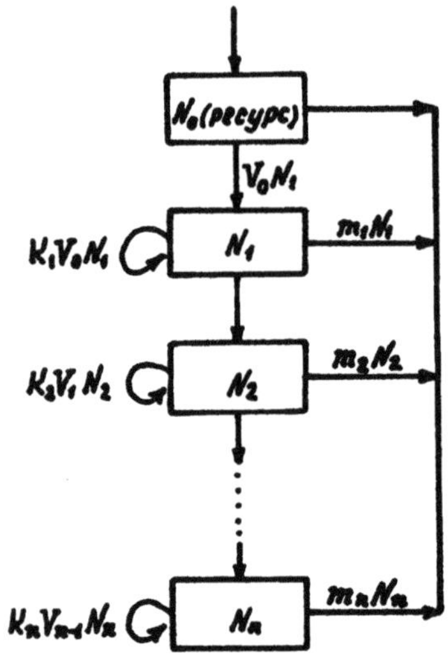

Fig. 4. Diagram of a closed trophic chain of length n

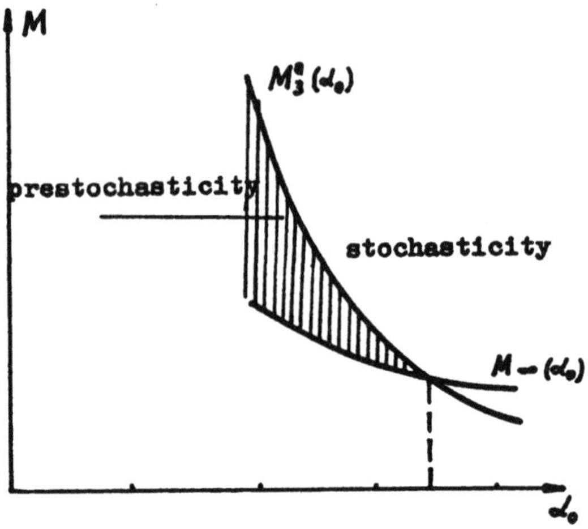

Fig 5. Regions of different dynamic regimes for the closed trophic chain

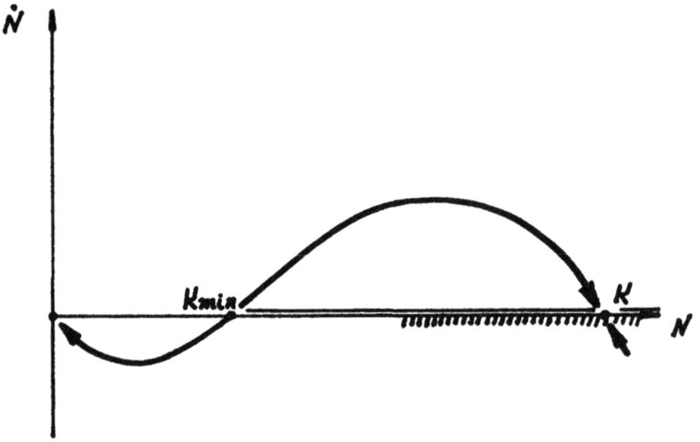

Fig. 6. The phase portrait for the population with a lower threshold in size:

▬▬ the domain of ecological stability;

/////// the estimate for that domain

Permanence in Lotka-Volterra and Replicator Equations

Erwin Amann and Josef Hofbauer [1)]

Let us start with the Lotka-Volterra equation

$$\dot{y}_i = y_i(a_{io} + \sum_{j=1}^{n} a_{ij}y_j) \quad i=1,\dots,n \qquad (*)$$

defined on the state space R^n_+, and the replicator equation

$$\dot{x}_i = x_i(\sum_{j=0}^{n} a_{ij}x_j - \Phi) \quad i=0,\dots,n \qquad (**)$$

with $\Phi = \sum_{i,j} x_i a_{ij} x_j$, defined on the probability simplex $S_n = \{x \in R^n, x_i \geq 0, \sum x_i = 1\}$. This equation was first introduced in 1930 by Fisher for symmetric matrices $a_{ij} = a_{ji}$ as selection equation. Eigen and Schuster [4] used it for modelling catalytic networks of selfreplicating macromolecules in an evolution reactor, their most prominent example being the hypercycle ($a_{ij} = k_i$ if $i = j+1$ (mod n), $= 0$ else). Maynard Smith [12] proposed (**) as a dynamics for an evolutionary game with payoff-matrix (a_{ij}).

Since it has been shown [6], that this replicator equation is equivalent by the simple change of coordinates $y_i = x_i/x_o$ or $x_i = y_i(1+\sum y_j)^{-1}$ to Lotka-Volterra equation, we may treat (**) without leaving the topic of this meeting.

In continuation of Sigmund's lecture on the last meeting [18] we want to discuss the <u>permanence</u> of these equations. The intuitive concept of "no species dies out" is precised to the

<u>Definition</u>: Equation (*),(**) is permanent
$\Leftrightarrow$ there is a compact set in the interior of the state space (R^n_+ resp. S^n) which contains the ω-limits of all orbits starting in the interior of the state space.

$\Leftrightarrow$ $\exists\, \delta > 0$: $\liminf_{t\to+\infty} x_i(t) \geq \delta$ whenever $x_i(0) > 0$ for all i and
for (*): $\limsup_{t\to+\infty} x_i(t) \leq 1/\delta$ whenever $x_i(0) > 0$ for all i.

Permanent systems run also under the name "cooperative" [4,7,8,15,17], "permanently coexistent" [9,10], "ecologically stable" [19]. In this paper we present a powerful technique for deciding whether a system is permanent or not. Applications are given to catalytic networks (generalizations of the hypercycle) and to 3-species ecological communities.

---

1) Institut für Mathematik, Universität Wien, Strudlhofgasse 4, A-1090 Wien

1. How to prove permanence of a system

For proving the permanence of the hypercycle, a special technique was developped in Schuster, Sigmund and Wolff [15], which was then refined [7] to the following theorem, applicable to more general problems:

Theorem 1: Let P be a function defined on S_n, such that $P(x)=0$ for $x \in \partial S_n$, $P(x)>0$ for $x \in \operatorname{int} S_n$, and $\psi = \dot{P}/P$ has a continuous extension to S_n. Then the condition

$$\text{(A)} \qquad \text{for all } x \in \partial S_n \ \exists T>1: \frac{1}{T}\int_0^T \psi(x(t))dt > 0$$

implies that (**) is permanent.

P is something like an "average Lyapunov function" for the boundary ∂S_n. In concrete examples we always use functions of the form $P = \prod_{i=1}^{n} x_i^{p_i}$. In order to fulfill condition (A), $\psi(x)>0$ must hold for all fixed points $x \in \partial S_n$. Sufficient to satisfy (A) would be that $\psi(x)>0$ on the whole ω-limit set of ∂S_n. Since $\psi>0$ just means $\dot{P}>0$, P(x), the "distance from the boundary" must then increase during the most time, and ∂S_n is a repelling set.

When applying this theorem one has first to study the behaviour on the boundary. But since in (A) only time averages are of importance, also qualitatively complicated behaviour can be handled with it - at least for Lotka-Volterra equations, since a limit cycle or even a chaotic motion averages to an equilibrium. So usually it is sufficient to test (A) at the fixed points.

Moreover we believe that our theorem is effective enough to characterize permanence for equations (*),(**), just taking $P = \prod x_i^{p_i}$ as an ALF.

Conjecture: (i) (**) is permanent
$\Leftrightarrow$ (ii) For some $p>0$, $P=\prod x_i^{p_i}$ satisfies condition (A)
$\Leftrightarrow$ (iii) For some $p>0$, $\psi = \dot{P}/P > 0$ at all fixed points on the boundary, i.e. $p.Ax > x.Ax$ for all fixed $x \in \partial S_n$.

(Observe the similarity of (iii) to the definition of an ESS [12,8]). The examples in § 4-6 of this paper support this conjecture. The results in the next two sections will be useful in order to disprove permance.

2. A necessary condition for permanence

A necessary condition for permanence is the existence of an interior fixed point. This is easily proved for (*),(**) using time averages [3,8,17].

For general systems it follows from the Brouwer fixed point theorem. For (*),(**) we can show even more:

<u>Theorem 2</u>: If a Lotka-Volterra equation (*) or a replicator equation (**) is permanent, then the <u>trace</u> of the Jacobian at the interior fixed point <u>cannot be positive</u>.

<u>Proof</u>:
$$\dot{x}_i = x_i(a_i + \sum_{j=1}^{n} a_{ij}x_j) \quad i = 1,\ldots,n \tag{2.1}$$

If this equation is permanent, then there is a unique stationary solution $\hat{x} > 0$ with $a_i + \sum a_{ij}\hat{x}_j = 0$ and the time averages of every interior orbit converge to $\hat{x}$:

$$\frac{1}{T}\int_0^T x(t)dt \to \hat{x} \quad \text{for} \quad T \to \infty$$

Since the Jacobian $J = (\partial\dot{x}_i/\partial x_j)$ is linear in the x_i's, also the time average of the Jacobian along every interior orbit converges to the Jacobian at $\hat{x}$, $J(\hat{x})$. In particular we have

$$\frac{1}{T}\int_0^T \text{div. at } x(t)dt = \frac{1}{T}\int_0^T \text{trace } J(x(t))dt \to \text{trace } J(\hat{x})$$

Now recall Liouville's theorem [1] for the volume of a small ball around $x(t)$:

$$\begin{aligned} \text{Vol }(x(t))^{\cdot} &= \text{Vol }(x(t)).\text{divergence of the vectorfield at } x(t) \\ \text{Vol }(x(t)) &= \text{Vol }(x(o)).\exp\int_0^T \text{trace } J(x(t)dt \sim \text{Vol}(x(o)).e^{T.\text{trace } J(\hat{x})} \end{aligned} \tag{2.2}$$

This holds along <u>every</u> interior orbit.
But since (2.1) was assumed to be permanent, there exists a compact set B in the interior of $\mathbb{R}^n_+$, homeomorphic to the n-dimensional ball, which is positively invariant. So its volume decreases (for some time). This implies trace $J(\hat{x}) \leq 0$ in (2.2).

<u>3. How to find attractors on the boundary</u>

Of course Theorem 1 can be dualized to prove the attractivity of ∂S_n (see [7,p.236]). In practice it is usually not possible to have $\phi < 0$ at all fixed points on the boundary. But if the set, where $\phi > 0$, is repulsive on ∂S_n, then this will do:

<u>Theorem 3</u>: Let M be an open subset of S_n with positfve invariant closure $\bar{M}$, and P, ϕ as in Theorem 1, but satisfying now condition

(B) $$\forall x \in M_o = \bar{M} \cap \partial S_n : \exists T > 1 : \frac{1}{T}\int_0^T \phi(x(t))dt < 0.$$

Then M_o contains an <u>attractor</u> for the flow on S_n (i.e. all orbit in S_n, starting near M_o, converge to M_o).

<u>Proof</u>: We consider for h> 0 the sets

$$\tau_h(x) = \{T > 1 : \frac{1}{T}\int_0^T \psi(x(t))dt < -h < 0\} \subseteq R_+$$

$$U_h = \{x \in S_n : \tau_h(x) \neq \emptyset\} \subseteq S_n$$

For $x \in U_h$ we can define $T_h(x) = \inf \tau_h(x)$. Note that U_h is an open set and $T_h(x)$ is an upper semicontinuous function (cf. [7]).
Condition (B) implies that M_o is contained in the union of all U_h's. Since M_o is compact, we will find one h (fixed for the following) with $M_o \subseteq U_h$. Now consider the sets $I(p) = \{x \in S_n : 0 < P(x) \leq p\}$, which are layers around the boundary. For some small $p > 0$ we have $\overline{I(p)} \cap \bar{M} \subseteq U_h$.

We now claim, that all orbits starting in $I(p) \cap \bar{M}$ converge to the boundary. Let $x \in I(p) \cap \bar{M}$. Then $x \in U_h$ and we will find some $T_1 > 0$ such that

$$\frac{P(x(T_1))}{P(x)} = \exp \int_0^T \psi(x(t)) < e^{-hT_1} \leq e^{-h}$$

Hence $P(x(T_1)) < e^{-h} P(x)$, so $x(T_1)$ is again $I(p) \cap \bar{M}$. Iteratin this argument we find a sequence of instants $0 = T_o < T_1 < T_2 < \ldots$ with

$$P(x(T_{i+1})) < P(x(T_i))e^{-h} \quad \text{and} \quad 1 \leq T_{i+1} - T_i < \bar{T} \tag{3.1}$$

where $\bar{T}$ is an upper bound of the upper semicontinuous function $T_h(x)$ on the set $I(p) \cap \bar{M}$.

Finally for $0 \leq t \leq T_{i+1} - T_i$ we have

$$\frac{P(x(T_i+t))}{P(x(T_i))} = \exp \int_{T_i}^{T_i+t} \psi(x(s))ds \leq e^{t\|\psi\|_\infty} \leq e^{\bar{T}\|\psi\|_\infty}.$$

This shows, that within one such time-interval $[T_i, T_{i+1}]$ the "distance from the boundary" $P(x(t))$ of the orbit can only increase by some fixed factor. Together with (3.1) this implies $P(x(t)) \to 0$ for $t \to +\infty$ for every $x \in I(p) \cap \bar{M}$. So all orbits in the neighbourhood $I(p) \cap M$ of M_o converge to M_o.

<u>4. A hypercycle of autocatalysists</u>

As a first example we consider the following catalytic network

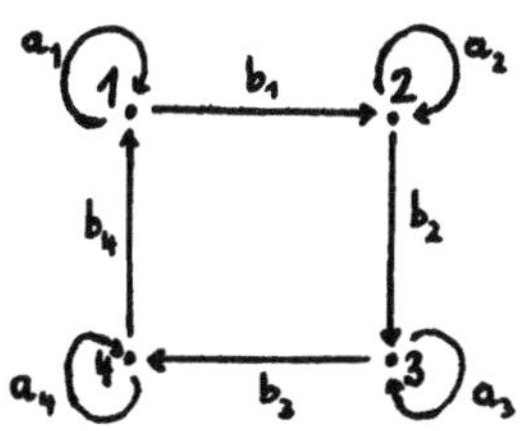

$$A = \begin{bmatrix} a_1 & 0 & 0 & \dots & 0 & b_n \\ b_1 & a_2 & 0 & \dots & 0 & 0 \\ 0 & b_2 & a_3 & \dots & 0 & 0 \\ \dots & \dots & \dots & \dots & \dots & \dots \\ 0 & 0 & 0 & \dots & b_{n-1} & a_n \end{bmatrix} \quad a_i, b_i > 0 \qquad (4.1)$$

If all $a_i = 0$, the system reduces to a hypercycle, which is a permanent system [15,7,8]. If all $b_i = 0$, we have a system of autocatalytic elements which are known to compete very strongly. In general only one species - depending on the initial values - can survive. Every vertex of the simplex S_n is a stable equilibrium. So the behaviour in both extreme cases is completely opposite and the question arises, what are the exact conditions on the rate constants a_i, b_i, for which the system is permanent.

(**) reads in this case

$$\dot{x}_i = x_i(a_i x_i + b_{i-1} x_{i-1} - \Phi) \quad i = 1, \dots, n \qquad (4.2)$$

Then

$$\left(\frac{x_1}{x_2}\right)^{\cdot} = \left(\frac{x_1}{x_2}\right) [(a_1 - b_1)x_1 + b_o x_o - a_2 x_2] \qquad (4.3)$$

If now $a_1 > b_1$ (i.e. species 1 does not cooperate will enough) then for large x_1 ($x_1 \approx 1$, all other $x_i \approx 0$), x_1/x_2 and all other x_1/x_i go to infinity. So the vertex $e_1 = (1,0,\dots,0)$ is a stable equilibrium and (4.2) cannot be permanent.

So we have to assume $a_i < b_i$ for all i as an (obvious) necessary condition for the permanence of (4.1).

Now let us study the behaviour on the boundary of the state space S_n: Assume $x_o = 0$.

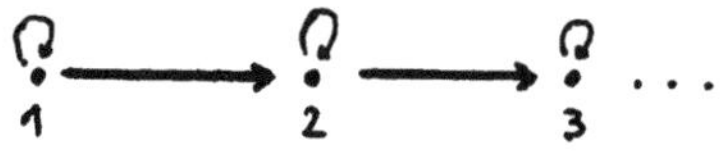

(4.3) then implies $(x_1/x_2)^{\cdot} < 0$, i.e. $x_1 \to 0$. Iteration shows that all species in this autocatalytic chain up to the endspecies will die out. Almost all orbits on the boundary converge to the vertices. (In fact a few orbits go to other fixed points, where $x_i x_{i-1} = 0$ for all i , but these do not seriously disturb the following arguments.)

Now look for an average Lyapunov function $P = \Pi x_i^{p_i}$ (with $\Sigma\, p_i = 1$):

$$\psi = \dot{P}/P = \Sigma\, p_i \dot{x}_i / x_i = \Sigma\, a_i p_i x_i - \Sigma\, b_{i-1} p_i x_{i-1} - \Phi$$

At the vertex e_k we obtain

$$\psi(e_k) = a_k p_k + b_k p_{k+1} - a_k > 0$$

So (4.1) will be permanent if we can find solutions $p_i > 0$ of the inequalities

$$
\begin{aligned}
& b_k p_{k+1} < a_k(1-p_k) \qquad k=1,\ldots,n \\
& \Sigma\, p_i = 1
\end{aligned}
\tag{4.4}
$$

Let us first simplify the problem by considering only equal rate constants

$$a_i = a < b_i = b$$

In view of the symmetry (4.4) reduces then to one inequality

$$b > a(n-1) \quad \text{or} \quad \frac{a}{b} < \frac{1}{n-1} \tag{4.5}$$

So Theorem 1 implies that (4.2) with equal rate constants is permanent, if (4.5) holds. But is this the <u>exact</u> answer? According to our conjecture, (4.2) should not be permanent if $b < a(n-1)$. However this is now a consequence of Theorem 2: Since the Jacobian at the interior equilibrium $(\hat{x}_i = \frac{1}{n})$ is circulant, the eigenvalues are easily computed to $a + \lambda^j b$, $j = 1,\ldots,n-1$ with $\lambda = \exp(2\pi i/n)$. So their sum is $(n-1)a - b$ which has to be ≤ 0, if (4.2) is permanent. This shows that in fact (4.5) is the precise condition for permanence (maybe up to the marginal case $b = a(n-1)$).

Let us now return to the general asymmetric case. Rewrite (4.4) as

$$p_1 + \ldots + p_{k-1} - (b_k/a_k - 1)p_{k+1} + \ldots + p_n < 0 \tag{4.6}$$

Denote

$$b_k/a_k - 1 = \alpha_k > 0 \tag{4.7}$$

We may then think of (4.6) as the (k+1)st equation in the matrix inequality

$$(Cp)_i > 0 \quad \text{with} \quad C = \begin{bmatrix} \alpha_1 & -1 & -1 & \ldots & -1 & 0 \\ 0 & \alpha_2 & -1 & \ldots & -1 & -1 \\ -1 & 0 & \alpha_3 & \ldots & -1 & -1 \\ \ldots & \ldots & \ldots & \ldots & \ldots & \ldots \\ -1 & -1 & -1 & \ldots & 0 & \alpha_n \end{bmatrix} \tag{4.8}$$

Now recall the following theorem on M-matrices, wellknown e.g. in economics:

<u>Theorem 4</u>: If C is a $n \times n$-matrix with nonpositive off-diagonal terms $c_{ij} \leq 0$ $(i \neq j)$, then the following conditions are equivalent: ($x > 0$ means $x_i > 0$ for all i)

(i) $\exists p > 0 : Cp > 0$

(ii) det $C > 0$ and all principal minors of C are positive.

(iii) All real eigenvalues of C are positive.

(iv) $\forall x > 0\ \exists i : (Cx)_i > 0$

The equivalence (i) ⇔ (ii) is sometimes called the Hawkins-Simon condition. For proofs seee e.g. [13] and [2] for some 50 other characterizations. Note that (i) ⇔ (iv) can be reformulated as:

$$\begin{array}{ll} \text{Either} & \exists p>0 : (Cp)>0 \\ \text{or} & \exists p>0 : (Cp)\leq 0 \end{array} \tag{4.9}$$

Theorem 1 and Theorem 4 (i,ii) now imply that (4.2) is permanent if the matrix C in (4.8) has positive determinant and all principal minors positive. For low dimensions this gives the explicit conditions

$$n=3 \qquad \alpha_1\alpha_2\alpha_3 > 1 \quad \Leftrightarrow \quad \prod_1^3 (b_i-a_i) > \prod_1^3 a_i$$

$$n=4 \qquad \alpha_1\alpha_2\alpha_3\alpha_4 > \alpha_1+\alpha_2+\alpha_3+\alpha_4+\alpha_1\alpha_3+\alpha_2\alpha_4$$

For $n \geq 5$ things become more and more complicated. But they are the <u>exact</u> conditions on the a_i, b_i which make (4.2) permanent. For if they are not satisfied, we can infer from (4.9) that at all vertices $\phi \leq 0$ holds (if we neglect the marginal case det $C=0$) and so Theorem 3 guarantees us the existence of an attractor on the boundary. (It is easy to find a repelling set which contains all the other boundary equilibria, different from the vertices, where ϕ will still be positive.)

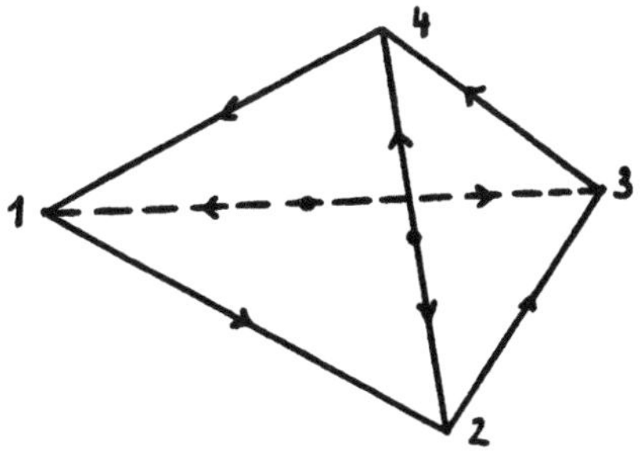

<u>Fig. 1</u>: System (4.2) with $a_i < b_i$ if $n=4$. Either the system is permanent or the heteroclinic boundary cycle 12341 is an attractor.

These arguments and results generalize directly to replicator equations (**) with $a_{ij}>0$ if $i=j+1 \pmod n$, $a_{ii}=0$, $a_{ij} \leq 0$ else; i.e. A is of the form

$$A = \begin{bmatrix} 0 & - & - & \dots & - & + \\ + & 0 & - & \dots & - & - \\ \dots & \dots & \dots & \dots & \dots & \dots \\ - & - & - & \dots & + & 0 \end{bmatrix} \tag{4.10}$$

<u>Theorem 5</u>: (**) with such a matrix A is permanent iff the matrix C with positive diagonal terms, obtained from A by a cyclic permutation of its rows, has positive determinant and all principal minors > 0.

This is a nice generalization of Zeeman's result [20, Theorem 6] on the "rock-scissors-paper" game:

Let

$$A = \begin{bmatrix} 0 & - & + \\ + & 0 & - \\ - & + & 0 \end{bmatrix}.$$

Then (**) is permanent, iff it has a globally stable interior equilibrium, iff $\det A > 0$.

5. Two counter-rotating hypercycles

Consider now networks of the form

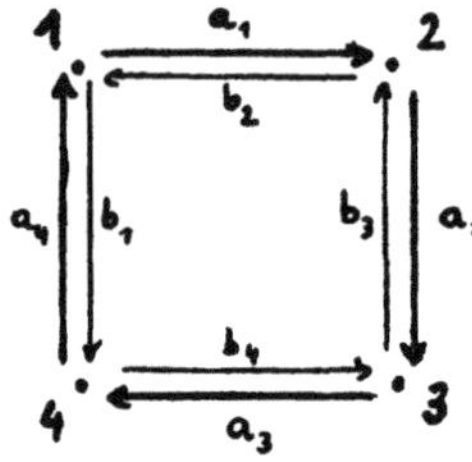

$$\dot{x}_i = x_i(a_{i-1}\, x_{i-1} + b_{i+1}\, x_{i+1} - \Phi)$$
$$i = 1, \ldots, n \qquad (5.1)$$
$$a_i, b_i > 0$$

Besides the vertices of the simplex we have now 2-species equilibria on the edges $\overline{i,i+1}$:

$$F_{i,i+1} = \left(0, \ldots, \frac{b_{i+1}}{a_i + b_{i+1}}, \frac{a_i}{a_i + b_{i+1}}, 0, \ldots\right) \qquad (5.2)$$

First observe that if (5.1) is permanent then either $a_i < b_i$ for all i or $a_i > b_i$ for all i. For otherwise there would be a change in the inequalities e.g. $a_1 > b_1$, $a_2 < b_2$. Then it is easy to see that the 1-2-subsystem

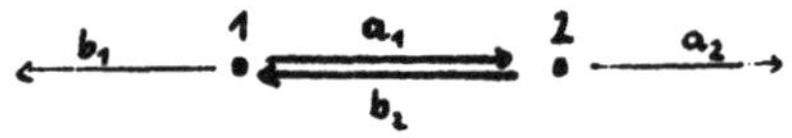

is uninvadable and F_{12} is a stable fixed point, which is incompatible with permanence.

So let us suppose $a_i > b_i$. We have to study the behaviour on the boundary. Let $x_o = 0$.

Since

$$\left(\frac{x_1}{x_3}\right)^{\cdot} = \left(\frac{x_1}{x_3}\right)[a_o x_o + (b_2 - a_2)x_2 - b_4 x_4] < 0,$$

x_1/x_3 is decreasing and $x_1 \to 0$. This shows that almost all orbits converge to the last edge equilibrium $F_{i,i+1}$ in the chain.

With the ansatz $P = \Pi x_i^{p_i}$ as an ALF we get

$$\psi = \dot{P}/P = \Sigma\, p_i a_{i-1} x_{i-1} + \Sigma\, p_i b_{i+1} x_{i+1} - \Phi$$

At the edge equilibria $F_{k,k+1}$ we obtain

$$p_{k-1}\frac{b_k}{a_k} + p_k + p_{k+1} + p_{k+2}\frac{a_{k+1}}{b_{k+1}} > 1 \qquad (5.3)$$

For the vertices and the other boundary equilibria $\psi > 0$ is then trivially satisfied. This system (5.3) of inequalities is easily seen to be solvable in $p_i > 0$, $\Sigma\, p_i = 1$, if $n \leq 4$. So far $n \leq 4$ (5.1) is always permanent if all $a_i > b_i$. This is no longer true for $n \geq 5$: In the symmetric case $a_k = a, b_k = b$, (5.3) reduces to

$$\frac{b}{a} + 2 + \frac{a}{b} > n$$

or

$$\frac{a}{b} > \frac{n-2+[n(n-4)]^{1/2}}{2}$$

In the general case we may again apply Theorem 4 to obtain the exact conditions for permanence. The corresponding matrix C is now given by

$$C = \begin{bmatrix} a_1/b_1 & -1 & -1 & \dots & -b_n/a_n & 0 & 0 \\ 0 & a_2/b_2 & -1 & \dots & -1 & -b_1/a_1 & 0 \\ 0 & 0 & a_3/b_3 & \dots & -1 & -1 & -b_2/a_2 \\ \dots & \dots & \dots & \dots & \dots & \dots & \dots \end{bmatrix}$$

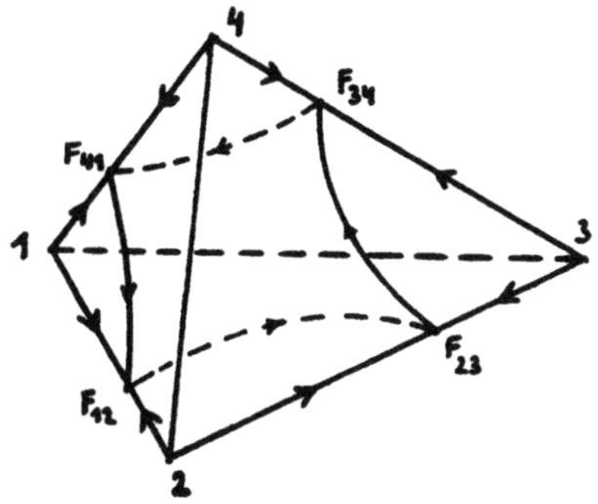

Fig. 2: System (5.1) with $a_i > b_i$ is either permanent or the heteroclinic connection of the saddle points $F_{12}, F_{23}, \dots$ is an attractor.

6. Applications to 3-dimensional Lotka-Volterra equations

We conclude this contribution by analyzing permanence (ecological stability) in 3-species communities governed by Lotka-Volterra dynamics. We restrict ourselves to two typical cases and present only the results omitting the concrete calculations.

I. One predator, two preys

This case was investigated by Hutson and Vickers [10], see also [9] for a related problem. (*) reads in this case

$$\begin{aligned} \dot{H}_1 &= H_1(a_1 - \epsilon_{11}H_1 - \epsilon_{12}H_2 + \alpha_1 P) \\ \dot{H}_2 &= H_2(a_2 - \epsilon_{21}H_1 - \epsilon_{22}H_2 + \alpha_2 P) \\ \dot{P} &= P\,(-c + \beta_1 H_1 + \beta_2 H_2 + \gamma P) \end{aligned} \tag{6.1}$$

Recall the possible behaviours on the boundary (cf. [14]).

Predator-Prey: 1) If there is an interior equilibrium, then it is globally stable: Prey and predator coexist. P—•—H

2) If there is no interior equilibrium, then the predator dies out. (P → H)

Two competitors: 1) There is an interior equilibrium which is globally stable: Both competitors coexist (H_1—•—H_2).

2) The interior equilibrium is a saddle point. Depending on the initial conditions one of the competitors dies out, the other survives. (Bistability).

3) No interior equilibrium: One of the species competes out the other, which one survives depends on the parameters. ($H_1 \to H_2$ or $H_2 \to H_1$)

Now we can give a list of all permanent cases

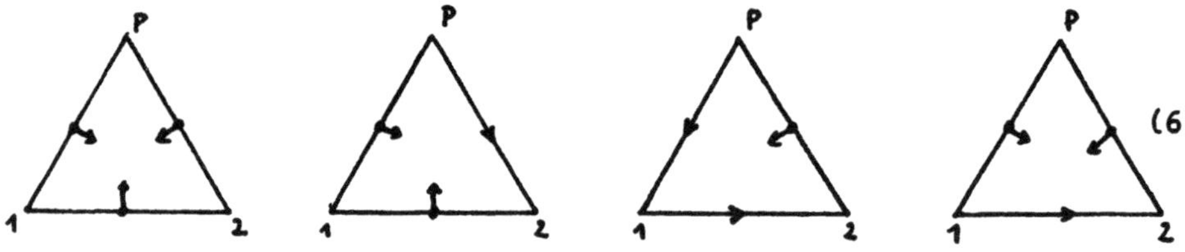

(6.2)

Taking $P = \prod x_i^{p_i}$ as an ALF, the condition $\Phi > 0$ reduces at the two-species equilibria to the simple fact, that the eigenvalue to the 3rd species is positive. This is indicated by an arrow pointing inwards the triangle. At all other fixed point this condition will be satisfied automatically for wellchosen p_i. All other possible cases not appearing in (6.2) cannot be permanent since one can find an interior orbit which converges to a boundary equilibrium.

In fact a more elegant characterization of permanence was given in [10]:

<u>Theorem 6</u>: System (6.1) is permanent ⇔

1) There is an interior equilibrium point ($P, H_1, H_2 > 0$)

and 2) The determinant of the community matrix is positive (⇔ the product of all eigenvalues at the equilibrium is negative)

and 3) The H_1-H_2-subsystem is not bistable.

We note that the interior equilibrium need not be stable in this case, limit cycles are possible [10].

II. Three competitors

$$\begin{aligned}\dot{x}_1 &= x_1(a_1 - e_{11}x_1 - e_{12}x_2 - e_{13}x_3)\\ \dot{x}_2 &= x_2(a_2 - e_{21}x_1 - e_{22}x_2 - e_{23}x_3)\\ \dot{x}_3 &= x_3(a_3 - e_{31}x_1 - e_{32}x_2 - e_{33}x_3)\end{aligned} \qquad (6.3)$$

This case has been treated by Hallam et al. [5], but with the weaker concept of "persistence", i.e. $\limsup_{t \to \infty} x_i(t) > \delta > 0$ for every $x \in \operatorname{int} R^n_+$ and every i.

Using their work we can list the permanent cases according to the possible behaviour on the boundary.

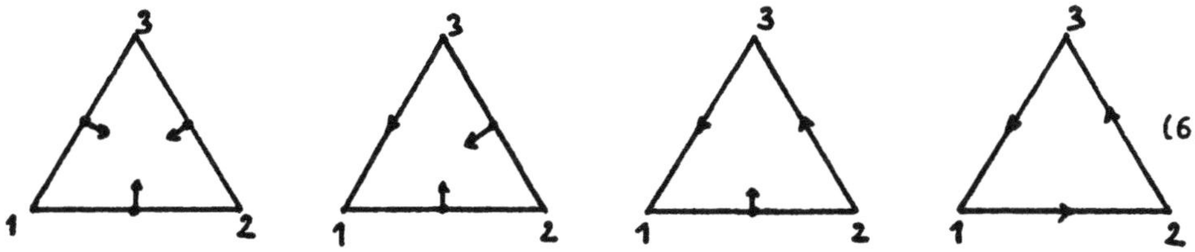

(6.4)

For the first three the situation is the same as in (6.3). The last one however is more interesting from the mathematical point of view, although it is of less biological relevance. Such a kind of "intransitive" competition systems was first considered by May and Leonard [11] and analysed further in [3 , 16]. This is the only case of 3-species communities which exhibit a nontrivial limit set on the boundary:

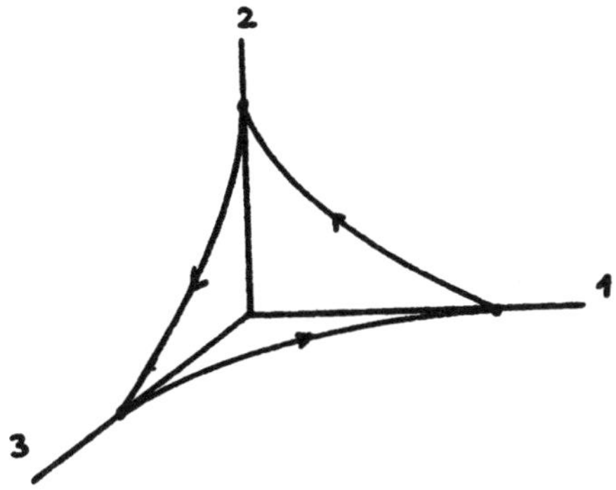

The heterodinic connection of saddle points is stable iff the product of the 3 "coming in"-velocities exceeds the product of the 3 "going off"-velocities. Otherwise the system is permanent.

Note that this case is always "persistent" in Hallam's sense, even when this heteroclinic cycle is an attractor.

If we neglect this "intransitive" case however, again the more concise characterization of permanence of (6.3) holds:

1) There is an interior (3-species) equilibrium.
2) The determinant of the community matrix (e_{ij}) is positive.
3) There is no "bistable" two-species subsystem.

References

[1] Amann, H.: Gewöhnliche Differentialgleichungen. De Gruyter, Berlin-New York 1938.

[2] Berman, A. and Plemmons, R.I.: Nonnegative Matrices in the Mathematical Sciences. Academic Press, New York-San Francisco-London 1979.

[3] Coste, J.; Peyra d, J. and Coullet, P.: Asymptotic behaviour in the dynamics of competing species. SIAM J. Appl. Math. 36 (1979), 516-542.

[4] Eigen, M. and Schuster, P.: The Hypercycle: A Principle of Natural Self-organization. Springer, Berlin-Heidelberg-New York (1979).

[5] Hallam, T.G.; Svoboda, L.I. and Gard, T.C.: Persistence and Extinction in three species Lotka-Volterra competitive systems. Math. Biosciences 46 (1979), 117-124.

[6] Hofbauer, J.: On the occurrence of limit cycles in the Volterra-Lotka equations. Nonlinear Analysis. TMA 5 (1981), 1003-1007.

[7] Hofbauer, J.: A general cooperation theorem for hypercycles. Monatsh. Math. 91 (1981), 233-240.

[8] Hofbauer, J. and Sigmund, K.: Evolutionstheorie und dynamische Systeme - Mathematische Aspekte der Selektion. Parey, Hamburg 1984.

[9] Hutson, V.: Predator mediated coexistence with a switching predator. Math. Biosciences, to appear.

[10] Hutson, V. and Vickers, G.T.: A criterion for permanent coexistence of species, with an application to a two-prey one-predator system. Math. Biosciences 63 (1983), 253-269.

[11] May, R.M. and Leonard, W.: Nonlinear aspects of competition between three species. SIAM J. Appl. Math. 29 (1975), 243-252.

[12] Maynard Smith, J.: Evolutionary Game Theory. Cambridge University Press 1982.

[13] Nikaido, H.: Introduction to Sets and Mappings in Modern Economics. North-Holland, Amsterdam-London 1970.

[14] Roughgarden, J.: Theory of Population Genetics and Evolutionary Ecology. Mac Millan, New York 1979.

[15] Schuster, P.; Sigmund, K. and Wolff, R.: Dynamical systems under constant organization. Part 3: Cooperative and competitive behaviour of hypercycles. J. Diff. Equat. 32 (1979), 357-368.

[16] Schuster, P.; Sigmund, K. and Wolff, R.: On ω-limits for competition between three species. SIAM J. Appl. Math. 37 (1979), 49-54.

[17] Schuster, P.; Sigmund, K. and Wolff, R.: Mass action kinetics of selfreplication in flow reactors. J. Math. Anal. Appl. 78 (1980), 88-112.

[18] Sigmund, K. and Schuster, P.: Permanence and uninvadability for deterministic population models. In: Stochastic Phenomena and Chaotic Behaviour in Complex Systems. Ed. P. Schuster, pp. 173-184, Springer.

[19] Svireshev, Iu.M.: The complex dynamics of simple ecological models, this volume.

[20] Zeeman, E.C.: Population dynamics from game theory. In: Global Theory of Dynamical Systems, Lecture Notes in Mathematics 819. Springer 1980.

DYNAMICAL REGRESSION IN COMPLEX SYSTEMS - ECOLOGICAL EXAMPLES

Jean COSTE*

INTRODUCTION

A fascinating problem, in the theory of complex dynamical systems (i.e. obeying many differential equations), concerns the dimensionality of their attractors. More precisely the question is : Can we find some (non trivial) classes of vector fields such that their attractors' dimension be much less than the phase space one ?

This problem is central in the physics of turbulence : numerous observations (experimental and numerical) seem indeed indicate that some non stationary flows may be described (on a "macroscopic scale") by relatively few variables.

In ecology, people obviously want to know if many species (biological or/and chemical) are able to coexist ; and this question has clear implications in the process of Darwinian evolution. Moreover, the ecologic version of the problem is perhaps simpler than in physics, one reason being that some situations of biological competition are reasonably described, in the absence of spatial diffusion, by ordinary differential equations (this also happens in physics, but it is usually less obvious). We shall essentially consider ecological models in the following, the evolutions being assumed continuous in time, and the interaction instantaneous (no delay).

Coming back to our first question, let us give a more precise formulation of the problem. Our basic assumption is that the solutions of the N differential equations defining the model are bounded for initial values taken in a convenient domain (boundedness must belong to any acceptable model). In other words, we suppose that there exists connex subspaces of $\mathbb{R}_N$ mapped into themselves by the dynamical transformation (such that the vector field vanishes or is inward at their boundaries). Then we shall say that the system suffers a strong "dynamical regression" if the attractors dimension (Hausdorff dimension) is much less than N , or if most of the variables are found arbitrarily small at large times (see below for a more precise definition). Actually the case where some of the variables asymptotically vanish is a particular case of dynamical regression. This happens when the models equations take the form :

$$\dot{x}_i = x_i \; \phi_i \left(\{x_j\} \right) \qquad i , j = 1 , \ldots , N \tag{1}$$

functions ϕ_i being regular at $x_i = 0$. Such a vector field has invariant planes passing by the origin, namely the coordinate planes $x_i = 0$. Many models of biological competition are described by equations of type (1), which guarantee the positivity of population variables at any time. We shall consider this type of vector field in the present lecture.

* Laboratoire de Physique de la Matière Condensée, L.A. 190, Mécanique Statistique, Parc Valrose, 06034 NICE Cedex, France.

A "principle of competitive exclusion" appeared in the very early years of ecological literature, Volterra (1927)[22], Gause (1934)[8], essentially as a conjecture. It was later reconsidered and developped along several ways in the sixties and seventies. However, few significant results were obtained and only on quite particular models (S. Levin [15], Mc Gehee-Armstrong [17]). A controversy about the problem "Does complexity favors (or not) coexistence" was developing during several years and did not end with any clear cut conclusion (May [18] , Robert[20] , Gilpin [9] , Gard, Hallam [7]) .

Why this interesting debate on the coexistence of many species faded out during the last years ? Perhaps because people were looking for an exclusion condition which would be satisfied by all members of a class of vector fields. This is a strong constraint, which shrinks very much the allowable classes. Less ambitiously one may look for "almost-always exclusive" classes, in the limit of large systems. I mean that, choosing at random the field coefficients inside some class, a necessary condition of coexistence will probably not be satisfied. We propose to show that some (non trivial) classes do exist, and even are such that most of the species asymptotically disappear.

TIME AVERAGING AND PERSISTENCE

Definition of persistence

Being given a vector field of type (1) defined on $\mathbb{R}_N$, the associated dynamical system will be said persistent if there exists an attractor in $\mathbb{R}_N^+$, stated equivalently : there exists ε_0 and initial values x_i^0 such that

$$|x_i(t, x_i^0)| > \varepsilon_0 \qquad \forall i \; t \; .$$

A necessary condition of persistence

Time averaging Eqs. (1) rewritten in the form $\frac{\dot{x}_i}{x_i} = \phi_i$, we obtain

$$\frac{1}{T} \operatorname{Log} \frac{x_i(T)}{x_i(0)} = \langle \phi_i(\{x_j\}) \rangle_T \qquad i, j = 1, \ldots, N \tag{2}$$

where

$$\langle \; \rangle_T = \frac{1}{T} \int_0^T dt$$

We see that the persistence of a N-species bounded solution implies that

$$\langle \phi_i(\{x_j\}) \rangle_{T \to \infty} = 0 \qquad i, j = 1, \ldots, N \tag{3}$$

The aim of this paper is to show, for two classes of vector fields, that satisfying the N equations (3) in which the x_i's must be positive, impose quite severe conditions on the field parameters. This makes the coexistence highly improbable, and usually causes a strong dynamical regression. The first case we shall consider is the celebrated Lotka-Volterra (L.V.) model.

THE LOTKA VOLTERRA MODEL [2, 3, 5]

In this model the ϕ_i's are linear functions of the x_j's

$$\phi_i = b_i - \sum_j a_{ij} x_j \qquad i, j \quad t[1, \dots N] \tag{4}$$

It is therefore the simplest description of competition, the species interactions being simply quadratic. This model was extensively studied on account of its simplicity although biologists usually prefer more sophisticated ones. It is also worth remarking that it describes, in Poincare variables, the dynamics of an ensemble of coupled oscillators near the threshold of linear unstability, provided their frequencies are not commensurable. We also want to mention the problem of Raighleigh-Benard convection at high Prandtl number : in a model where the velocity field may be analyzed as a set of discrete hydrodynamic modes with amplitudes x_i, and neglecting the slow diffusion of these modes, the x_i's obey L.V. equations [4].

As was said above, we assume the boundedness of the solutions of Eqs. (1, 4). It is easily proved when coefficients b_i, a_{ij}, are all positive. For simplicity we shall assume these coefficients to be positive, although this assumption is not essential in the argument. Let us first observe that the L.V. field exhibits 2^N fixed points defined by

$$\xi^k = 0 \quad , \quad k \in \{p\} , \quad \text{any set of p integers}$$

$$\sum_j a_{ij} \xi^j = b_i \quad , \quad i, j \in [1, \dots, N] - \{p\} \tag{5}$$

Now Eqs. (3) give here :

$$\sum_{j=1}^{N} a_{ij} \langle x_j \rangle = b_i \tag{6}$$

ξ_N being the unique fixed point in $\mathbb{R}_N^+$, and $\langle x \rangle$ the time averaged point associated with an orbit (coordinates $\langle x_j \rangle$), we conclude that the condition of N-species coexistence is $\langle x \rangle = \xi_N$. It can be satisfied only if ξ_N belongs to the positive octant $(\xi_N^j > 0, \forall_j)$, and this sets up the problem : is it difficult to find a positive solution to the system of N linear equations :

$$\sum_{j=1}^{N} a_{ij} \xi^j = b_i \qquad ? \tag{7}$$

The answer is yes in the trivial case where the inequalities $a_{ij} \ll a_{ii} \; \forall i,j$ hold, i.e. if the interactions between individuals of the same species are dominant compared to the interaction between individuals belonging to different species. The species are then weakly coupled, and we expect that they evolve toward an equilibrium state in the neighborhood of $x_i = \frac{b_i}{a_{ii}}$; therefore the coexistence is trivially realized. In the non trivial case, most of the coefficients a_{ij} are of the same order of magnitude. It is then clear that studying the existence of a positive solution of Eqs. (7) is only meaningful in a statistical sense : choosing at random the field coefficients in a convenient domain, which is the probability $\mathcal{P}(N)$ of finding a positive solutions ? This problem of probability is by no means obvious, and only partial results are available at present time. They are the following :

1° Considering a system of similar species, i.e. with coefficients of the form :

$$
\begin{aligned}
a_{ij} &= a\,(1+\varepsilon_{ij}) \\
b_i &= b\,(1+\varepsilon_i) \qquad \left\{\begin{matrix}\varepsilon_{ij}\\ \varepsilon_i\end{matrix}\right. < \varepsilon \ll 1
\end{aligned}
\tag{8}
$$

it is shown that, in the limit of large N

$$\mathcal{P}(N) < (N-1)\,e^{-\frac{(N-1)}{\varepsilon}}$$

2° It is also possible to show, in the general case of positive a_{ij} and b_i, and under rather weak assumptions on distribution probabilities, that $\mathcal{P}(N)$ decreases faster than $\frac{1}{N}$.

3° We have performed an ensemble of numerical calculations, solving large sets of linear equations whose coefficients are random numbers in (0,1) and (-1,1), and observe that $\mathcal{P}(N) \leqslant 2^{-N}$ for large N.

Let us remark, at this stage, that the assumption of strong coupling between all species (or of a non empty a_{ij} matrix) may seem quite strong, and difficult to be verified in real ecosystems. The following example show that such ecosystems are conceivable and rather natural : consider the off-springs of one initial bacteria or virus after many successive generations (asexual reproduction), and suffering random mutations : this is called a"clone" by biologists. It is reasonable to assume that most of the mutations (which modify the genome and therefore create new "species") are weak in the sense that the phenotype of new born species are only slightly modified. Then it is clear that the interactions between individuals of the same species or belonging to different species are very similar. Therefore, we have $a_{ij} \simeq a_{ii} \quad \forall i, j$, and we are in the case of strong coupling, where the probability $\mathcal{P}(N)$ can be evaluated and is indeed very small.

Now, we give a first rough argument showing the dynamical regression. If a positive solution is not found to Eqs. (7), relation $\langle x \rangle = \xi_N$ cannot be satisfied and at least one population variable, say x_k , must become small after a large enough time. Therefore, the orbit in phase space approaches $(\mathbb{R}^+)^{N-1}_k$ subspace (i.e. that subspace from which variable x_k is excluded). When x_k has become small enough we are left with a (N-1) dimension problem and we look again for a fixed point in $(\mathbb{R}^+)^{N-1}_k$. In the absence of such a point, one more species must disappear, and so on ... and this regressive process continues until some subspace $\{\mathbb{R}^+\}^{N-p}$ is reached, which contains a fixed point (even then of course the coexistence is not garanteed since this last fixed point may be unstable).

Finally, strong dynamical regression is expected (i.e. the number of coexisting species is much less than N) due to the rapid decay of $\mathcal{P}(N)$ with N. Indeed, admitting that $\mathcal{P}(N)$ decreases faster than 2^{-N} (on the basis of above arguments), we see that the probability of finding a positive fixed point in one of the (2^N-1) subspaces is very small, unless the dimension p of this subspace is small (say $p < 10$) as a rough estimate). Therefore, most of the species must disappear.

Actually, the above argument is insufficient, because one cannot assume that the populations decrease monotically. It is even known that the evolution may be highly irregular, as this happens in the May-Leonard model [9] : some population get nearly extincted at some time, then starts growing up to a finite value, and later on decreases again. This fact led us to introduce the concept of "non persistence of the system of order $>$ N-p" by the

Definition

A system will be said "not persistent of order $>$ N-p" if, along any orbit and in the limit $t \to \infty$, one can almost always find a set of p variables $\{x_i\}$ such that $|x_i| < \varepsilon$, ε given arbitrarily small. Then we shall say that p variables are "almost always extincted" or "A.A. extincted".

Now it is possible to show (see [5]) that the non positivity of fixed points in successive subspaces $\mathbb{R}^+_N, \mathbb{R}^+_{N-1}, \dots, \mathbb{R}^+_{N-m}$ implies non persistence of order $> N-m$. This statement makes precise the concept of strong dynamical regression in the case of the L.V. field.

Two very simple examples of L.V. regression

The first one is the above mentionned model of Raileigh-Benard convection. In this case $b_i = b > 0$ and the a_{ij}'s are positive and symmetrical $(a_{ji} = a_{ij})$. Here the regression can be shown even without the use of previous theorem. Two properties are needed :

i- There exists a Liapunov function [16] $\left(H = \sum_i b x_i - \frac{1}{2} \sum a_{ij} x_i x_j\right)$ which garantees that the attractor is a fixed point

ii- If $a_{ij} > a_{ii}$ $(\forall ij, i \neq j)$, and this condition is actually verified in the hydrodynamic model, then it can be shown (see Appendix) that all the fixed points are unstable, except the one-component fixed points.

We conclude that the attractor must be a one-component equilibrium, which is in agreement with the fact that roll solutions (characterized by one hydrodynamic mode) appear at convection's threshold.

The second example is the simplest version of Eigen's "flow reactor" (see [21]), whose model equations are of the form:

$$\dot{x}_i = x_i \left(E_i - \frac{1}{c} \sum_{j=1}^{N} E_j x_j \right) .$$

The fixed points of this L.V. field are given by :

$$\sum_{j=1}^{p} \frac{E_j}{c} \xi_j = E_i$$

p being any integer in [1,N] and $i,j \in [1,2,\dots,p]$. One immediately sees that these equilibrium equations only admit a one-component solution.

Remarks on the dynamics of the regression. Effect of fluctuations

How much time does it take for most of the species to get extincted (in the case of strong regression) ? If field's coefficients are of the same order of magnitude, "extinction time" t_c is expected to be of the order of the average inverse growth rate, which is confirmed by numerical integrations of the equations of many species motion. Things are different in the interesting case of a system made of similar species (when inequalities (8) are obeyed). Then, it may be conjectured that t_c is a decreasing function of parameter ε of Eq. (8). Indeed, for $\varepsilon = 0$, any set $\{x_j\}$ such that $\sum_j x_j = \frac{b}{a}$ is an equilibrium solution. Then, for ε small but finite, this solution becomes quasi stationary and its life time goes to infinity when $\varepsilon \to 0$. Let us also remark that, in the case of small ε, the selection process is determined by small differences between the field parameters. Therefore, the fluctuations of environment could likely make these parameters to fluctuate. How these fluctuations modify the dynamical regression ? We have performed a numerical experiment where we choose the a_{ij} 's and b_i's randomly in [0,1] at random time intervals t_i. A non trivial situation occurs when average time interval $\langle t_{i+1} - t_i \rangle$ is much smaller than the average selection time in the non fluctuating field. The numerical results suggest that dynamical regression is a robust phenomenon with respect to fluctuations.

PREDATORS-PREYS MODEL WITH MICHAELIS-MENTEN INTERACTIONS

Some predators-preys models

The following models have been used to describe experimental situations such as the "chemostat" [10] [23]. Two types have been studied (in addition to the L.V. model), in which the predator-prey interaction is of the Michaelis-Menten kind, and the predator-predator interactions are ignored. The predators equations are :

$$\frac{\dot{x}_i}{x_i} = -d_i + \frac{m_i s}{a_i + s} \qquad (i \in [1, N]) \tag{9}$$

where m_i, a_i, d_i are positive coefficients.

In model A, the prey s is renewable, and its evolution described by

$$\frac{\dot{s}}{s} = \gamma\left(1 - \frac{s}{k}\right) - \sum_{i=1}^{N} \frac{\alpha_i x_i}{a_i + s} \qquad (\alpha_i > 0) \tag{10}$$

In model B, the ressource is limited and obeys equation

$$\dot{s} = (s_0 - s) D - \sum_{i=1}^{N} \frac{\alpha_i x_i s}{a_i + s}$$

The models have been the object of several mathematical studies [11] [12] [13] [1].
The many predators-case has been solved for model B. Using a method based on Lasalle's theorem [14], it was shown by S.B. Hsu [12] that, generically, all predators get extincted except one : an extreme case, indeed of dynamical regression ! Model A was only studied in the two-predators case [13], [1]. It is interesting to mention that such a system violates our L.V. persistence criterium. Indeed, G. Butler and P. Waltman proved [1] the existence of a periodic orbit, in the absence of any fixed point in the positive octant. We shall consider here model A in the case of many predators (N large).
Equations (9) will be rewritten in the form :

$$\frac{\dot{x}_i}{x_i} = (m_i - d_i)\left(\frac{s-\lambda_i}{s+a_i}\right) \qquad \left(\lambda_i = \frac{a_i d_i}{m_i - d_i}\right) \tag{11}$$

which define the model, together with Eq. (10). We assume bounded positive coefficients λ_i, a_i, and

$$\begin{cases} \lambda_i \neq \lambda_j & i \neq j \qquad (\lambda_i > 0) \\ 0 < a_{min} < a_i < a \end{cases} \tag{12}$$

Study of model A : Many predators competing for one renewable prey [6]
First of all, the model is biological relevant, i.e. it may be shown that the solutions of Eqs. (10,11) are bounded, and $s(t \to \infty) \leq k$ (k finite). Therefore, we need only consider $\lambda_{min} < \lambda_i < k$ (since any predator with $\lambda_i > k$ get extincted). Persistence conditions (3) imply here that

$$\left\langle \frac{s-\lambda_i}{s+a_i} \right\rangle = 0 \qquad i = 1, \ldots N \tag{13}$$

Satisfying Eq. (13) imposes to parameters λ_i, μ_i conditions given in the following theorem.
We put

$$\varepsilon = 1 - \frac{a_{min}}{a} \qquad \mu_i = \frac{a}{a_i + \lambda_i}$$

and call μ the point $\{\mu_i\}$ in μ_i-space, $\mathcal{C}$ the dense set of μ points such that, for given $\{a_i\}$, Eqs. (13) are satisfied. Then the following theorem can be proved :

Theorem
The Lebesgue measure of $\mathcal{C}$ goes to zero for large N faster than $\frac{\varepsilon^{N-1}}{\sqrt{N}}$.
We only give here the starting point of the demonstration (see [6])
Let $A_i = \frac{s-\lambda_i}{s+a_i}$. A_i can be put in the form

$$A_i = \frac{1-\frac{g_i}{\sigma}}{1-\frac{f_i}{\sigma}} \quad \text{with} \quad \sigma = \frac{s+a}{a}, \; f_i = 1-\frac{a_i}{a}, \; g_i = 1+\frac{\lambda_i}{a}.$$

Since $0<\frac{f_i}{\sigma}<1$, $0<f_i<\varepsilon$ and $0<\sigma^{-1}$, we can expand A_i in powers of σ^{-p}, which gives

and Eqs. (13) are rewritten as

$$\sum_{p=1}^{\infty} f_i^{p-1} y_p = \mu_i \qquad \left(\frac{a}{a+k} < \mu_i < \frac{a}{a_{min} + \lambda_{min}} \right) \tag{14}$$

with $y_p = \langle \sigma^{-p} \rangle$ obeying conditions :

$$0 < y_p < 1 \tag{15}$$

Now

$$\eta_i = \sum_{N+1}^{\infty} f_i < \left(\frac{\sigma}{f_i}\right)^p > \; < \frac{\varepsilon^{N+2}}{1-\varepsilon} \quad ,$$

and therefore goes to zero for large N. Finally Eq. (14) takes the form

$$\sum_{p=1}^{N} f_i^{p-1} y_p = \tilde{\mu}_i \quad .$$

with $\tilde{\mu}_i \to \mu_i$ for large N.

Conditions (15) together with $0 < f_i < \varepsilon$ impose on solutions of Eqs. (16) a strong constraint which can be shown to force parameters μ_i to lie in a vanishing volume of $\{\mu_i\}$ space in the limit of large N.

Corollaries of the previous theorem are :

i- Choosing at random λ_i , μ_i respectively in (λ_{min}, k) and (μ_{min}, μ_{max}) the probability of finding a "good set" of field parameters (i.e. such that Eqs. (13) are satisfied) goes to zero for large N.

ii- It can be shown [6] that most of the species populations are expected (in a probabilistic sense) to be found arbitrarily small at large times.

CONCLUSION

We have considered two types of complex systems in which coexistence was shown to be highly improbable, and where strong dynamical regression was expected. Our method of time averaging and momenta expansion seems well fitted for studying more general systems, at least those where many predators feed on finitely many preys.

APPENDIX

The proof is given in the general case of arbitrary growth rates b_i $(b_i > 0)$. Let us linearize the L.V. equations near fixed point $\{\bar{x}_i\}$ (with non zero components).

Putting $y_i = x_i - \bar{x}_i$, we obtain

$$\dot{y}_i = -\bar{x}_i \sum_{j=1}^{N} a_{ij} y_j \qquad i \in \{p\} \tag{A1}$$

$$\dot{y}_k = y_k \left\{ b_k - \sum_{j \in \{p\}} a_{ij} \bar{x}_j \right\} \qquad k \in \{N-p\} \tag{A2}$$

The a_{ij} and b_i are all assumed positive, and $p>1$.

Looking for a sufficient condition of fixed point's unstability, we assume that there does not exist any b_k such that $b_k - \sum_{j \in \{p\}} a_{ij} \bar{x}_j > 0$. Therefore, the y_k's are exponentially vanishing, and we are left with the study of Eqs. (A1) in which i, j will be taken in set $\{p\}$. Let $-s_0, s_1, \ldots s_{p-1}$ be the eigen values of Eqs. (A1). We have :

$$-s_0 + \sum_{i=1}^{p-1} s_i = -\sum_{i=1}^{p} a_{ii} \bar{x}_i \qquad \text{(A3)}$$

Now Frobenius theorem says that, for matrix $\{a_{ij} \bar{x}_i\}$ with positive elements, there exists an eigenvalue s_0 such that $\mathrm{Re}\, s_0 > 0$ and $\mathrm{Re}\, s_0 > s_i \ \forall i$, and whose eigenvector has all its components positive. Let $\mu^{(0)}$ the eigen vector of the stability matrix (associated to $-s_0$). We have :

$$s_0 \sum_i \mu_i^{(0)} = \sum_{i,j} a_{ij} \bar{x}_c \mu_j^{(0)} \geqslant \left(\operatorname*{Inf}_{k \in \{p\}} \sum_k a_{ik} \bar{x}_c \right) \left(\sum_i \mu_i^{(0)} \right)$$

Therefore : $s_0 \geqslant \operatorname*{Inf}_{k \in \{p\}} \sum_k a_{ik} \bar{x}_c$. Now Eq. (A3) implies :

$$\sum_{i \neq 0} s_i \geqslant \operatorname*{Inf}_k \left\{ \sum_k (a_{ik} - a_{ii}) \bar{x}_c \right\} ,$$

and condition $a_{ik} > a_{ii} \ \forall i, k$ implies $\sum_{i \neq 0} \mathrm{Re}\, s_i > 0$ which proves the unstability of the fixed point. It is immediately verified that, among one-component equilibria, those satisfying $b_i > \frac{a_{ii}}{a_{ik}} b_k$ are stable.

REFERENCES

[1] Butler, G. and Waltmann, P. : J. Math. Biol. **12** (1981) 295-310.

[2] Coste, J., Peyraud, J. and Coullet, P. : J. Theor. Biol. **73** (1978) 359-362.

[3] Coste, J., Peyraud, J. and Coullet, P. : SIAM J. Appl. Math. **36** (1979) 516-543.

[4] Coste, J., Peyraud, J. and Coullet, P. : J. Stat. Phys. **21** (1979) n° 1, 33.

[5] Coste, J. : Lectures Notes in Bio. Math. **52** (1983) 218.

[6] Coste, J. : submitted for publication.

[7] Gard. T. and Hallam, T. : Bull. Math. Biol. **41** (1979) 877-891.

[8] Gause, C.F. : **The struggle for existence**, Williams & Wilkins Co., Baltimore (1934).

[9] Gilpin, M. : Nature **254** (1975) 137.

[10] Hansen, S. and Hubbell, S. : Science **207** (1980) 1491-1493.

[11] Hsu, S., Hubbell, S. and Waltmann, P. : SIAM J. Appl. Math. **32** (1977) 366-383.

[12] Hsu, S. : SIAM J. Appl. Math. **34** (1978) 760-763.

[13] Hsu, S., Hubbell, S. and Waltmann, P. : SIAM J. Appl. Math. **35** (1978) 617-625.

[14] Lasalle, J. : IRE Trans. Circuit Theory **7** (1960) 520-527.

[15] Levin, S. : American Naturalist **104** (1970) 413.

[16] Mac Arthur, R.H. : Proc. Nat. Acad. Sci. **64** (1969) 1369.

[17] Mc Gehee, R., Armstrong, R. : J. Diff. Equations **23** (1977) 30-52.

[18] May, R. : Nature **238** (1972) 413.

[19] May, R., Leonard, W. : SIAM J. Appl. Math. **29** (1975) 243.

[20] Roberts, A. : Nature **254** (1975) 137.

[21] Schuster P. : **Biochemical evolution**, H. Gutfreund Ed., Cambridge University Press (1980).

[22] Volterra, V. : Memorie del R. Comitato Talassografico Italiano **131** (1927) 1-142.

[23] Waltmann, P., Hubbell, S. and Hsu, S. : **Modeling and Differential Equation in Biology**, T. Burton ed., M. Dekker, New York, 1980.

CHAIN DECOMPOSITION OF GROWTH FUNCTIONS AND STRUCTURE MODELS OF ECOLOGICAL SYSTEMS

M.Peschel [1], W.Mende [2], W.Jansen [1], F.Albrecht [1]

INTRODUCTION

In [1] it was shown, that usually systems described by ordinary differential equations can be transformed into a description by Lotka-Volterra equations. This transformation corresponds to a structure in form of a network with the logarithmic derivative

$$dx/dt = K\,x\,y$$

as basic module.

On the other hand growth phenomena are frequently bound to highly aggregated socalled soft systems consisting of a large number of homogenous parts (particles, species).

Therefore it seems to be necessary to look for a structure decomposition of a given large system into higher aggregated modules which already possess autocatalytic properties and expose certain growth phenomena.

The aim of this paper is the generalisation of the STRUCTURE DESIGN PRINCIPLE [2] on the case with higher aggregated basic modules.

Instead of a Lotka-Volterra network we get for basic modules of the form

$$dx/dt = K\,x^k\,y$$

a realization by a socalled Riccati - network demonstrating at the same time the extraordinary role of Cobb-Douglas production functions.

1. CHAIN DECOMPOSITION, SIGNAL-GENERATORS, GROWTH MODELS

Chains (for energy, food, matter) are dominating structures in real systems.

Let us assume that on an aggregated consideration level such a chain consists of modules with a functional law of the following kind

$$d\,\phi(x_i)/dt = K_i\,x_{i+1} \quad \text{for} \quad i = 0,1,2,\ldots$$

with $\phi(u)$ being a continuously differentiable strict monotonous function.

---

1) Central Institute of Cybernetics and Information Processes of Academy of Sciences of the GDR.

2) Institute of Geography and Geoecology of Academy of Sciences of the GDR.

In this description $\phi(u)$ expresses the 'richness' of the local growth.

If we expand a given signal $x_0 = x(t)$ into such a ϕ-chain, then

$$d\,\phi(x_i)/dt = K_i x_{i+1} \quad \text{for} \quad i = 0,1,2\ldots$$

is a socalled signal generator model of $x(t)$.
For any given signal $x(t)$ we have to specify the initial values $x_i(0)$ and the chain parameters K_i.
For initial values $x_i(0) = 1$ the K_i can be identified by

$$K_i = d\,\phi(x_i)/dt \Big|_{t = 0}$$

We can easily construct a state-variable description by substituting

$$x_i(t) = \phi^{-1}\left(\phi(x_i(0)) + \phi(\xi_i)\right)$$

with the normalisation condition

$$\phi(\xi_i(0)) = 0$$

Then we get the following state-space description

$$d\,\phi(\xi_i)/dt = K_i \phi^{-1}\left(L_{i+1} + \phi(\xi_{i+1})\right)$$

In [1] we considered in detail the following special cases

1. $\phi(u) = u \qquad d\xi_i/dt = K_i + M_i \xi_{i+1}$ and $\xi_i(0) = 0$

   Especially for $M_i = 1$ we get the socalled Taylor-chain

   $$d\xi_i/dt = K_i + \xi_{i+1}$$

   corresponding to the Taylor-Expansion of $x_0(t) = x(t)$.

2. $\phi(u) = \ln u$

   Now we get with $\xi_i(0) = 1$ the socalled EXPONENTIAL CHAIN

   $$d\xi_i/dt = K_i \xi_i \xi_{i+1} \quad , \quad K_i = d \ln \xi_i /dt \Big|_{t = 0}$$

In this paper we are more interested in inhomogenous chains of the form

$$d\,\phi_i(\xi_i)/dt = K_i \phi_{i+1}^{-1}\left(L_{i+1} + \phi_{i+1}(\xi_{i+1})\right)$$

We have studied the following special cases

1. $\phi_i(u) = u^{-k_i + 1}$

   $$dx_i/dt = K_i x_i^{k_i} x_{i+1} \qquad \text{Riccati Chains}$$

2. $dx_i/dt = K_i x_i (B_i - x_i)\, x_{i+1}$ Logistic Chains

3. $dx_i/dt = K_i x_i^{k_i} (B_i - x_i^{w_i})^{l_i}\, x_{i+1}$ Hyperlogistic Chains

It can be shown by simulation and sometimes it can be proofed that these chains usually converge very fast and are robust against equally distributed random disturbances, this was studied for example in [7] .

2. RICCATI CHAINS AS A GENERALISATION OF EXPONENTIAL CHAINS

Riccati chains proof to be an interesting signal expansion possibility.

$x_0 = x(t)$ may be a given growth process in the form of a s-formed transition (Fig. 1)

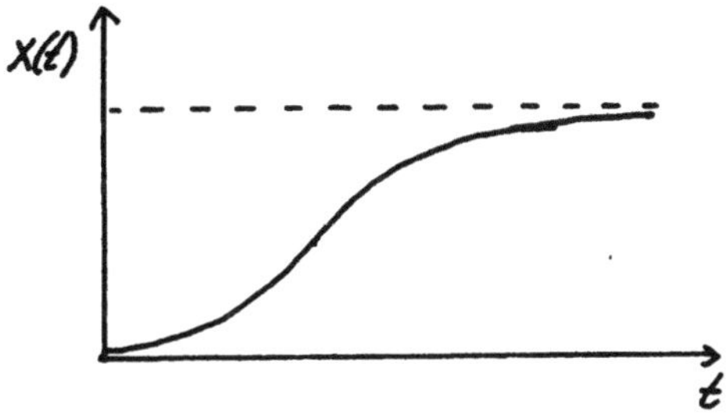

Fig.1. s-formed transition (EVOLON)

We want to generate $x_0(t)$ by a finite Riccati chain

$$dx_i/dt = K_i\, x_i^{k_i}\, x_{i+1} \quad \text{with} \quad x_i(0) = 1 \quad (i=1,2,\ldots,n)$$

$$\text{for } i = 0,1,\ldots,n$$

$$x_{n+1}(t) \equiv 1, \quad K_i = dx_i(0)/dt$$

We have to distinguish two cases:

- with given values for the exponents k_i
- with optimally chosen values for the exponents k_i (Chain Design Principle)

The Riccati Chain Design is based on the following identification procedure of the cooperativity exponents k_i:

From $dx_i/dt = K_i\, x_i^{k_i}\, x_{i+1}$ $i = 0,1,2,\ldots$

we derive the expression

$$\mathcal{X}_i(t) = \frac{d \ln \dot{x}_i /dt}{d \ln x_i /dt} = k_i + \lambda_i(t)$$

with

$$\lambda_i(t) = \frac{d \ln x_{i+1}/dt}{d \ln x_i /dt}$$

Obviously $\lambda_i(t)$ has an invariant sign.

For the identification of the exponents k_i we get the following criterion:

If sign $\lambda_i(t) > 0$ and there exists a t_o with $\lambda_i(t_o) = 0$ then

$$k_i = \text{Min } \mathcal{X}_i(t)$$

If sign $\lambda_i(t) < 0$ and there exists a t_o with $\lambda_o(t_o) = 0$ then

$$k_i = \text{Max } \mathcal{X}_i(t)$$

3. SOME EXAMPLES FOR RICCATI CHAIN DESIGN

Riccati chain for the EVOLON

$$dx/dt = K\, x^k (B - x^w)^l \qquad x(0) = xA$$

We get the following Riccati chain

$$dx/dt = K_o\, x^k\, x_1 \qquad K_o = K(B - xA^w)^l \qquad x(t_o) = 0$$

$$dx_1/dt = - \frac{w\, l\, xA^{(w-1+k)} K_o}{(B - xA^w)} \; x_1^{(2l-1)/l} \; x_2, \qquad x(t_o) = B^{1/w}$$

$$dx_2/dt = \frac{(w-1+k)\, K_o}{xA^{1-k}} \; x_2^{(2k+w-2)/(k+w-1)} \; x_1 \qquad x(t_o) = 0$$

This is obviously a hypercycle of 2-nd order. It turns out that frequently the first 3 steps (putting $x_3 \equiv 1$) already give a good approximation of the EVOLON.

This example offers an identification procedure for the EVOLON parameters after the following scheme

$$K_o,\ k \longrightarrow x_1(t) \longrightarrow K_1,\ l \longrightarrow x_2(t) \longrightarrow K_2,\ w$$

From K_1 we then get B assuming the value of xA, and K_2 gives us the possibility to get a better estimation for the initial value xA.

Riccati chain models for some probability distributions

$dx/dt = K\, e^{\varkappa(t-a)^k}$ with integer k gives

$$dx/dt = K\, e^{\varkappa(-a)^k}\, x_1$$

$$dx_1/dt = \varkappa k(-a)^{k-1}\, x_1 x_2$$

$$dx_2/dt = \frac{(k-1)}{(-a)}\, x_2^{(k-2)/(k-1)}$$

$dx/dt = K\, t^k e^{-at}$ Γ-distribution gives

$$dx/dt = K\, x_1$$

$$dx_1/dt = -\, a x_1 x_2$$

$$dx_2/dt = \frac{k}{a}\, x_2^2\, x_3$$

$$dx_3/dt = -\, 2\, x_3^{3/2}$$

$dx/dt = \dfrac{\lambda}{\pi(\lambda^2 + (t-\mu)^2)}$ Cauchy density gives

$$dx/dt = \frac{\lambda}{\pi(\lambda^2 + \mu^2)}\, x_1$$

$$dx_1/dt = \frac{2\mu}{(\lambda^2 + \mu^2)}\, x_1^{3/2}\, x_2$$

$$dx_2/dt = \frac{\lambda^2}{(\lambda^2 + \mu^2)}\, x_2\, x_1$$

$$dx/dt = \frac{d}{dt}\,(1 - (1-t)^w\,)^k$$

$$dx/dt = k\, w\, x^{(1-1/(kw))}\, x_1$$

$$dx_1/dt = -\,(w-1)\, x_1^{(w-2)/(w-1)}\, x_2$$

$$dx_2/dt = -\,\frac{(w+1)}{(kw)}\, x_2(1 + kw/(w+1))$$

4. STRUCTURE MODELS OF ECOLOGICAL SYSTEMS

Basic structures of ecological systems are chains and cycles (hypercycles) as it is schematically shown in Fig. 2 with a relatively small number of interconnections and feedbacks.

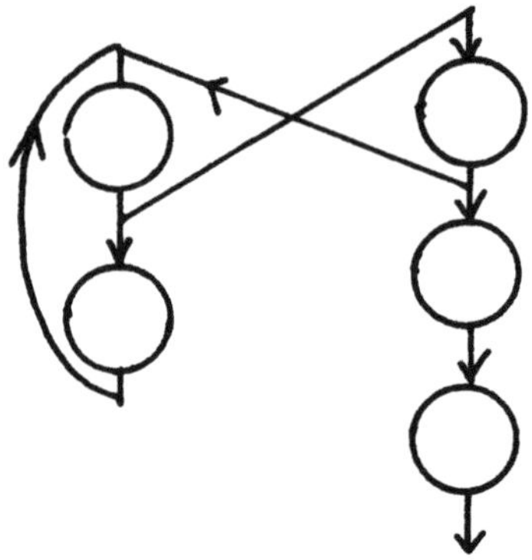

Fig.2. Scheme of the structure of an ecological system

The growth potential concentrated in the single modules depends on the functional integration within the corresponding system and on the aggregation of consideration (observation) of the system. Some mixture of autocatalysm and rate-coupling seems to be obligatory.
Examples of possible module-functions

$dx/dt = K\,x\,y$	Logarithmic derivative, Rate-coupling, EXPONENTIAL CHAINS
$dx/dt = K\,x^k y$	Riccati-module, RICCATI - CHAINS
$dx/dt = K\,x\,(B-x)y$	Logistic module
$dx/dt = K\,x^k(B-x^w)^l y$	Hyperlogistic module

Structure Design

We are interested in a decomposition of a complex function into a network based on an agreed basic module and taking into account substructures in the form of chains and cycles.
In this paper we refer on Structure Design based on

- Logarithmic module
- Riccati module

Structure Design with the Logarithmic Derivative

For a given set of ordinary differential equations we determine a structure decomposition applying the following.
Structure Design Rules

1. The logarithmic derivative $F = d\ \ln/dt$ is applied on every intermediate state variable x_i

$$F\,x_i = A_i$$

In general A_i is an arithmetic expression containing signals of type I, which already appeared previously and signals of type II (new signals).

2. For all signals of type I in A_i we construct feedback loops, for all signals of type II we fix names and consider these new signals as additional state-variables (Return to statement 1).

3. The structure design process stops when all of its branches bear arithmetic expressions A_i containing only known signals.

In this process we apply the following reduction rule

If $F\, x_i = A_i = A_{i_1} \pm A_{i_2}$ then we agree $x_{i_1} = A_{i_1}$, $x_{i_2} = A_{i_2}$

If $F\, x_i = A_i = A_{i_1} \dot{\dot{x}} A_{i_2}$ then we agree $x_{i+1} = A_{i_1} \dot{x} A_{i_2}$

By applying these structure design rules the structure design process based on the logarithmic derivative leads always to a system description in the form of the famous

Lotka-Volterra Equations

$F\, x_i = \sum G_{ij} x_j + \sum H_{is} y_s$ Lotka-Volterra network [3]

The area of application of the Structure Design Principle based on the Logarithmic Derivative is determined by the socalled

admissable differential equations

$p^n x$ = arithmetic differential expression

= rational function ($p^j x$, $p^r y$, $p(t)$, $p(x)$, $j < n$)

$p(t)$, $p(x)$ are themselves admissable parameter functions depending on one variable and obeying an admissable differential equation.

As example we consider the Lotka-Volterra network for the EVOLON

For $dx/dt = K\, x^k (B-x^w)^l$

we get the following Lotka-Volterra equations by application of the Structure Design Principle

$$F\, x_0 = K\, x_1 \quad x_0 = x,\ x_1 = x^{k-1}(B-x^w)^l$$

$$F\, x_1 = (k-1)K\, x_1 - w\, l\, K\, x_2 \quad x_2 = x^{k+w-1}\,(B-x^w)^{l-1}$$

$$F\, x_2 = (k+w-1)K\, x_1 - w(l-1)K\, x_2$$

Fig. 3 shows the corresponding structure

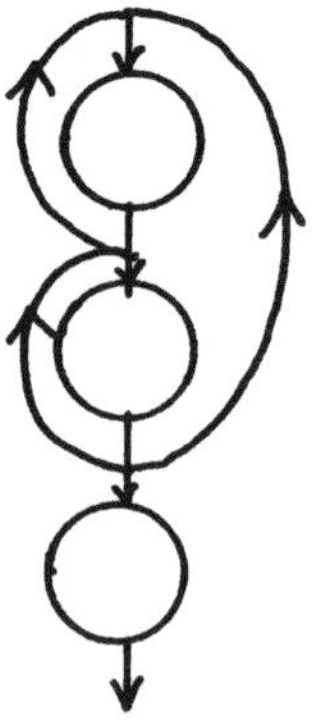

Fig.3. Lotka-Volterra structure of the EVOLON

Riccati - representation of Lotka-Volterra Equations

Under the restriction $x_i \geqslant 0$ we substitute $x_i = K_i \; F \; u_i$ in the autonomous Lotka-Volterra Equations

$$F \; x_i = \sum_j G_{ij} x_j$$

and get then the Riccati-representation

$$du_i / \; dt = k_i \prod_j u_j^{k_{ij}}$$

Of special importance is that Riccati-representation corresponding to the removal of all autocatalytic terms from the Lotka-Volterra Equations.

In this case we have

$$x_j = \frac{-1}{G_{jj}} \; F \; u_j \qquad \text{for} \quad G_{jj} \neq 0$$

$$x_j = F \; u_j \qquad \text{for} \quad G_{jj} = 0$$

Structure Design with the Riccati-module

For modelling systems on a higher aggregated level we are interested in a modified Structure Design Principle based on the Riccati-module instead of the logarithmic basic module.

It turns out that this is possible under the same assumptions as for the Structure Design based on the logarithmic module.

This follows from the following considerations.

For a given set of admissable differential equations we get after the application of the Structure Design Principle based on the Logarithmic Derivative a description by the Lotka-Volterra Equations

$$F \; x_i = \sum_j G_{ij} x_j$$

On the positive cone $x_i \geqslant 0$ the Lotka-Volterra representation can be transformed into the Riccati-representation

$$du_i/dt = k_i \prod_j u_j^{k_{ij}}$$

We rewrite this representation in the following form

$$du_i/dt = k_i u_i^{k_{ii}} \prod_j z_{ij}$$

with the new state-variables

$$z_{ij} = u_j^{k_{ij}} \quad \text{for} \quad j \neq i$$

For these new state-variables we get the following differential equations

$$dz_{ij}/dt = k_{ij} k_j z_{ij}^{1 + (k_{jj}-1)/k_{ij}} \prod_r z_{jr}$$

Altogether we arrive with the following representation in form of a socalled Riccati-network

$$dz_i/dt = K_i z_i^{k_i} \prod_{r \in I_i} z_r$$

This does not mean that in general the Riccati-network has more nodes than the original Lotka-Volterra network.
If we apply the Riccati-module immediately in the design process we get usually a network with a remarkably reduced number of basic modules. The possibility to transform a complex interaction structure into a Riccati representation proofs the universality of Cobb-Douglas production functions concept.

5. Literature

[1] Peschel, M., W. Mende: Leben wir in einer Volterra-Welt? Mathematical Research, vol. 14, Akademie-Verlag, Berlin 1983

[2] Mende, W., Peschel, M.: Strukturentwurf für instationäre und nichtlineare Systeme, msr 10(1982), 581-583

[3] Volterra, V.: Lecons sur la theorie mathematique de la lutte pour la vie, Gauthier-Villars, Paris 1931

[4] Mende, W., M. Peschel: Structure-Building Phenomena in Systems with Power-Product Forces, In: Haken, H.: Chaos and Order in Nature, Springer-Verlag, Berlin 1981, 196-206

[5] Peschel, M., W. Mende: Ecological approach to systems analysis based on Volterra Equations, Cybernetics and Systems, 13(1982), 175-186

[6] Peschel, M., W. Mende, M. Grauer: Qualitative analysis of non-linear systems by the Lotka-Volterra approach, collaborative paper CP-83-60, IIASA 1983

[7] Albrecht, K.F., V. Chetverikov, W. Ebeling, R. Funke, W. Mende, M. Peschel: Random phenomena in nonlinear systems in connection with the Volterra approach In: Schuster, P.: Stochastic Phenomena and Chaotic Behaviour in Complex Systems, Springer-Verlag, Berlin 1984, 250-270

CONTINUOUS LOTKA-VOLTERRA MODELS FOR EVOLUTION PROCESSES

W. Ebeling and R. Feistel[1)]

1. Introduction

Evolution processes are possibly the most complicated processes we observe in nature. Evolution may be defined as an infinite series of selforganization processes [1] . It seems to be at time starting from the well founded theory of selforganization [2,3,4] to develop a theory of evolution processes on this basis. There exist already several attempts pointing in this direction [5-8] .
However it should be underlined that all these approaches may be considered only as a beginning. There is still much to do before we may say that the essence of evolutionary processes is well understood.
The approach presented here is devoted to the development of a new formal model of evolution processes: the continuous Lotka-Volterra model. In this model the evolution of species is considered as a stochastic process on a many-dimensional space where the coordinates denote the phenotypic properties. Over this space a complicated value function with many minima and maxima - like stochastic potentials in disordered semiconductors - is defined which forms an adaptive landscape [6,7,9,10,11]. Evolution is considered as hill-climbing in this rich-structured landscape, which itself depends on the distribution and therefore is time-dependent due to internal and external processes.
This model may be considered as a continuous generalization of the discrete Lotka-Volterra schema with mutations which reads [1,12,13,14]

$$\partial_t x_i(t) = x_i(a_i - b_{ij}x_j) + (a_{ij}x_j - a_{ji}x_i) \qquad (1.1)$$

Here $x_i(t)$ is the density of the population of the species numbered by i, further a_i is tis reproduction rate b_{ij} the community matrix and a_{ij} the mutation matrix. Silently it has been assumed in this traditional approach that the set of species is countable (i = 1,2, ...). Since in reality the number of species is very high the treatment of the model (1.1) leads to quite complicated problems. The alternative model developed by us [11,15-17] works with a continuous set of species represented by a d-dimensional vector space

$$q = \{ q_1 q_2 \dots q_d \} \quad ; \quad q_i \in R \qquad (1.2)$$

---

1) Sektion Physik der Humboldt-Universität, Bereich 04, 1040 Berlin Invalidenstr. 42, DDR

We may associate e.g. q_1 with the weight of the species, q_2 with the height, q_3 with the length etc. and q_{d-2}, q_{d-1}, q_d with the position in space. The continuous generalization of $x_i(t)$ is a density over the state space $x(q,t)$. In the simplest case we may imagine a colony of bacteria distributed with the density $x(q_1,q_2)$ over a two-dimensional Petri dish (d = 2). The generalization of eq. (1.1) is

$$\partial_t x(q,t) = x(q,t)\cdot W(q,t;x)$$

$$+ \int dq'(a(q,q')x(q',t) - a(q',q)x(q,t)) \qquad (1.3)$$

$$W(q;x) = a(q,t) - \int dq' b(q,q')x(q',t) \qquad (1.4)$$

Instead of eq. (1.3) we shall use in the following a diffusion approximation which takes into account only undirected mutation or transport to states near in space

$$\partial_t x(q,t) = x(q,t)\cdot W(q,t;x) + \nabla\cdot D(q)\nabla x(q,t) \qquad (1.5)$$

The mathematical problems connected with the solution of the integro-differential equation (1.5) are of course not simpler than those connected with the solution of the traditional discrete Lotka-Volterra scheme (1.1). We will show however that there are special problems which are more easily solved in the continuous Lotka-Volterra scheme. Some of the advantages of the continuous scheme are connected with:
(i) the possibility to use the concepts of metric spaces as e.g. the distance concept and several results of integral equation theory.
(ii) the close relation to modern physical theories of disordered materials.

The latter point is based on the close analogy of eq. (1.5) to the Schrödinger equation with imaginary time and also to the reaction-diffusion equation which is one of the basic mathematical tools in the theory of selforganization.

2. The Strategies of Evolution

The basic elements of the picture developed so far are two time-dependent (moving) landscape

$$x(q,t) \quad , \quad W(q,t;x(t)) \qquad (2.1)$$

which characterize the whole process (Fig. 1). These landscapes are connected by a feedback loop

$x(q,t) \longrightarrow W(q,t;x)$

$\uparrow$ ← evolution equation ← $\downarrow$

Let us discuss first the time evolution neglecting diffusion i.e.

$$\partial_t x = x W \qquad (2.2)$$

The neutral surfaces $W = 0$ divide the phenotype space into regions where $W > 0$ i.e. the population density increases and regions where $W < 0$ i.e. the population density decreases (Fig. 2)

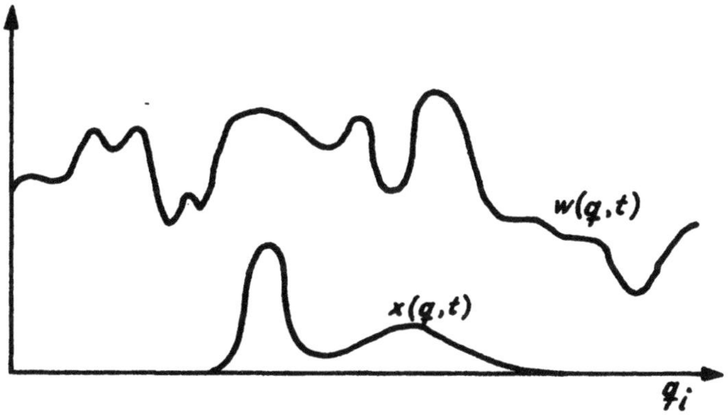

Fig. 1 One-dimensional cross-section of the occupation landscape $x(q,t)$ and the valuation landscape $W(q,t)$.

The decomposition of the area into islands may be studied by means of percolation theory [18] . A similar role as the islands of increasing population play the islands of maximal population density which may be approximated by means of Gaussian functions centered around certain maxima q^i i.e. the density is given by

$$x(q,t) = \sum_i x_i(t) \exp\left(-(q-q^i)^2/2\cdot\sigma_i^2\right) \qquad (2.3)$$

The time behaviour of the maxima which correspond to the "species" is given by [11]

$$d_t q^i(t) = \langle qW\rangle_i - \langle q\rangle_i \cdot \langle W\rangle_i \approx \sigma_i^2 \, \partial_q W(q,t)_{q=q^i} \qquad (2.4)$$

which is nothing else than Fisher's law. This law expresses the tendency that evolution is hill-climbing. Therefore hill-climbing is the first and the main strategy of evolution.

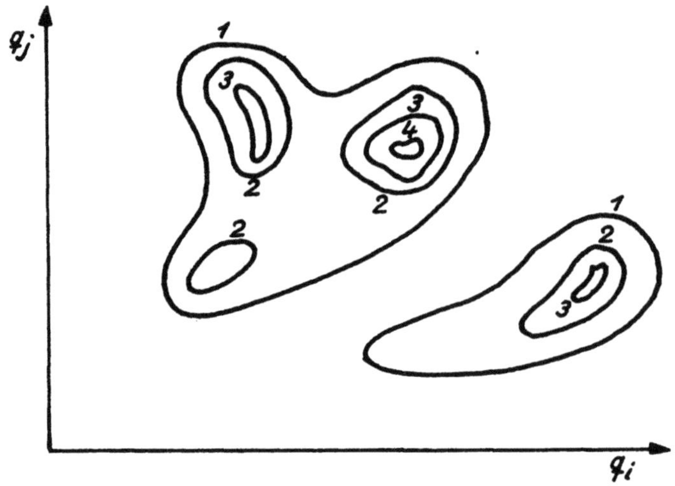

Fig. 2. The islands of increasing population density and their (deterministic) change with increasing time ($t_1 < t_2 < t_3 < t_4$).

Now we shall discuss other more subtly strategies of evolution which are connected with the interaction of the species. The most efficient way to proceed is to use Hilbert-Schmidt expansions of the interaction kernel.

$$b(q,q') = \sum_{k=1}^{\infty} \mu_k^{-1} \; \bar{\Phi}_k(q) \; \bar{\Psi}_k(q') \tag{2.5}$$

where the two sets of eigen functions are defined by

$$\mu_k \int b(q,q') \, \bar{\Phi}_k(q')dq' = \bar{\Psi}_k(q) \tag{2.6}$$

$$\mu_k \int b(q',q) \, \bar{\Psi}_k(q')dq' = \bar{\Phi}_k(q)$$

In this way we get for the growth function

$$W(q,t;x) = a(q,t) - \sum \mu_k^{-1} \; \bar{\Phi}_k(q)z_k(t) \tag{2.7}$$

$$z_k(t) = \int \bar{\Psi}_k(q')x(q',t) \, dq' \tag{2.8}$$

Each term in the expansion of the growth function corresponds to an eigen strategy. Since in evolutionary systems the ressources are always finite it seems to be natural to assume that the number of existing strategies is also finite. In other words we shall assume that the community kernel is degenerated i.e.

$$W(q,t;x) = a(q,t) - \sum_{k=1}^{m} \mu_k^{-1} \bar{\Phi}_k(q) z_k(t) \tag{2.9}$$

Here m is the number of independent niches and the predator-prey strategies in the niche i is characterized by the functions

$$\bar{\Phi}_i(q) \quad , \quad \bar{\Psi}_i(q)$$

In the stationary state $W = 0$ the distribution between the niches is given by the solutions of the linear equations

$$\sum_{l=1}^{m} (\mathcal{B}_{kl}/\mu_l) z_l^0 = A_k \tag{2.10}$$

$$\mathcal{B}_{kl} = \int \bar{\Phi}_k \, \bar{\Psi}_l dq \quad , \quad A_k = \int a(q) \, \bar{\Psi}_k(q) dq$$

Let us assume for simplicity that the eigen functions $\bar{\Phi}_k(q)$ are well separated i.e. the niches are nonoverlapping (Fig. 3). In the niche k the reproduction function may be approximated by

$$\begin{aligned} W(q,t);x) &= a(q,t) - b(t) \\ b(t) &= (\bar{\Phi}_k / \mu_k) z_k(t) \\ \bar{\Phi}_k &= \max \bar{\Phi}_k(q) \end{aligned} \tag{2.11}$$

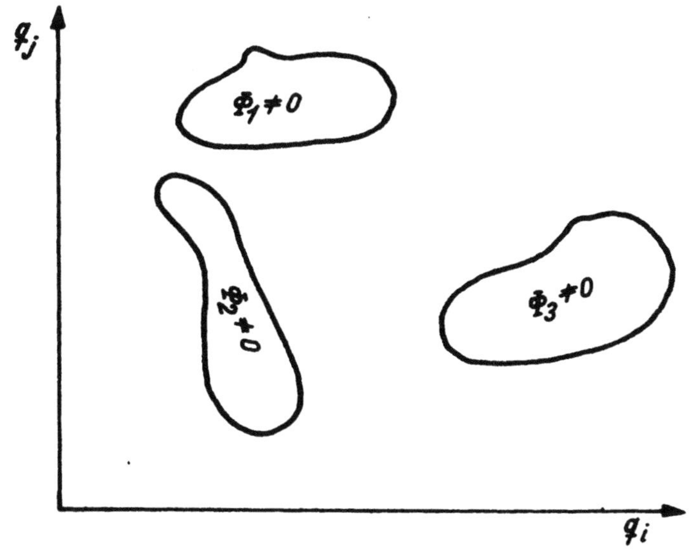

Fig. 3 The separation of niches in the phenotype space

3. The Evolution in one Niche

Including diffusion (mutation) the evolution in one of the niches may be described by

$$\partial_t x(q,t) = x(q,t)\ (a(q) - b(t)) + D\Delta\ x \tag{3.1}$$

with the ansatz

$$x(q,t) = y(q,t)\ \exp\ (- \int_0^t dt' b(t')) \tag{3.2}$$

we get the Schrödinger-type equation (with imaginery time)

$$\partial_t y(q,t) = a(q)y + D\Delta\ y \tag{3.3}$$

Defining an eigen function system by

$$(\hbar^2/2m)\ \Delta \psi_n + (\varepsilon_n - U(q))\ \psi_n = 0 \tag{3.4}$$

$$U(q) = -a(q)\quad ,\quad D = \hbar^2/2m$$

The solution may be represented by the expansion

$$y(q,t) = \sum c_n \exp\ (-\ \varepsilon_n t)\ \ \psi_n(q) \tag{3.5}$$

In the following we shall assume that the reproduction rate a(q) has a very complicated shape with many maxima and minima (see Fig. 1). the spectrum will contain a discrete part corresponding to localized states an a continuous part corresponding to extended states

In the developed island regime which will occur after some time of evolution mainly the localized states will be populated. We may assume that each of them has the approximately shape

$$\psi_n(q) \sim \exp \left\{ -\ |q-q^n|\ /l(\varepsilon_n) \right\} \tag{3.6}$$

centered around q^n. Then the mean time necessary for a transition from a maximum near to q^n corresponding to ε_n to a maximum near to q^m corresponding to ε_m is given by [17, 19]:

$$t(q^m \rightarrow q^n) \sim |q^n - q^m|\ /(l(\varepsilon_n)\ |\varepsilon_n - \varepsilon_m|\ ) \tag{3.7}$$

A direct consequence of this expression is that the evolution in one niche has hopping character. The system is hopping from populated hills of a(q) (minima of U(q)) to other hills at finite distance which correspond to lower eigen values [19] .

4. Conclusions

In the model presented here the main idea is to replace the ordinary discrete Lotka-Volterra scheme by a continuous one, turning from the vector space of population numbers to a function space of population density. Darwinian selection appears now as a collapse process, tending to localize the species in the space of possible phenotypic properties. As a counterpart, mutations (diffusion) spread the distributions; in combination of both evolution is displayed as a motion of the species through the phenotype space. This drift is governed by a self-consisting valuation landscape, i.e. the species move hill-climbing in the landscape which depends itself on the distribution of species. Where the landscape changes softly in space, evolution is a continuous drift following Fisher's law, but where it is very cleft, a hopping character of phenotypic change dominates, i.e. an "evolution quantum" becomes visible. The way the species influence and deform the fitness landscape can be decomposed uniquely into a set of orthogonal real functions, each in one-to-one correspondence to an ecological niche as well as to a species interaction strategy, called here eigenstrategy. Discovering and occupying these discrete strategies yields again a hopping process, now on a qualitatively higher level.

References

[1] Ebeling, W., R. Feistel: Theorie der Selbstorganisation und Evolution. Akademie-Verlag Berlin 1982

[2] Romanovsky, Yu.M. et al.: Mathematical Modelling in Biophysics (in Russian). Nauka Moscow 1975

[3] Nicolis, G., I. Prigogine: Selforganization in Nonequilibrium Systems. Wiley New York 1977

[4] Haken, H.: Synergetics - An Introduction. Springer Berlin u.a. 1978; Advanced Synergetics Springer Berlin u.a. 1983.

[5] Eigen, M., P. Schuster: Naturwiss. 64 (1977) 541;

[6] Conrad, M.: Biosystems 11 (1979) 167; Adaptability. The Significance of Variability from Molecule to Ecosystem. Plenum Press New York 1983

[7] Bremermann, H.J.: J. theor. Biol. 87 (1980) 671

[8] Anderson, P.W.: Proc. Nat. Acad. Sci. USA 80 (1983) 3386

[9] Rechenberg, I.: Evolutionsstrategie. Fromman-Verlag Stuttgart 1973

[10] Landauer, R.: Helv. Phys. Acta 56 (1983) 847

[11] Feistel, R., W. Ebeling: Biosystems 15 (1982) 291-299, Stochastic Models of Evolutionary Processes. In: A.I. Zotin (Ed.) Thermodynamics and Regulation in Biological Processes. Nauka Moscow 1984

[12] Svireshev, Yu.M., D.O. Logofet: Stability of Biological Communities (in Russian). Nauka Moscow 1978.

[13] Peschel, M., W. Mende: Leben wir in einer Volterra-Welt? Akademie-

Verlag Berlin 1983
[14] Allen, P., W. Ebeling: Biosystems 16 (1983) 113
[15] Ebeling, W., R. Feistel: Proc. Conf. Autowaves in Biology, Chemistry and Physics (Puscino 1983). Springer Series in Synergetics Springer-Verlag Berlin u.a. 1984
[16] Esser, B. et al.: Localization Phenomena in the Theory of Evolution Processes. Teubner Texte für Physik. Teubner Leipzig 1984
[17] Ebeling, W., A. Engel, B. Esser, R. Feistel: J. Stat. Phys., in press
[18] Zeldovich, Ya. B.: Dokl. Akad. Nauk 270 (1983) 1369
[19] Engel, A: Discretness in a Continuous Model of Biological Evolution. Contribution to this Volume.

THE MAXIMUM PRINCIPLE FOR REPLICATOR EQUATIONS

Karl Sigmund [1)]

1. Introduction

The notion of adaptive landscape is a familiar one in theoretical biology. Evolution is often pictured as an uphill movement leading to higher and higher fitness under the driving force of selection. We mention Wright [19] and Simpson [18] for explicit descriptions of the concepts of adaptive genotypic resp. phenotypic landscapes. The general idea of evolution as progressive optimization is so pervasive, however, that it is difficult to give a precise account of its origin.

The sternest version of an uphill movement is that of steepest ascent. In this case, the adaptive landscape itself determines the path, and the dynamics is given by the gradient of the slope. Gradient systems are well-behaved: in fact, they are often too tame for a realistic description of the antics of biological evolution. Even if random drift and other stochastic influences are excluded, the effects of time-dependence, frequency dependence, developmental constraints, genetic linkage, coevolutionary interactions etc. will lead to phenomena incompatible with the existence of a potential function. Nevertheless, gradient systems play an important conceptual role in basic dynamical models, both in micro- and in macro-evolution (cf. Akin [1] and Lande [11]). They lead to highly suggestive extremum principles and provide a link between the methods of population genetics and mathematical physics.

In this paper we shall consider gradient systems within the framework of replicator equations. Such equations model a rich diversity of phenotypic and genotypic evolution. The state of the system is described by relative frequencies within a population, and hence by a point on the unit simplex. The basic idea of Shahshahani [17] was to replace the Euclidean metric by a Riemannian metric. The gradient systems with respect to this metric occur as important examples of replicator equations, both in classical population genetics and in the chemical kinetics of polynucleotide replication. The corresponding maximum principles have been stated by Kimura [8] and Küppers [9]. Their rigorous proof was an immediate consequence of Shahshahani's introduction of the appropriate metric.

2. Replicator equations

Let

$$S_n = \{\underline{x} = (x_1, \ldots, x_n) \in R^n : \Sigma x_i = 1,\ x_i \geq 0\}$$

1) Institut für Mathematik der Universität Wien, Strudlhofgasse 4, A-1090 Wien, Österreich

denote the unit simplex and let $\underline{f}$ be a vector field defined in a neighborhood of S_n. We associate with $\underline{f}$ the vector field $\hat{\underline{f}}$ on S_n with coordinates

$$\hat{f}_i(\underline{x}) = x_i(f_i(\underline{x}) - \Phi(\underline{x})) \quad i = 1,\dots,n \tag{1}$$

where

$$\Phi(\underline{x}) = \Sigma x_i \; f_i(\underline{x}) \tag{2}$$

A differential equation of the type

$$\dot{\underline{x}} = \hat{\underline{f}}(\underline{x}) \tag{3}$$

is called a replicator equation. Two simple properties are easily checked:

(a) The unit simplex S_n, as well as its faces, are invariant under (3);

(b) If $\underline{f}$ and $\underline{g}$ are two vectorfields which are equivalent, then $\hat{f} = \hat{g}$ on S_n. Here, $\underline{f} \sim \underline{g}$ if $f_i(\underline{x}) - g_i(\underline{x})$ is independent of i for all $\underline{x} \in S_n$.

Replicator equations are very common in mathematical biology. They describe the action of selection on many different levels of biological organization. We refer to Schuster and Sigmund [16] and Hofbauer and Sigmund [6] for a short resp. long survey and only point out two of the simplest examples.

(a) If $\underline{f}$ is constant, then (3) becomes

$$\dot{x}_i = x_i(a_i - \Phi) \tag{4}$$

This equation describes the evolution of gene frequencies for frequency-independent asexual reproduction, and in particular the relative concentrations of selfreproducing macromolecules in the absence of mutations and chemical interactions (see Eigen and Schuster.[3] and Küppers [10].

(b) If $\underline{f}$ is linear, then (3) becomes

$$\dot{x}_i = x_i(\Sigma \; a_{ij}x_j - \Phi) \tag{5}$$

This equation occurs in at least four different fields of evolutionary biology. First of all, it is equivalent to the general Volterra-Lotka equation in mathematical ecology

$$\dot{y}_i = y_i(b_{in} + \sum_{j=1}^{n-1} b_{ij} \; y_j) \quad i = 1,\dots,n-1 \tag{6}$$

with $b_{ij} = a_{ij} - a_{nj}$. This equivalence (obtained by setting $y_i = \frac{x_i}{x_n}$) has been pointed out by Hofbauer [5]. Secondly, (5) describes the evolution of gene frequencies for asexual reproduction, if the fitnesses are (linearly) dependent, and of phenotype frequencies in game theoretic models of animal behaviour (cf. Maynard Smith [14]). Thirdly, it plays an important role in the chemical kinetics of catalytically interacting polynucleotides, hypercycles etc. (see Eigen and Schuster [3]). Finally, (5) describes the action of selection in a one locus viability model, under the assumption of Hardy-Weinberg equilibrium (cf. Hadeler [4]). In this case,

x_i is the frequency of allele A_i in the gene pool and a_{ij} is the probability for survival, from zygote to adult age, of the genotype A_iA_j. This yields a special case of (5), namely

$$\dot{x}_i = x_i(\Sigma\, a_{ij}\, x_j - \Phi) \quad \text{with} \quad a_{ij} = a_{ji} \tag{7}$$

(7) is the so-called selection equation of Fisher-Haldane-Wright.

It is easy to show that for both (4) and (7), the "average fitness" Φ is always increasing (see [4]). Kimura [8] claimed that the orbits of (7) always point in the direction of maximal increase of Φ, and Küppers [9] stated that this same property of "steepest ascent" holds for (4).

At first glance, this seems to be wrong. Indeed, maximal increase implies that the direction of the orbits is orthogonal to the constant level sets of Φ (as every hiker intuitively knows). This is not the case in general.

It turns out, however, that with another notion of orthogonality, the orbits do cross the constant level sets at right angles. Thus the maximum principles become valid if one modifies the notion of inner product (see Shahshahani [17] and Akin [1]).

3. Shahshahani gradients

The relevant state space for replicator equations is S_n. One is interested therefore in angles between vectors belonging to T_pS_n, the tangent space to S_n at the point $\underline{p} \in \text{int } S_n$: these vectors are characterized by the property that the sum of their components is 0. For two vectors $\underline{x}$ and $\underline{y}$ in $T_{\underline{p}}S_n$, we define, following Jacquard [7] and Shahshahani [17],

$$\langle \underline{x}, \underline{y} \rangle_{\underline{p}} = \sum_{i=1}^{n} \frac{1}{p_i} x_i y_i \tag{8}$$

and check that this is indeed an inner product. It differs from the "usual" Euclidean inner product

$$\langle \underline{x}, \underline{y} \rangle = \sum_{i=1}^{n} x_i y_i \tag{9}$$

by the factors $\frac{1}{p_i}$. The i-th term in the sum gains in importance if p_i is small. (8) leads to a notion of orthogonality which depends on $\underline{p}$, and induces a distance which differs from the Euclidean one by attaching more weight to changes which occur near the boundary of S_n. (We refer to Akin [1] for details).

Let V be a differentiable function from some neighborhood U of S_n (in $\mathbb{R}^n$) into $\mathbb{R}$. For each $\underline{p} \in \text{int } S_n$, the derivative $DV(\underline{p})$ is a linear map from the tangent space into $\mathbb{R}$. There exists a unique vector grad $V(\underline{p})$ such that

$$\langle \text{grad } V(\underline{p}), \underline{y} \rangle = DV(\underline{p})(\underline{y}) \tag{10}$$

holds for all $\underline{y} \in T_p\mathbb{R}^n$. This "Euclidean" gradient grad $V(\underline{p})$ has components

$\frac{\partial V}{\partial x_i}(\underline{p})$. Similarly, there is a unique vector Grad $V(\underline{p})$ such that

$$\langle \text{Grad } V(\underline{p}), \underline{y} \rangle_{\underline{p}} = DV(\underline{p})(\underline{y}) \tag{11}$$

holds for all $\underline{y} \in T_{\underline{p}}S_n$. This vector is called the Shahshahani-gradient of V.

4. Replicator equations and Shahshahani gradients.

It is easy to characterize those replicator equations (3) which are Shahshahani gradients:

<u>Theorem</u>: $\hat{\underline{f}}$ = Grad V iff $\underline{f} \sim$ grad V

Indeed, suppose that $\underline{f} \sim$ grad V. We know that $\underline{g} \sim \underline{h}$ implies $\hat{\underline{g}} = \hat{\underline{h}}$ on S_n. We may therefore assume, without restricting generality, that $\underline{f}$ = grad V. For $\underline{y} \in T_{\underline{p}}S_n$, one gets

$$\langle \hat{\underline{f}}(\underline{p}), \underline{y} \rangle_{\underline{p}} = \Sigma \frac{1}{p_i} p_i(f_i - \Phi)y_i = \Sigma f_i y_i - \Phi \Sigma y_i = \Sigma f_i y_i$$

since $\Sigma y_i = 0$. Thus

$$\langle \hat{\underline{f}}(\underline{p}), \underline{y} \rangle_{\underline{p}} = \Sigma \frac{\partial V}{\partial x_i} y_i = DV(\underline{p})(\underline{y}) \tag{12}$$

Hence, by (11), $\hat{f}$ = Grad V.

If conversely $\hat{\underline{f}}$ = Grad V, then (12) implies

$$\Sigma f_i y_i = \Sigma \frac{\partial V}{\partial x_i} y_i$$

for all $\underline{y} \in T_{\underline{p}}S_n$. With $y_i = 1$, $y_n = -1$ and $y_j = 0$ for all $j \neq i,n$, this implies

$$\frac{\partial V}{\partial x_i}(\underline{p}) - \frac{\partial V}{\partial x_n}(\underline{p}) = f_i(\underline{p}) - f_n(\underline{p}).$$

It follows that

$$\frac{\partial V}{\partial x_i}(\underline{p}) - f_i(\underline{p})$$

does not depend on i, and hence that $\underline{f} \sim$ grad V.

Thus if $\underline{f}$ is an Euclidean gradient, i.e. if

$$\frac{\partial f_i}{\partial x_j} = \frac{\partial f_j}{\partial x_i}$$

holds (for all i and j) in some small neighborhood U of S_n which is simply connected, then the orbits of $\dot{\underline{x}} = \hat{\underline{f}}(\underline{x})$ are orthogonal, in Shahshahani's sense, to the constant level sets of V (in S_n).

5. Homogeneous potentials

If the potential function V is a homogeneous function of degree $s>0$, i.e. if

$$V(\alpha x_1,\dots,\alpha x_n) = \alpha^s\, V(x_1,\dots,x_n) \tag{13}$$

holds for all $\alpha \in R$, then the "average fitness" Φ satisfies (by Euler's theorem)

$$\Phi(\underline{p}) = \Sigma\, p_i f_i(\underline{p}) = \Sigma\, p_i \frac{\partial V}{\partial x_i}(\underline{p}) = s\, V(\underline{p}).$$

Hence Φ increases at a maximal rate, in the sense of Shahshahani. The rate of increase is

$$\dot{\Phi}(\underline{x}) = s\, \dot{V}(\underline{x}) = s\, \Sigma \frac{\partial V}{\partial x_i}\dot{x}_i = s \sum_i \frac{\partial V}{\partial x_i} x_i\Big(\frac{\partial V}{\partial x_i} - \Sigma\, x_j \frac{\partial V}{\partial x_j}\Big)$$

$$= s\Big[\sum_i x_i (f_i(\underline{x}))^2 - (\Sigma\, x_i f_i(\underline{x}))^2\Big] \geq 0 \tag{14}$$

The rate of increase can be viewed (up to the factor s) as variance of a random variable taking the value $f_i(\underline{x})$ with probability x_i, $i=1,\dots,n$.

If, for example, $\underline{f} = \text{grad}\, V$ with

$$V(\underline{x}) = \Sigma\, a_i x_i \tag{15}$$

then (3) becomes (4) and one obtains the (modified) maximum principle by Küppers. If

$$V(\underline{x}) = \frac{1}{2} \sum_{ij} a_{ij} x_i x_j \tag{16}$$

then

$$\frac{\partial V}{\partial x_i}(\underline{x}) = \frac{1}{2} \sum_j (a_{ij}+a_{ji}) x_j$$

If in particular $a_{ij} = a_{ji}$, then (3) becomes (7) and one obtains the (modified) maximum principle by Kimura. In this case (14) is just Fisher's Fundamental Theorem of Natural Selection (see, e.g. Hadeler [4]): the rate of increase of the average fitness is proportional to the variance of the fitness in the gene pool.

It is obvious that the orbits of any gradient system converge to the set of fixed points. Does every orbit converge to an equilibrium? This need not always be the case, as Takens has shown. But for (4), it is obviously true. For (7), it is also valid, but demands an elaborate proof (see Akin and Hofbauer [2], and Losert and Akin [12]). It would be interesting to know whether for any Shahshahani-gradient system with homogeneous potential function V, every orbit converges to an equilibrium.

If V is not homogeneous, then Φ need not always increase. For example $V = x_1^2 + x_2$ has its minimum at $x_1 = \frac{1}{2}$, but the corresponding $\Phi = 2x_1^2 + x_2$ at $x_1 = \frac{1}{4}$. For a state with x_1 between $\frac{1}{4}$ and $\frac{1}{2}$, the average fitness decreases.

6. First order replicator equations and Shahshahani gradients.

Let us now characterize those linear replicator equations (5) which are Shahshahani-gradients

Theorem (5) is a Shahshahani gradient iff

$$a_{ij} + a_{jk} + a_{ki} = a_{ji} + a_{ik} + a_{kj} \tag{17}$$

holds for all i,j,k between 1 and n.

Condition (17) states that the sum of the coefficients of the matrix $A = (a_{ij})$ over all three-cycles $i \to j \to k \to i$ of indices is independent of the orientation. The same holds, then, for all p-cycles, $p > 3$, as one can see by "triangulation".

Indeed, if (17) holds, one has only to set $c_k = a_{kn} - a_{nk}$ to see that

$$b_{ij} = a_{ij} + c_j$$

satisfies $b_{ij} = b_{ji}$. The equation

$$\dot{x}_i = x_i(\Sigma\, b_{ij}x_j - \Phi) \tag{18}$$

is therefore a Shahshahani gradient, and so is (5), since it coincides on S_n with (18).

Conversely, if $\hat{\underline{f}} = \text{Grad } V$, then $\underline{f} \sim \text{grad } V$, i.e. there is some function Ψ such that

$$f_i(\underline{p}) - \frac{\partial V}{\partial x_i}(\underline{p}) = \Psi(\underline{p})$$

holds for all i (and all $\underline{p} \in S_n$). From this follows

$$\frac{\partial f_i}{\partial x_j}(\underline{p}) - \frac{\partial f_j}{\partial x_i}(\underline{p}) = \frac{\partial \Psi}{\partial x_j}(\underline{p}) - \frac{\partial \Psi}{\partial x_i}(\underline{p}) \tag{19}$$

If $f_i = \sum_i a_{ij}x_j$ then $\frac{\partial f_i}{\partial x_j} = a_{ij}$. Thus

$$a_{ij} - a_{ji} = \frac{\partial \Psi}{\partial x_j}(\underline{p}) - \frac{\partial \Psi}{\partial x_i}(\underline{p}) \tag{20}$$

From this, (17) follows immediately.

Let us call two $n \times n$ matrices A and B equivalent ($A \sim B$) if there exist constants c_j s.t. $a_{ij} = b_{ij} + c_j$ for all i and j between 1 and n. A and B are equivalent iff the functions $\underline{x} \to A\underline{x}$ and $\underline{x} \to B\underline{x}$ are equivalent in the sense of section 2. The theorem implies that (5) is a Shahshahani gradient iff one of the following conditions is satisfied:

(a) there is a symmetric matrix within the equivalence class of the matrix A;

(b) there exist constants c_i such that $a_{ij} - a_{ji} = c_i - c_j$ for all i and j.

7. Further remarks

A) For (5), Φ is homogeneous. From this follows that Φ increases at a maximal rate, in Shahshahani's sense, iff A is symmetric.

On the other hand it is easy to check that Φ is an invariant of motion for (5) (i.e. constant along every orbit) iff for each pair (i,j) of indices,

$$\text{(a)}\ a_{ij}+a_{ji} = 2a_{ii} = 2a_{jj} \quad \text{or} \quad \text{(b)}\ a_{ij}-a_{jj} = a_{ji}-a_{ii} = 0 \tag{21}$$

If (5) is interpreted as the dynamics of a game (see, e.g. Schuster et al [15]), this means that the average payoff increases at a maximal rate iff the game is a partnership game, and is constant iff every 2×2 submatrix either corresponds to a constant sum game or is equivalent to the zero matrix.

It would be interesting to characterize these equations (5) for which Φ is monotonically increasing along every orbit.

B) Game dynamics between two populations leads to equations of the type

$$\begin{aligned} \dot{x}_i &= x_i\,\Big(\sum_{j=1}^{n} a_{ij}x_j + \sum_{j=1}^{m} b_{ij}y_j - \Phi\Big) \qquad i = 1,\dots,n \\ \dot{y}_i &= y_j\,\Big(\sum_{i=1}^{n} c_{ji}x_i + \sum_{j=1}^{m} d_{ji}y_i - \Psi\Big) \qquad j = 1,\dots,m \end{aligned} \tag{22}$$

with

$$\Phi = \sum_{ij} a_{ij}x_ix_j + \sum_{ij} b_{ij}x_iy_j \qquad \Psi = \sum_{ij} c_{ji}y_jx_i + \sum_{ij} d_{ji}y_jy_i \tag{23}$$

(see, e.g., Schuster et al [15]). This equation lives on the product space $S_n \times S_m$ of two simplices. One may introduce in an obvious way a Shahshahani-type inner product in the corresponding tangent spaces. Equation (22) is a gradient system with respect to this metric if the matrices (a_{ij}) and (d_{ji}) satisfy (17) and if there exist constants c_i and d_j such that with $g_{ij} = c_{ij} - b_{ji}$

$$g_{ij} = c_i - d_j \tag{24}$$

for $i=1,\dots,n$ and $j=1,\dots,m$. This the case iff

$$g_{sj} + g_{ti} = g_{tj} + g_{si} \tag{25}$$

holds for all indices i,j,s and t. (Condition (17) means that $g_{ij} = a_{ij} - a_{ji}$ satisfies (25)).

C) As mentioned in section 2, the first order replicator equation (5) is equivalent to the Volterra-Lotka-equation (6). The mapping

$$(x_1,\dots,y_1) \to (y_1,\dots,y_{n-1})$$

transforms the Shahshahani inner product on S_n into an inner product on R_+^{n-1}. For $\underline{q} \in \operatorname{int} R_+^{n-1}$ and two vectors Y and Z in $T_{\underline{q}} R_+^{n-1}$, this yields

$$\langle Y,Z\rangle_{\underline{q}}' = \sum_{i=1}^{n-1} \frac{1}{q_i} Y_i Z_i - \left(\sum_{i=1}^{n-1} Y_i\right)\left(\sum_{i=1}^{n-1} Z_j\right)$$

A more natural inner product would be

$$\langle Y,Z\rangle_{\underline{q}} = \sum_{i=1}^{n-1} \frac{1}{q_i} Y_i Z_i \tag{26}$$

With this metric,

$$\dot{y}_i = y_i f_i(y_1,\ldots,y_{n-1}) \qquad i = 1,\ldots,n-1$$

is a gradient iff

$$\dot{y}_i = f_i(y_1 \ldots y_{n-1})$$

is a gradient with respect to the Euclidean metric. In particular, the Volterra-Lotka equation (6) is a gradient system with respect to the metric defined by (26) iff $b_{ij} = b_{ji}$ for $1 \leq i,j \leq n-1$. Volterra-Lotka equations of this type have been investigated in Mac Arthur [13].

Literature

[1] Akin, E.: The geometry of population genetics. Lecture notes in biomathematics, vol. 31. Springer, Berlin-Heidelber-New York 1979.

[2] Akin, E. and Hofbauer, J.: Recurrence of the unfit. Math. Biosci. 61 (1982), 51-63.

[3] Eigen, M. and Schuster, P.: The hypercycle: A principle of natural selforganization, Springer Verlag, Berlin-Heidelberg 1978.

[4] Hadeler, K.P.: Mathematik für Biologen. Springer Verlag, Berlin-Heidelberg-New York 1974.

[5] Hofbauer, J.: On the recurrence of limit cycles in the Volterra-Lotka equation, Nonlinear Analysis, TMA 5 (1981), 1003-1007.

[6] Hofbauer, J. and Sigmund, K.: Evolutionstheorie und dynamische Systeme, Parey, Hamburg-Berlin 1984.

[7] Jacquard, A.: The genetic structure of populations, Springer Verlag, Berlin 1974.

[8] Kimura, M.: On the change of population fitness by natural selection, Heredity 12 (1958), 145-167.

[9] Küppers, B.: Some remarks on the dynamics of molecular selforganization, Bull. Math. Biol. 41 (1979), 803-809.

[10] Küppers, B.O.: Molecular theory of evolution, Springer Verlag, Berlin 1983.

[11] Lande, R.: Natural selection and random genetic drift in phenotypic evolution, Evolution 30 (1976), 314-334.

[12] Losert, V. and Akin, E.: Dynamics of Games and Genes. J. Math. Biol. 17 (1983), 241-251.

[13] Mac Arthur, R.: Species packing and competitive equilibrium for many species, Theor. Pop. Biol. 1 (1970), 1-11.

[14] Maynard Smith, J.: Evolutionary game theory. Cambridge U.P. 1982.

[15] Schuster, P., Sigmund, K., Hofbauer, J. and Wolff, R.: Selfregulation of behaviour in animal societies. Biol. Cybern. 40 (1981), 1-25.

[16] Schuster, P. and Sigmund, K.: Replicator dynamics. J. Theor. Biol. 100 (1983), 333-338.

[17] Shahshahani, S.: A new mathematical framework for the study of linkage and selection. Memoirs AMS 211, 1979.

[18] Simpson, G.G.: The major features of evolution. Columbia U.P., New York 1953.

[19] Wright, S.: Evolution in Mendelian population. Genetics 16 (1931), 97-159.

BILINEAR AND HAMILTONIAN SYSTEMS, MACRODYNOMICS

A.M. Molchanov [1)]

Multiparticle systems,

$$\frac{dx_i}{dt} = a(x_i) + \sum_{k=1}^{k=N} b(x_i, x_k), \quad (1)$$

with number N,

$$N \gg 1 \quad (2)$$

of components x_i and not small interactions $b(x,y)$ is very important in physics and biology. Its study is a very difficult one, because computer is useless (if $N \sim 10^6$ for example) and the common perturbation theory is not valid, because interaction term $b(x,y)$ is not small.

Luckily, there is an important class of systems, which admits detailed investigation. These are the systems with linear $a(x)$ and bilinear $b(x,y)$.

Introducing a new variable z,

$$z = x_1 + \dots + x_N \quad (3)$$

we obtain a much more simple system

$$\frac{dx_i}{dt} = a(x_i) + b(x_i, z) \quad (4)$$

The sum of all equations gives an equations for z,

$$\frac{dz}{dt} = a(z) + b(z,z) \quad (5)$$

The difference between the equations for x_i reduces to different initial conditions.

Therefore we obtain the Macrodynamical System

$$\begin{aligned} \frac{dz}{dt} &= a(z) + b(z,z) \\ \frac{dx}{dt} &= a(x) + b(x,z) \end{aligned} \quad (6)$$

---

1) Research Computer Centre USSR Academy of Sciences Pushchino Moscow Region

The particular case of the stationary point,

$$\frac{dz}{dt} = 0, \qquad z = z_0$$

gives the very important notion of the Thermodynamical limit.

We can now interprete the phase portrait of the equations for x. It is the macrodynamical limit of the original system, because every trajectory of this system generates many ($N \gg 1$) different trajectories of the second equation of the Macrodynamical System.

<u>Generalisation</u>

The notion of the summatory function

$$F(x) = f(x_1) + \ldots + f(x_N) \tag{7}$$

permits an important generalisation of the macrodynamical approach.

Let us introduce a few functions of x:

$$f_1(x), \ldots, f_k(x) \tag{8}$$

and suppose that they satisfy the structural relations,

$$\frac{df_\alpha}{dx} I \left(\frac{df_\beta}{dx} \right)^* = R^\gamma_{\alpha\beta} f_\gamma(x) \tag{9}$$

where I obeys

$$I^2 = - E \tag{10}$$

Let H,

$$H = H(F_1, \ldots, F_k), \tag{11}$$

be an arbitrary function of its arguments.

Construct the Hamiltonian system with Hamiltonian

$$\mathcal{H}(x_1, \ldots, x_N) = H(F_1, \ldots, F_k) \tag{12}$$

$$\frac{dx_i}{dt} = I \left(\frac{\partial \mathcal{H}}{\partial x_i} \right)^* \tag{13}$$

From (13) follow equations for F_α,

$$\frac{dF_\alpha}{dt} = \sum \frac{df_\alpha}{dt} \frac{dx_i}{dt} = \sum \frac{df_\alpha}{dx_i} I \left(\frac{\partial \mathcal{H}}{\partial x_i} \right)^* = \left\{ \sum \frac{df_\alpha}{dx_i} I \left(\frac{df_\beta}{dx_i} \right)^* \right\} \left(\frac{\partial H}{\partial F_\beta} \right)^*$$

Structural relations permit to express the right parts in terms of $\{F_\alpha\}$ only:

$$\frac{dF_\alpha}{dt} = R^{\gamma}_{\alpha\beta} F_\gamma \left(\frac{\partial H}{\partial F_\beta}\right)^{*} \tag{14}$$

It is a principial distinction between (13) and (14). Last one don't depend on N and is the Macrodynamical equation for the multi-particle system (13).

Macrodynamical approach permits to formulate the important hypothese:

<u>Bifurcation processes in the macrodynamical equation correspond to the general phase transitions in the multiparticle system.</u>

In supplement is demonstrated how to enclose an arbitrary system (with polynomial right parts) into a system with square right parts.

<u>Conclusion</u>

Systems with square right parts and especially Lotka-Volterra systems can be regarded as macrodynomical equations for multiparticle systems.

БИЛИНЕЙНЫЕ И ГАМИЛЬТОНОВЫ СИСТЕМЫ. МАКРОДИНАМИКА.

А.М.Молчанов

Предложены системы, допускающие точный предельный подход / $N \to \infty$ / при увеличении числа частиц без предположения о малости взаимодействия.

п°1. Парное взаимодействие

Динамика системы одинаковых компонент описывается системой дифференциальных уравнений вида:

$$\frac{dx_i}{dt} = a(x_i) + \sum_{k=1}^{N} b(x_i, x_k) \qquad /1/$$

Вектор-функция $a(x)$ задает "внешнее поле", а вектор-функция $b(x,y)$ "воздействие" компоненты на компоненту /"частицу"/ x.

Для задач естествознания характерны большие значения N,

$$N \gg 1 \qquad /2/$$

и существенная роль взаимодействия $b(x,y)$.

Первое обстоятельство делает неперспективным чисто вычислительный /"компьютерный"/ подход, а второе приводит к неприменимости обычных методов теории возмущений /типа "разложения по параметру связи"/.

п°2. Билинейные системы

Существует, к счастью, важный класс систем - билинейные системы - когда изучение может быть проведено весьма полным образом.

Это случай, когда функции $a(x)$ и $b(x,y)$ линейны по своим аргументам:

$$a(\lambda x + \lambda' x') = \lambda a(x) + \lambda' a(x') \qquad /3/$$

$$\left.\begin{aligned} b(\lambda x + \lambda' x', y) &= \lambda b(x,y) + \lambda' b(x', y) \\ b(x, \mu y + \mu' y') &= \mu b(x,y) + \mu' b(x,y') \end{aligned}\right\} \qquad /4/$$

В этом случае система /1/ резко упрощается.

Вводя новое переменное z,

$$z = x_1 + \ldots + x_N \qquad /5/$$

получаем, используя линейность b по второму аргументу

$$\frac{dx_i}{dt} = a(x_i) + b(x_i, z) \qquad /6/$$

Складывая теперь все полученные уравнения и используя линейность и линейность b по первому аргументу, получаем уравнение для :

$$\frac{dz}{dt} = a(z) + b(z,z) \qquad /7/$$

Систему /6/ и /7/ удобно записать вместе:

$$\frac{dz}{dt} = a(z) + b(z,z) \qquad /8/$$

$$\frac{dx}{dt} = a(x) + b(x,z) \qquad /9/$$

Таким образом вместо N уравнений с N^2 членами мы получили всего два уравнения, причем уравнение для z интегрируется независимо от x, а уравнения для x_i одинаковы и отличаются друг от друга лишь начальными данными.

п°3. Макродинамический предел

Сравнение /6/, /7/ с системой /8/, /9/ позволяет понять, что происходит в предельном случае $N \to \infty$. Увеличение числа компонент $x_1, \ldots, x_N$ приводит к тому, что одна единственная траектория системы /6/ проектируется на N различных /вообще говоря/ траекторий уравнения /9/. Предельный переход $N \to \infty$ порождает фазовый портрет уравнения /9/. Уравнение /8/ сохраняет одну траекторию.

Важный частный случай, когда эта траектория является стационарной точкой уравнения /8/, соответствует термодинамическому пределу

$$\frac{dz}{dt} = 0 \quad , \quad z = z_0 \qquad /10/$$

Возникающие в уравнении макродинамики нетривиальные стационарные режимы - предельные циклы или еще более сложные стационарные режимы с перемешиванием - соответствуют нетривиальным обобщениям понятия термодинамического равновесия.

п°4. Гамильтоновы системы

Принадлежащее А.Я.Хинчину понятие сумматорной функции,

$$F(x) = f(x_1) + \ldots + f(x_N) \qquad /11/$$

позволяет существенно обобщить идею билинейных систем на гамильтоновы системы.

Рассмотрим некоторый /пока произвольный/ набор функций

$$f_1(x),\ f_2(x),\ldots,\ f_K(x) \qquad /12/$$

и построим соответствующий набор сумматорных функций

$$F_\alpha(x) = f_\alpha(x_1) + \ldots + f_\alpha(x_N) \qquad /13/$$

$$\alpha = 1,2,\ldots,K$$

Рассмотрим теперь произвольную функцию ,

$$\mathcal{H}(x_1,\ldots,x_N) = H(F_1,\ldots,F_K) \qquad /14/$$

и примем ее за гамильтониан системы:

$$\frac{dx_i}{dt} = I\left(\frac{\partial\mathcal{H}}{\partial x_i}\right)^* \qquad /15/$$

Учитывая формулу /14/, получаем:

$$\frac{dx_i}{dt} = I\left(\frac{\partial H}{\partial F_\beta}\frac{df_\beta}{dx_i}\right)^* = I\left(\frac{df_\beta}{dx_i}\right)^*\left(\frac{\partial H}{\partial F_\beta}\right)^* \qquad /16/$$

Найдем уравнения для F_α , вытекающее из /16/

$$\frac{dF_\alpha}{dt} = \sum_i \frac{df_\alpha}{dx_i}\frac{dx_i}{dt} = \left\{\sum_i \frac{df_\alpha}{dx_i} I\left(\frac{df_\beta}{dx_i}\right)^*\right\}\left(\frac{\partial H}{\partial F_\beta}\right)^* \qquad /17/$$

Начиная с этого места, существенно ограничим возможные наборы вводя <u>структурные соотношения</u>

$$\frac{df_\alpha}{dx} I\left(\frac{df_\beta}{dx}\right)^* = R^\gamma_{\alpha\beta}\, f_\gamma(x) \qquad /18/$$

Здесь $R^\gamma_{\alpha\beta}$ структурные коэффициенты.

Условия /18/ выполнены для наборов функций $f_\alpha(x)$, квадратичных по аргументу x .

Однако множество таких наборов значительно богаче и поэтому системы вида /16/ являются существенным обобщением класса билинейных систем.

Используя структурные соотношения, получаем уравнения макродинамики для гамильтоновых систем:

$$\frac{dF_\alpha}{dt} = R^{\gamma}_{\alpha\beta} F_\gamma \left(\frac{\partial H}{\partial F_\beta}\right)^* \qquad /19/$$

Уравнения для F_α снова не содержат x_i , которые могут быть затем найдены из системы /16/

$$\frac{dx}{dt} = I \left(\frac{d f_\beta}{dx}\right)^* \left(\frac{\partial H}{\partial F_\beta}\right)^* \qquad /20/$$

И снова предельный переход $N \to \infty$ приводит к построению фазового портрета векторного уравнения /20/.

Термодинамический предел соответствует частному случаю макродинамического предела /19/, когда эта система имеет стационарную точку:

$$\frac{dF_\alpha}{dt} = 0 \ , \quad F_\alpha = F_{\alpha 0} \qquad /21/$$

Система /16/ является обобщением билинейных систем /6/ еще и в том смысле, что вывод систем уравнений макродинамики /19/ не использует предположением о парности взаимодействия.

Гамильтоновы системы с парным взаимодействием возникают в предположении квадратичности функции H :

$$H = \sum_\alpha \sum_\beta H^{\alpha\beta} F_\alpha F_\beta \qquad /22/$$

п^{0}5. Фазовый переход

Развиваемый макродинамический подход позволяет сопоставить идею фазового перехода с понятием бифуркации в качественной теории обыкновенных дифференциальных уравнений.

Если уравнения макродинамики /19/ содержат параметры, то понятно, что их изменение может привести к потере устойчивости стационарного режима.

Возникающий при этом переходный процесс естественно сопоставить с фазовым переходом. Однако более подробное обсуждение этого важного вопроса выходит за рамки обсуждаемой темы.

Дополнение. Полиномиальные системы.

Пусть дана система с полиномиальными правыми частями степени $\ell + 1$,

$$\frac{dy_i}{dt} = \sum_\beta A_{i\beta}\, x_1^{\beta_1} \dots x_n^{\beta_n} \qquad /Д.1/$$

Здесь $\beta = (\beta_1, \dots, \beta_n)$ неотрицательные целые числа, причем

$$\beta_1 + \dots + \beta_n \leq l + 1 \qquad /Д.2/$$

Покажем, что увеличением числа переменных любую такую систему можно точно включить в квадратичную.

На первом шаге сделаем систему однородной степени $l+1$, введя всего одно дополнительное переменное y_0,

$$\frac{dy_0}{dt} = 0, \qquad y_0 = 1, \qquad /Д.3/$$

Новую систему запишем в виде:

$$\left.\begin{aligned} &\frac{dy_0}{dt} = 0 \\ &\frac{dy_i}{dt} = \sum_{\beta} A_{i\beta}\, y_0^{\beta_0} y_1^{\beta_1} \dots y_n^{\beta_n} \end{aligned}\right\} \qquad /Д.4/$$

определив неотрицательную степень β_0 из условия однородности

$$\beta_0 + \beta_1 + \dots + \beta_n = l + 1 \qquad /Д.5/$$

На следующем шаге введем большое количество новых переменных u_α;

$$u_\alpha = y_0^{\alpha_0} y_1^{\alpha_1} \dots y_n^{\alpha_n} \qquad /Д.6/$$

где α всевозможные наборы неотрицательных целых числе, но в сумме на единицу меньше, чем β :

$$\alpha_0 + \alpha_1 + \dots + \alpha_n = l \qquad /Д.7/$$

Введение u_α позволяет систему /Д.4/ в квадратичной форме:

$$\frac{dy_i}{dt} = \sum_{\kappa,\gamma} B_{i\kappa\gamma}\, y^\kappa u_\gamma \qquad /Д.8/$$

Однако наше цель еще не достигнута, ибо система /Д.8/ незамкнута - она содержит "лишние" переменные u_γ.

Увеличим размерность нашей системы, написав уравнения для всех u_α:

$$\frac{du_\alpha}{dt} = \sum_i y_0^{\alpha_0} \dots y_i^{\alpha_i - 1} \dots y_n^{\alpha_n} \frac{dy_i}{dt}$$

Подставляя вместо $\frac{dy_i}{dt}$ его выражение /Д.8/ получаем:

$$\frac{du_\alpha}{dt} = \sum_{i,k,\gamma} B_{ik\gamma} u_\gamma \left(y_0^{\alpha_0} \dots y_i^{\alpha_i - 1} \dots y_k^{\alpha_k + 1} \dots y_n^{\alpha_n} \right) \qquad /Д.9/$$

Заметим, что член, заключенный в скобки имеет степень ровно ℓ , ибо единица, на которую уменьшилась степень y_i компенсируется единицей, на которую увеличилась степень y_k . В результате имеем:

$$\frac{du_\alpha}{dt} = \sum_{\gamma s} B_{s\gamma} u_\gamma u_s \qquad /Д.10/$$

Система /Д.8/, /Д.10/ уже замкнута, квадратична и даже однородна. Она имеет большую серию первых интегралов

$$u_\alpha - y_0^{\alpha_0} y_1^{\alpha_1} \dots y_n^{\alpha_n} = C_\alpha \qquad /Д.11/$$

а все решения исходной системы /Д.1/ лежат в пространстве (y_i, u_α) на пересечении поверхностей

$$\left. \begin{aligned} x_0 &= 1 \\ C_\alpha &= 0 \end{aligned} \right\} \qquad /Д.12/$$

Теорема о включении доказана.

PHASE TRANSITIONS AND JULIA SETS

H.-O. Peitgen, M. Prüfer, P. H. Richter*)

I. Introduction

The renormalization group approach to phase transitions and the classical Yang-Lee theory have recently been brought together in a remarkable observation by Derrida, De Seze, and Itzykson [1]. For a hierarchical q state Potts model they indicated that the zeroes of the partition function are the Julia set corresponding to the renormalization group transformation. The Yang-Lee theory describes a physical phase as a domain of analyticity of the free energy, viewed as a function of complex temperature [2]. The boundaries of these domains are given by the zeroes of the partition function; in terms of the free energy this implies a phase separation made up of logarithmic singularities. The Wilson approach interprets each phase as the basin of attraction of a stable fixed point of the renormalization transformation. A phase transition is viewed as a separatrix between such basins of attraction.

In cases where the renormalization operates on the temperature only it should not be surprising that both approaches, though different in perspective, lead to the same set of points in the complex temperature plane. After all, they both attempt to describe the same physical object. The finding of Derrida et al. has nevertheless somewhat unexpected consequences because of the high degree of complexity of typical Julia sets.

The present note is a report on some preliminary work that has been done in Bremen following Derrida et al. In Sec. II the basic ideas are presented in connection with a one-dimensional Ising model where both the Yang-Lee zeroes and the renormalization transformation are easily determined. The nature of the Yang-Lee distribution is investigated for two hierarchical models in Sec. III. The emphasis is not on physical results but rather on a demonstration of the bewildering complexity that the analytic properties of even the simplest phase transition models may exhibit. The appendix contains a general discussion of the morphology of Julia sets.

*) Forschungsschwerpunkt 'Dynamische Systeme'
Universität Bremen, 2800 Bremen 33, Fed. Rep. of Germany

II. One-dimensional Ising model

1. The partition function and its zeroes.

Consider a chain of N spins, each with 2 possible states $\sigma_i = 1,2,$ and interaction energy between nearest neighbors:

$$E(\sigma_i,\sigma_{i+1}) = - J\delta(\sigma_i,\sigma_{i+1}) \equiv \begin{cases} -J & (\sigma_i = \sigma_{i+1}) \\ 0 & (\sigma_i \neq \sigma_{i+1}) \end{cases} . \quad (2.1)$$

The partition function is the sum of Boltzmann factors over all configurations $\{\sigma_i\}$,

$$Z_N = \sum_{\{\sigma_i\}} \exp\left\{\frac{J}{k_BT} \sum_i \delta(\sigma_i,\sigma_{i+1})\right\} \equiv \sum_{\{\sigma_i\}} y^{\sum_i \delta(\sigma_i,\sigma_{i+1})} . \quad (2.2)$$

In the last equality we have introduced the variable

$$y = e^{J/k_BT} \quad (2.3)$$

which can be viewed as a convenient representation of temperature T. The coupling constant J is characteristic of the material, $J > 0$ for ferromagnetic, and $J < 0$ for antiferromagnetic coupling. From Z_N we obtain the free energy per particle

$$f_N = - \frac{k_BT}{N} \ln Z_N , \quad (2.4)$$

and the internal energy per particle

$$u_N = - \frac{Jy}{N} \frac{\partial}{\partial y} \ln Z_N . \quad (2.5)$$

The partition function $Z_N(y)$ is a polynomial in y as is obvious from (2.2). Its precise form depends on the choice of boundary conditions for the chain of spins. To be specific we shall discuss two cases, free ends and closed loop. For the case of free ends an elementary calculation shows that

$$Z_N^f(y) = 2(y+1)^{N-1} \tag{2.6}$$

while the result for periodic boundary conditions is [3]

$$Z_N^c(y) = (y+1)^N + (y-1)^N \quad . \tag{2.7}$$

For the physical temperature range $y > 0$ these two partition functions are equivalent in the thermodynamic limit although their distributions of zeroes are quite different. Z_N^f has an $(N-1)$-fold zero at $y = -1$ while Z_N^c has N zeroes on the imaginary axis:

$$y_k = i \cdot \cot \frac{(2k+1)\pi}{2N} \quad , \quad k = 0,1,\ldots, N-1 \quad . \tag{2.8}$$

Parametrizing the imaginary axis by $y = is$ we find that the density of these zeroes, in the limit $N \to \infty$, is

$$\mu(s) = \frac{1}{N}\frac{dk}{ds} = \frac{1}{\pi}\frac{1}{1+s^2} \quad . \tag{2.9}$$

Using the product representation $Z_N^c(y) = 2\Pi(y-y_k)$ and (2.5) we can express the internal energy $u = \lim_{N\to\infty} u_N$ in terms of this density

$$u = -Jy \int_{-\infty}^{\infty} ds\, \mu(s) \frac{1}{y-is} \quad . \tag{2.10}$$

The integral is easily evaluated and found to agree with the result from (2.6),

$$u = -Jy/(y+1) \quad . \tag{2.11}$$

The well known fact that the one-dimensional Ising model does not exhibit a phase transition at finite temperature corresponds to the finding that the distribution of zeroes does not approach the real axis at positive y. If this were the case, however, the critical index α could be derived from $\mu(s)$ [4]. For example assume that the zeroes are located on the line $y_c + is$ $(-\infty<s<\infty)$, with density $\mu(s) \propto |s|^x$ for $s \to 0$. Then the internal energy is

$$u \propto \int ds \; |s|^x \; \frac{1}{y-y_c-is} \propto |y-y_c|^x \tag{2.12}$$

so that the specific heat diverges with critical exponent $\alpha = 1-x$. Before the advent of renormalization group theory, strong efforts were made to determine the location and density of zeroes in the neighborhood of critical points. It was hoped, and in some special cases proven, that the zeroes were distributed along lines. The recent work of Derrida et al., however, shows that this hope was very naive. In order to see why we next discuss Wilson's approach which has been so successful in providing critical indices. We shall stick with the one-dimensional Ising model to illustrate the connection between the Yang-Lee picture and renormalization group theory. From that connection, however, it will be apparent that in the general case much more complex things than line distributions are to be expected.

2. The RG transformation and its Julia set

The renormalization group (RG) approach to the one-dimensional Ising problem is as follows [5,6]. Consider a chain of $N = r^n$ spins, and in the partition function (2.2) perform a partial summation such that only every r-th spin survives. This relates the N-particle problem to the N/r-particle problem:

$$Z_N(y) = h^N(y) \; Z_{N/r}(y') \tag{2.13}$$

where y' is a renormalized temperature variable,

$$y' = T_r \, y \; , \tag{2.14}$$

and $h^r(y)$ absorbs a shift in the energy zero. The RG transformation is easily determined; for $r = 2$ we find

$$y' = T_2 \, y = \frac{y^2+1}{2y} \; , \qquad h^2(y) = 2y \; , \tag{2.15}$$

and for $r = 3$

$$y' = T_3\, y = \frac{y^3+3y}{3y^2+1} , \quad h^3(y) = 3y^2+1 . \tag{2.16}$$

The general form of T_r in terms of binomial coefficients is obvious but will not be needed in the following. The case $r = 3$ has an advantage over $r = 2$ when antiferromagnetism is of interest [7]; T_2 eliminates antiferromagnetic coupling $(0<y<1)$ in the first step while T_3 preserves antiferromagnetism.

The only attractors of both T_2 and T_3 are $y = \pm 1$. Since these attractors happen to be critical points as well they are superstable; there are no other critical points. The Julia set is the separatrix between the domains of attraction. That it coincides with the imaginary y - axis can best be seen from the following conjugacies:

$$\begin{aligned} g \circ T_2 \circ g^{-1}(z) &\equiv \tilde{T}_2(z) = z^2 \\ g \circ T_3 \circ g^{-1}(z) &\equiv \tilde{T}_3(z) = z^3 \end{aligned} \tag{2.17}$$

where $g(z) = (z+1)/(z-1)$ maps the attractors $+1, -1$ into ∞ and 0 respectively, and the imaginary axis onto the unit circle. The unit circle is the Julia set of both $\tilde{T}_2$ and $\tilde{T}_3$, so T_2 and T_3 have the imaginary axis as their Julia set.

Now what is the connection of these Julia sets to the Yang-Lee distribution of zeroes? Consider (2.13) and imagine $Z_N(y)$ as well as $Z_{N/r}(y')$ expressed in terms of their zeroes. It is then obvious that the zeroes y_K of $Z_N(y)$ are the preimages under T_r of the zeroes y_k' of $Z_{N/r}(y')$. By iteration of (2.13) it follows that the zeroes of $Z_{r^n}(y)$ are obtained from those of $Z_r(y)$ by application of $T_r^{-(n+1)}$. The zeroes of $Z_r(y)$ are easily determined. For the closed chain (periodic boundary conditions) we have

$$\begin{aligned} Z_2^c(y) &= 2(y^2+1) \Rightarrow \text{zeroes} \quad \pm i \\ Z_3^c(y) &= 2(y^3+3y) \Rightarrow \text{zeroes} \quad 0, \pm i\sqrt{3} . \end{aligned} \tag{2.18}$$

In both cases the zeroes are preimages of repulsive fixed points of T_r , and therefore (see (4.5)(3)) are contained in the Julia set of T_r . The zeroes of $Z_{r^n}(y)$, as $n \to \infty$, are therefore dense in the

Julia set of T_r .

The situation is different, however, for the chain with free ends. The RG equations (2.13) - (2.16) still hold if $N = r^n$ is interpreted as the number of bonds (i.e. N+1 is the number of particles). The partition functions $Z_r^f(y)$ are then

$$Z_2^f = 2(y+1)$$
$$Z_3^f = 2(y+1)^2 \qquad (2.19)$$

with zeroes at $y = -1$ which is an attractive fixed point of T_r . Application of T_r^{-n} to a stable fixed point does not lead away from it. Therefore in this particular case the zeroes of $Z_{r^n}(y)$ are not related to the Julia set of T_r .

For general boundary conditions the zeroes of $Z_r(y)$ may neither be stable fixed points of T_r nor belong to its Julia set. They may or may not approach the Julia set, as $n \to \infty$. Their delicate dependence on the parameters that represent the boundary conditions will be clarified in connection with (4.6). Cases like (2.19) where the zeroes get stuck in a stable fixed point are certainly quite exceptional.

The one-dimensional Ising model is a convenient example for discussing the connection of Yang-Lee zeroes and Julia sets. It is not representative, however, in that the Julia set is confined to a simple line. As a rule, Julia sets are much more complicated objects of fractal dimension [8] . Some typical features of Julia sets are reviewed in the appendix.

III. Two hierarchical Potts models

Hierarchical models have been studied as models for phase transitions because they allow for an exact treatment in terms of renormalization group theory and because their RG transformation is relevant as an approximate description of two-dimensional Ising-Potts systems [9-13] Derrida et al. [1] discovered the connection between Yang-Lee zeroes and Julia sets of RG transformations in a study of a diamond hierarchical model. We present here some results of a numerical investigation into the morphology of the corresponding Julia sets, and in addition analyze a second hierarchical model which allows for antiferromagnetic phase transitions.

1. <u>The diamond hierarchical lattice.</u>

Consider the lattice constructed according to the scheme of fig. 1. There are 4^{n-1} bonds on the n-th level of the construction. On each lattice site we assume a spin with

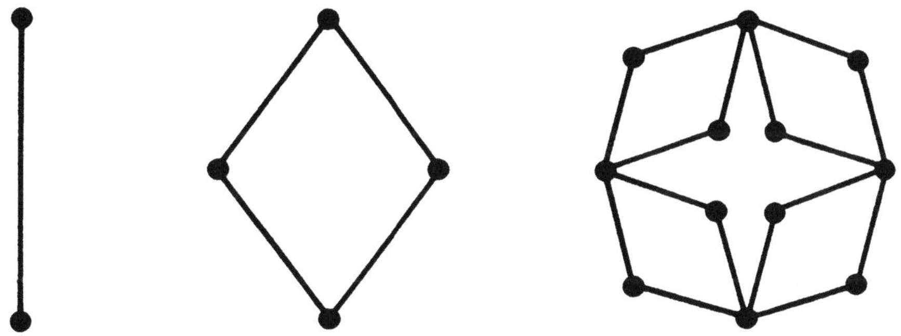

<u>Figure 1</u>

q possible states $\sigma_i = 1,2,\ldots,q$ and interaction energy across a bond

$$E(\sigma_i,\sigma_j) = - J\delta(\sigma_i,\sigma_j) \equiv \begin{cases} -J & (\sigma_i = \sigma_j) \\ 0 & (\sigma_i \neq \sigma_j) \end{cases} \tag{3.1}$$

The partition function (2.2) is most easily evaluated in terms of RG theory. Denoting by $Z_{(n)}$ the n-th level partition function (number of spins is $(2(4^{n-1}+2)/3)$, the recursion relation is

$$Z_{(n)}(y) = (h(y))^{2\cdot 4^{n-2}} \cdot Z_{(n-1)}(y') \tag{3.2}$$

with RG transformation

$$y' = T(y) = \left(\frac{y^2+q-1}{2y+q-2} \right)^2 \tag{3.3}$$

and function $h(y)$ given by

$$h(y) = 2y + q - 2 \ . \tag{3.4}$$

The two particle partition function

$$Z_{(1)}(y) = q(y + q - 1) \tag{3.5}$$

has just one zero at $y = 1-q$. Using the same arguments as in sec. II it can be seen that the zeroes of $Z_{(n)}(y)$ are obtained by $(n-1)$ fold application of T^{-1} to this initial zero. This generates 4^{n-1} zeroes of $Z_{(n)}(y)$, and figure 2 shows limit sets resulting from the corresponding backward iteration process, for some choices of q . Whether and how these limit sets are related to the Julia sets of T_q is a delicate question in general. This will be addressed in the appendix (cf. 4.7 and 4.8). However, when $1-q$ is contained in the basin of attraction of a fixed point of T it is easy to see that the limit set approaches the Julia set. This has been made sure in all four cases of fig. 2.

The initial zero $1-q$ happens to be a critical point of T . This is, according to (4.7) and (4.8), crucial for an understanding of the nature of the Julia sets. Analyzing (3.3) we find that T has two superstable attractive fixed points, independent of q , and six critical points:

$$\begin{aligned} &\text{attractive fixed points:} && y = 1 \ , \ \infty \\ &\text{critical points:} && y = 1 \ , \ \infty \ , \ 1-q \ , \ \pm\sqrt{1-q} \ , \ \frac{2-q}{2} \end{aligned} \tag{3.6}$$

From (3.3) it is obvious that both $\pm\sqrt{1-q}$ are mapped to zero and $(2-q)/2$ is mapped into ∞ . Therefore, in view of (4.7) we have to analyze the orbits of two critical values $(1-q)^2$ and zero only. According to theorem 4.7 the fate of these critical values may be one of the five possible types characterized by D. Sullivan [9] (see 4.6).

The fate of $y = 0$ and $y = 1-q$ has been investigated in the spirit of B. Mandelbrot's [8] fundamental experiment for the mapping $z \to z^2+\lambda$.

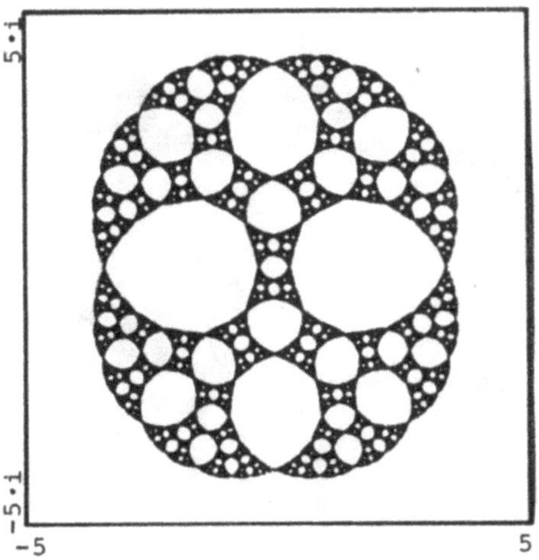

(a) q = 2 , $q \in A_2$

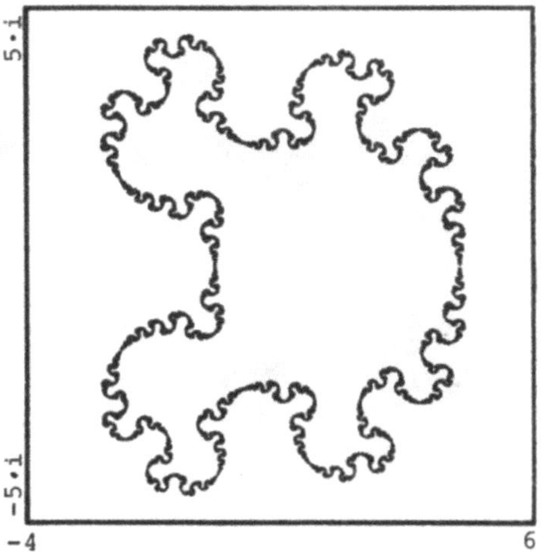

(b) q = 4 , $q \in A_1$

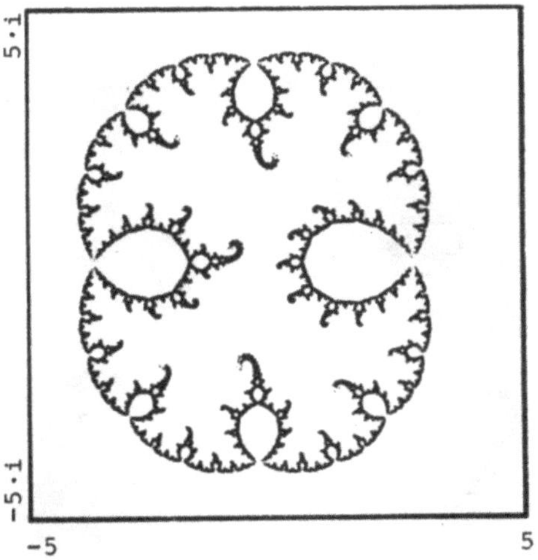

(c) q = 1.21 + 0.01·i , $q \in D$

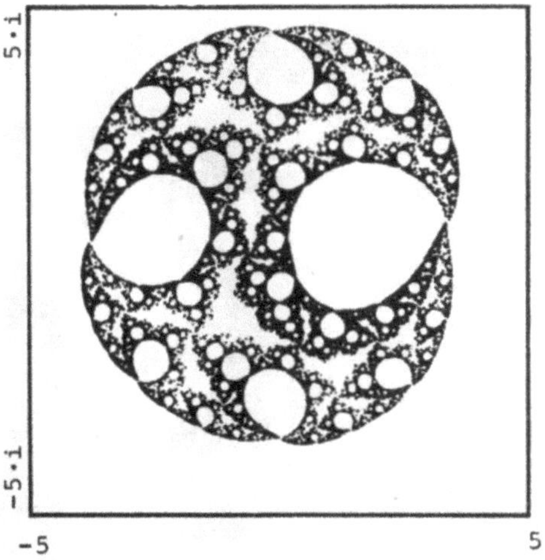

(d) q = 1.79 + 1.09·i

Figure 2 : Julia sets for T (3.3) and four values of q .

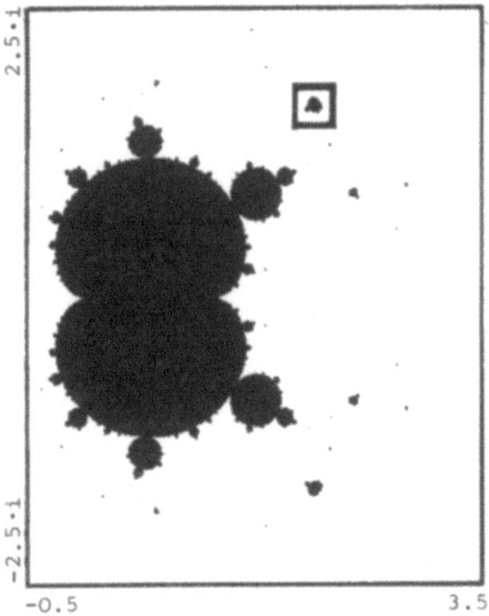

(a) $A_3 = A_6$

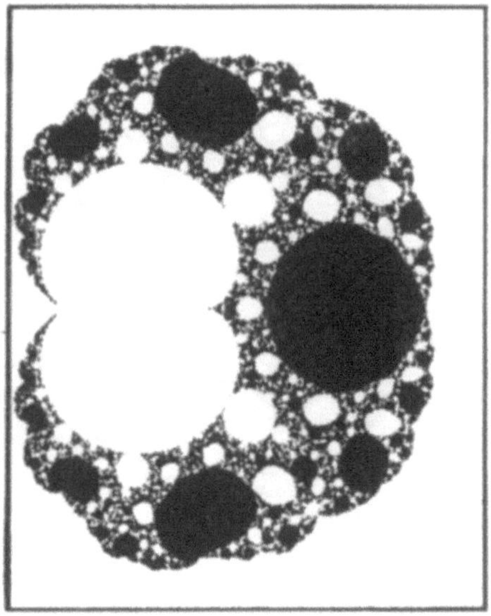

(b) $A_2 = A_4$

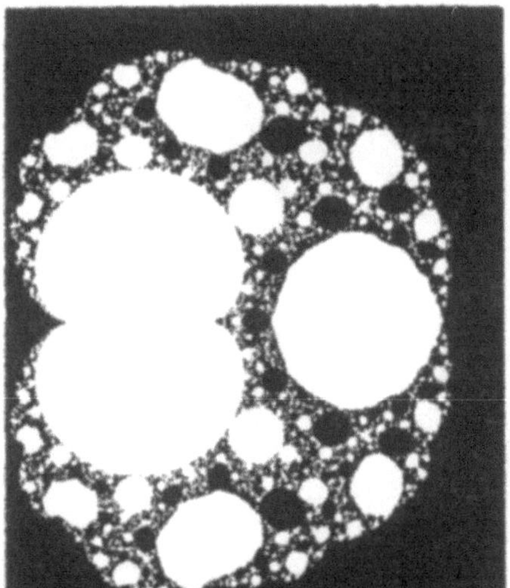

(c) $A_1 = A_5$

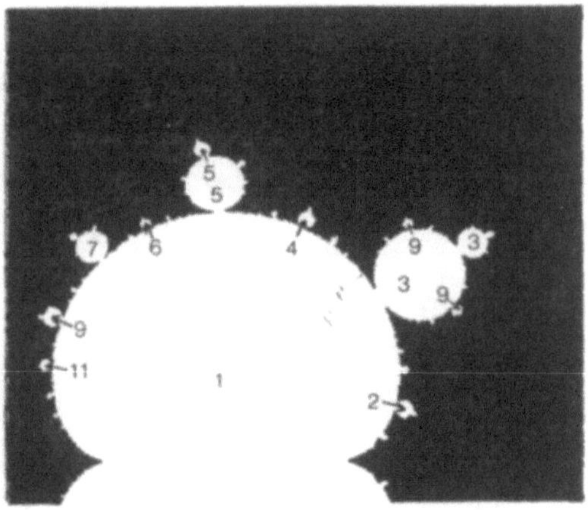

(d) Bifurcation dynamics in the set D

Figure 3 : Fate of the critical points $y = 0$, $y = 1 - q$ in system (3.3) .

We introduce the following notation:

$$A_1 = \{c \in C : T^n(0) \to 1 , n \to \infty\} ,$$
$$A_2 = \{c \in C : T^n(0) \to \infty , n \to \infty\} , A_3 = C \setminus (A_1 \cup A_2) ,$$
$$A_4 = \{c \in C : T^n(1-q) \to 1 , n \to \infty\} , \tag{3.7}$$
$$A_5 = \{c \in C : T^n(1-q) \to \infty , n \to \infty\} , A_6 = C \setminus (A_4 \cup A_5) .$$

Experimentally it turns out that $A_1 = A_5$, $A_2 = A_4$, $A_3 = A_6$, and these sets are shown in fig. 3a-c . Indeed, one shows that $T = \varphi \circ S^2 \circ \varphi$, where $\varphi(y) = (y+q-1)/(y-1)$ and $S(y) = (y^2+q-1)/(y^2-1)$. Thus, it follows that $T^n(0) \to 1$ for $n \to \infty$ iff $T^n(1-q) \to \infty$ for $n \to \infty$ and $T^n(0) \to \infty$ for $n \to \infty$ iff $T^n(1-q) \to 1$ for $n \to \infty$. Moreover, it can be seen that $2 \in A_2$ and $4 \in A_1$, so in the two cases of fig. 2a,b the superstable fixed points 1 , ∞ are the only attractors of the mapping T . Fig. 2b ($q = 4$) is an example for case (2) in (4.8) (see appendix). We have that $A^*(1) = A(1)$, $A^*(\infty) = A(\infty)$, and therefore the Julia set J separating the basins is a Jordan curve.

For $|q|$ sufficiently large an inspection of (3.3) shows that 0 is attracted by 1 and 1-q by infinity. Therefore, the sets $A_3 = A_6$ and $A_2 = A_4$ are bounded. However, it is evident from fig. 3 that all three subsets A_1 , A_2 , A_3 are intimately interwoven. We shall mainly be interested in A_3 which contains those q-values where the critical points have a q-dependent asymptotic fate. We now give a brief description of the set $A_3 = A_6$, i.e. the set shown in fig. 3a. It is certainly reminiscent of the Mandelbrot set [8] for the family $T_\lambda : z \to z^2 + \lambda$ (cf. fig. 8 in the appendix). Let us denote the solid main part of A_3 by D . Experimentally we find the following n-furcation systematics in D (cf. fig. 3d) . When q is taken from the central part of D , the orbits $\{T^n(0)\}_{n \in \mathbb{N}}$, $\{T^n(1-q)\}_{n \in \mathbb{N}}$ converge towards an attractive fixed point a_q of T , $1 \neq a_q \neq \infty$. When q is taken from the buds that are attached to D , the attractors for 0 and $1-q$ are periodic orbits, i.e. fixed points of T^k , the power k being a characteristic of the respective bud. Fig. 3d indicates these periods for the most prominent portions of D .

As q passes from one part of D to another we observe one of three types of n-furcations.

- Pitchfork bifurcation: An attractive fixed point of T^k , $k > 1$, splits into one repulsive fixed point of T^k plus two attractive fixed points of T^k . Before this bifurcation takes places the orbits of 0 and 1-q converge towards the same periodic attractor. Beyond the bifurcation, they converge towards the two different periodic attractors of the same length. This happens where the large buds of period 3, 5, 7, ... develop their own major buds, see fig. 3d.

- Period n-tupling bifurcation: An attractive fixed point of T^k , $k > 1$, becomes repulsive, and an attractive fixed point of $T^{n \cdot k}$ appears instead. This occurs, e.g., where the large buds of period 3, 5, 7, ... grow out of the main body of D , and at other places shown in fig. 3d.

- Combined bifurcation: An attractive fixed point of T^k , $k > 1$, bifurcates into two attractive fixed points of $T^{n \cdot k}$. The orbits of the two free critical points converge towards the two different attractors. This takes place where the buds of period 2, 4, 6, ... appear on the main body of D .

This bifurcation scheme differs from that of the Mandelbrot set which only exhibits the n-tupling bifurcations.

The Julia sets of figs. 2c, d correspond to q-values taken from D . The value $q = 1.21 + 0.01 \cdot i$ from fig. 2c is taken from the main body; both critical points are attracted by the same fixed point a_q , $1 \neq a_q \neq \infty$. By closer inspection of the basins of attraction we find $A(a_q) = A^*(a_q)$. We also find that no critical point of T lies in J . Therefore the assumptions of case (3) in (4.8) are fulfilled, and J must contain an infinite number of Jordan curves. These Jordan curves appear as closed loops in fig. 2c. An instructive picture is given in fig. 4 which is a close-up from fig. 2c with basins A(1) , $A(\infty)$ shown in solid black.

The parameter $q = 1.79 + 1.09 \cdot i$ in fig. 2d is taken from the main bud where two distinct periodic attractors of period 3 exist, attracting 0 and 1-q respectively.

The set D is not the only part of $A_3 = A_6$, as can be seen in fig. 3a. Fig. 5 shows a close-up (see inserted square in fig. 3a) of a region where a typical set of Mandelbrot type appears. The only difference to Mandelbrot's set [8] is that the main body corresponds to two attractors of period 2, each one attracting a different critical

Figure 4

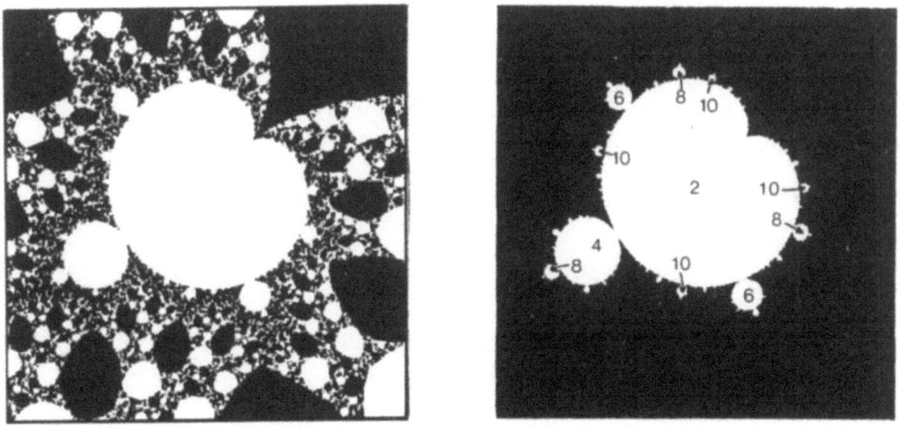

(a) $A_1 = A_5$

(b) Bifurcation dynamics

Figure 5 : Close-up from figure 3.

point of $\{0,1-q\}$. Otherwise the n-furcation scheme is the familiar one, as fig. 5b illustrates.

2. A hierarchical model for antiferromagnetism.

The diamond hierarchical model is not appropriate for a discussion of antiferromagnetism since elimination of every other layer of spins eliminates antiferromagnetism altogether [7] . We therefore looked into the model of Fig. 7 where each step of the RG transformation takes care of two out of three layers. There are

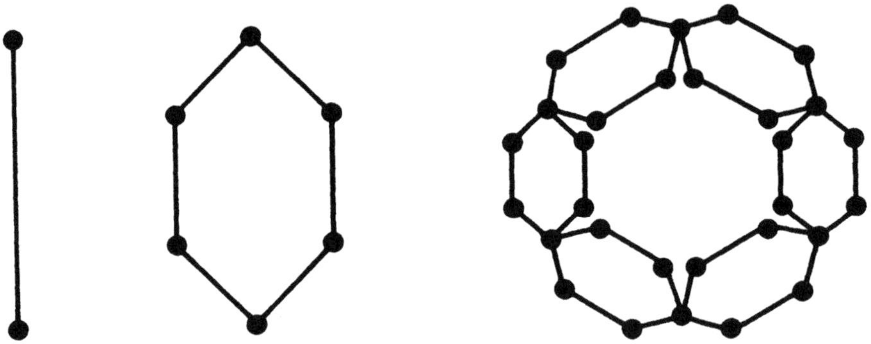

n = 1 n = 2 n = 3

Figure 6

6^{n-1} bonds on the n-th level and $(4\cdot 6^{n-1}+6)/5$ particles. The RG transformation is easily worked out:

$$Z_{(n)}(y) = h(y)^{2\cdot 6^{n-2}} \; Z_{(n-1)}(y') \tag{3.8}$$

with

$$y' = T(y) = \left(\frac{y^3+3(q-1)y+(q-1)(q-2)}{3y^2+3(q-2)y+q^2-3q+3} \right)^2 \tag{3.9}$$

and

$$h(y) = 3y^2 + 3(q-2)y + q^2 - 3q + 3 \quad . \tag{3.10}$$

By (3.8) and (3.5), the zeroes of $Z_{(n)}(y)$ are again given by $T^{-(n-1)}(1-q)$. For large n this set converges towards the Julia set of T provided $1-q$ is contained in a basin of attraction.

The morphology of these Julia sets depends again on the fate of the two free points $y = 0$ and $y = 1-q$. This follows from (3.9) and

$$\frac{dy'}{dy} = \frac{6(y-1)^2(y+q-1)^2(y^3+3(q-1)y+(q-1)(q-2))}{(3y^2+3(q-2)y+q^2-3q+3)^3} \tag{3.11}$$

We find, as in the diamond model, that $y = 1$ and $y = \infty$ are superstable attractors, and that $y = 1-q$ and the three preimages of $y = 0$ are the remaining critical points. Their story is told in Fig. 8. In contrast to the model of Derrida et al. this gives distinctly different fates for the two free critical points. We shall not go into more detail here and leave the analysis to a forthcoming publication.

3. The physics of these models

The choice of hierarchical lattices in the foregoing analysis was motivated by feasibility rather than physical interest. The models are constructed so as to allow for an exact treatment in terms of RG theory. Their only physical justification seems to be that they represent a certain approximation scheme of two-dimensional Ising-Potts models [11]. But even there it appears that the approximations are not terribly good. The critical index α , at least, does not come out satisfactorily.

To calculate α in the framework of RG theory [6], recall that phase transitions are represented by unstable fixed points of the RG transformation, and critical indices derive from the linearization of T at these points. For the diamond hierarchical model we find from (3.2) that the specific heat diverges with critical index

$$\alpha = 2 - \frac{\ln 4}{\ln\frac{dy'}{dy}\Big|_{y_c}} \tag{3.12}$$

where y_c is the Curie "temperature". An unstable fixed point $y_c > 1$ of T exists for all $q > 0$ and has been determined numerically. For $q = 2$, we find $y_c = 3.38$ and $\alpha = -0.677$ which is quite different from the two-dimensional Ising result $\alpha = 0(\log)$. Both y_c and α increase with increasing q . α turns positive at $q \approx 7$

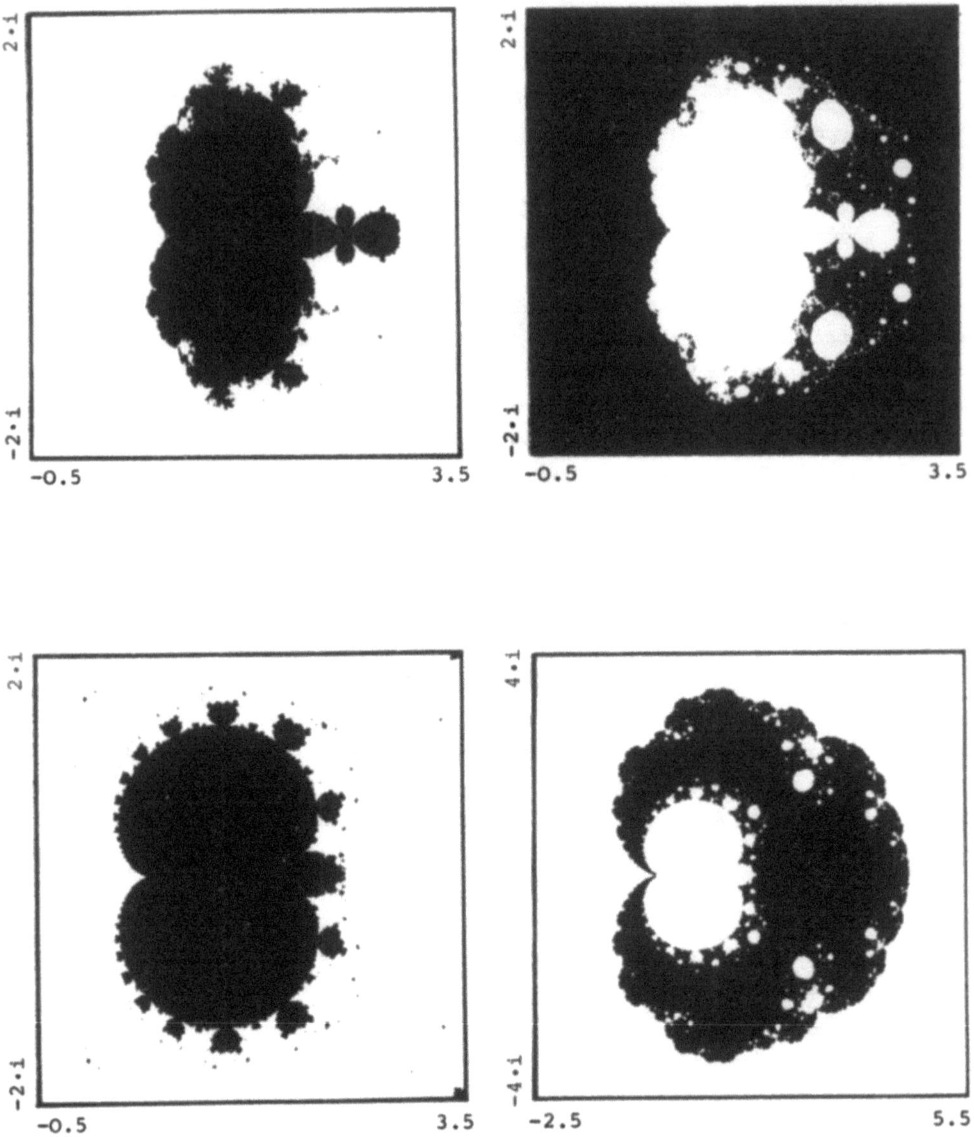

(c) A_6

(d) A_5

Figure 7 : Fate of the points $y = 0$, $y = 1 - q$ in system (3.9). The sets A_i are defined as in (3.7).

and for $q = 10$ assumes the value 0.14.

Our second hierarchical model has a corresponding ferromagnetic phase transition for $q > -0.2$, and in addition exhibits a Néel point for $0 < q < 2.1$ (this Néel point turns into a Curie point for $q<0$ but we do not discuss negative q here). For the Ising case $q = 2$ the Curie point is found to be at $y_c = 8.35$ and the Néel point at $y_c = 0.120$. Both transitions have critical index $\alpha = -0.902$, calculated from

$$\alpha = 2 - \frac{\ln 6}{\ln \frac{dy'}{dy}\Big|_{y_{c,N}}} \tag{3.13}$$

These poor results for α, however, do not affect the main point of this paper: the intimate connection between RG theory and the Yang-Lee distribution of zeroes. The old assumption of line distribution can definitely not be upheld any more, and it becomes an interesting challenge to study the Julia sets of more realistic RG transformations.

IV. Appendix

In this chapter we give a brief survey of some of the fundamental results on Julia sets due to G. Julia [15] and P. Fatou [16] as well as of some of the recent achievements due to D. Sullivan [14], A. Douady and J.H. Hubbard [17] and B. Mandelbrot [8].

1. Julia sets of rational mappings

Let R be a rational mapping of the extended complex plane $\Sigma = \mathbb{C} \cup \{\infty\}$ of degree $\deg(R) \geq 2$. The Julia set of R is the set

$$J = \{z \in \Sigma : R \text{ is not normal in } z\} . \tag{4.1}$$

Recall that R is said to be normal in z provided there is a neighborhood V of z, such that the family $\{R^k_{|V}\}_{k\in\mathbb{N}}$ of mappings from V to Σ is equicontinuous.

A first fact is (cf. [16]):

$$J \neq \emptyset \text{ and } R(J) = J = R^{-1}(J) . \tag{4.2}$$

In view of this fact a natural question is: What can be said about J and $\Sigma \setminus J$? Though G. Julia and P. Fatou had given numerous answers in their pioneering works it seems that only now we begin to have a more complete understanding. This is certainly due to the contributions of A. Douady and J. H. Hubbard [17], D. Sullivan [14]. Moreover it seems to be noteworthy that many of the recent investigations heavily rely on sophisticated numerical and computergraphical experiments.

In the context of our discussion of phase transition models we have the following facts: A periodic point p of period k (i.e. $R^k(p)=p$) is called

$$\left\{\begin{matrix} \text{attractive} \\ \text{repulsive} \\ \text{indifferent} \end{matrix}\right\} \Leftrightarrow |(R^k)'(p)| \left\{\begin{matrix} < 1 \\ > 1 \\ = 1 \end{matrix}\right\} . \quad (4.3)$$

If p is indifferent, i.e. $(R^k)'(p) = \exp(2\pi i \alpha)$, one distinguishes <u>rationally indifferent</u>, i.e. α is rational, <u>from irrationally indifferent</u>, i.e. α is irrational. If p is a fixed point of R then

$$A(p) = \{z : R^n(z) \to p \text{ as } n \to \infty\} \quad (4.4)$$

is called the <u>basin of attraction of p</u>. The connected component of $A(p)$, which contains p is denoted by $A^*(p)$ and is called the <u>immediate basin of attraction</u>.

Now the following fundamental results are well known ([15],[16],[19]):

(4.5)

(1) J is closed and perfect (i.e. dense in itself).

(2) The set of repulsive periodic points is dense in J.

(3) Given a repulsive periodic point p, then $J = cl\{z \in \Sigma : R^n(z) = p \text{ for some } n \in \mathbb{N}\}$

(4) The Julia sets of R and R^n are the same for all $n \in \mathbb{N}$.

(5) If a is an attractive periodic point of period k and $A(p) = \{z \in \Sigma : R^{n \cdot k}(z) \to p \text{ as } n \to \infty\}$ then $\partial A(p) = J$.

(6) J is selfsimilar, i.e. if V is any open set and $V \cap J \neq \emptyset$, then there is an $n \in \mathbb{N}$ such that $R^n(V \cap J) = J$.

In particular the third and fifth property of J in (4.5) allow to construct the Julia set in computergraphical experiments. The fifth property also has another remarkable consequence: If R has several attractive fixed (periodic) points, then the boundaries of all basins of attraction are the same.

2. Components of $\Sigma \setminus J$

Recently D. Sullivan in a series of three papers [14] gave a complete classification of $\Sigma \setminus J$ by answering the long standing question: Do there exist wandering domains (a component in $\Sigma \setminus J$ whose forward images are disjoint)? His result was negative. Thus, every component in $\Sigma \setminus J$ of a rational map of Σ is eventually periodic, i.e. for a sufficiently high iterate R^k its image becomes periodic. As a consequence he obtained, building on the works of G. Julia, P. Fatou, C. L. Siegel [20], J. Moser [21], V. Arnold [22] and M. Herman [23]:

> Each component $C \subset \Sigma \setminus J$ is eventually one of five types (i.e. for a sufficiently high iterate $R^n(C)$ one has):
>
> (1) (attractive case) The immediate basin of attraction of an attractive periodic point p of period k with $0 < |(R^k)'(p)| < 1$.
>
> (2) (parabolic case) The immediate basin of attraction of a rationally indifferent periodic point p of period k with $|(R^k)'(p)| = 1$.
>
> (3) (Siegel disk) A maximal periodic disk on which R^k is analytically equivalent to an irrational rotation.
>
> (4) (Herman ring) A maximal periodic anulus in which R^k is analytically equivalent to an irrational rotation.
>
> (5) (superattractive case) The immediate basin of attraction of a superattractive periodic point p of period k, i.e. $(R^k)'(p) = 0$.

(4.6)

Note, however, that $\Sigma \setminus J$ may be empty. For example, if

$$R(z) = \left(\frac{z-2}{z} \right)^2$$

one has that $J = \Sigma$. In fact the only critical points are 0 and 2. Now $2 \to 0 \to \infty \to 1 \to 1$ and $R'(1) = -4$. Hence, all critical points are absorbed by a repulsive fixed point.

Types (3) and (4) are "small denominator results". Given a concrete rational mapping of Σ it remains to understand which of the five types actually occur. It turns out that this question can essentially be reduced to an analysis of the orbits of all critical points.

3. The fate of critical points

A point $c \in \Sigma$ is called a critical value of R provided $R(z) - c = 0$ has a double root. The critical points of R are the preimages of the critical values. Note that if R has degree $\deg(R) = d$, then R has at most $2d-2$ critical points. The following are important results of P. Fatou [16]:

(1) Any component of type (4.6)(1) or (4.6)(2) contains at least one critical point. (2) Forward orbits of critical points in J (the Julia set) are dense in the frontiers of Siegel disks and Herman rings.	(4.7)

Note that a component of type (4.6)(5) contains a critical point by definition. Another famous long standing open problem was this: Does the frontier of a Siegel disk or Herman ring always contain a critical point? M. Herman recently announced to have solved this problem in a positive sense.

4. Morphology of Julia sets

Given a one-parameter family, say $\{R_\lambda\}_{\lambda \in C}$, of rational mappings of Σ we denote by J_λ their respective Julia sets. As λ is changed we may observe dramatic changes in J_λ such as J_λ is connected for one parameter of λ and totally disconnected for another parameter of λ. Also, in view of (4.5)(5) we expect to see changes in J_λ as we vary λ in such a way that we pass from a situation with n attractive periodic points to a situation with m $(n \neq m)$ attractive periodic points. Here is a collection of useful results (cf. [19]):

Let C denote the set of all critical points of R^k, $k \in \mathbb{N}$ and suppose that R is a rational mapping of Σ:

(1) Let a be the only attractive fixed point of R and suppose that $\mathrm{cl}\, C \subset A^*(a)$. Then J is totally disconnected. (2) If R has two attractive fixed points a and b and if $A^*(a)=A(a)$, $A^*(b)=A(b)$, then J is a Jordan curve. (3) Suppose that the number of attractive fixed points is ≥ 2, and that one and only one of them, say a, has $A^*(a)=A(a)$ and finally $\mathrm{cl}\, C \cap J = \emptyset$, then J contains an infinite number of Jordan curves.	(4.8)

(4) Let R be a polynomial. Then J is connected if and only if $A(\infty)$ contains no finite critical points of R. (Note that ∞ is a superattractive fixed point of R.)

(5) Let R be a polynomial and suppose that all finite critical points of R are contained in J, then J is connected.

The first investigations of a morphology of Julia sets go back to B. Mandelbrot [8]. He considered the polynomial $R_\lambda(z) = z^2+\lambda$. Since ∞ is a superattractive fixed point for all $\lambda \in C$, there is only one "free" critical point $c = 0$. For a given $\lambda \in C$ we have the alternative that either $0 \in A(\infty)$ (hence J_λ is totally disconnected (see (4.8)(1))) or $0 \notin A(\infty)$ (hence J_λ is connected (see (4.8)(4))). Plotting black dots for values of $\lambda \in C$ for which $0 \notin A(\infty)$ (i.e. $\{R_\lambda^k(0)\}$ bounded as $k \to \infty$) one obtains a picture like in figure 8 :

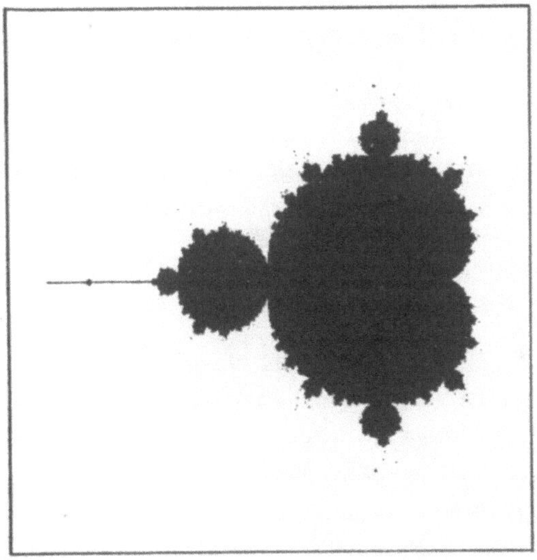

Figure 8

This important experiment was first carried out by B. Mandelbrot [8] The black set

$$M = \{\lambda \in C : \{R_\lambda^k(0)\} \text{ bounded for } k \in \mathbb{N}\}$$

is called the Mandelbrot set of R_λ. In view of (4.7) and Sullivans classification a parameter value $\lambda \in M$ may indicate one of the five types described in (4.6). In fact, it is not hard to show that types (1), (2), (3) and (5) occur. Meanwhile the Mandelbrot set M has attracted a lot of attention (see P. Cvitanović and J. Myrheim [24], A. Douady and J. H. Hubbard [17]). Computer graphical studies have revealed an unbelievable structure near the boundary of M (see [18]).

A. Douady and J. H. Hubbard have shown that M is connected. They have furthermore succeeded in characterizing Julia sets associated with various parts of M by unique combinatorial "trees". A particularly interesting Douady-Hubbard result is their determination of exterior angles associated with the boundary of M. Recently, in a remarkable computer graphical study L.Garnett, J. Curry and D. Sullivan [25] have found analogues of the Mandelbrot set in the investigation of Newton's method applied to the family $p_\lambda(z) = z^3 + (\lambda-1)z - \lambda$.

REFERENCES

[1] B. Derrida, L. De Seze, C. Itzykson,J. Statist.Phys.33, 559 (1983)

[2] C. N. Yang, T. D. Lee, Phys. Rev. 87, 404, 410 (1952)

[3] H. E. Stanley, Introduction to Phase Transitions and Critical Phenomena, Oxford Univ. Press 1971

[4] S. Grossmann, in: Festkörperprobleme IX, Ed. O. Madelung, Vieweg 1969

[5] M. Nauenberg, J. Math. Phys. 16, 703 (1975)

[6] H. J. Maris, L. P. Kadanoff, Amer. J. Phys. 46, 652 (1978)

[7] D. R. Nelson, M. E. Fisher, Ann. Phys. 91, 226 (1975)

[8] B. B. Mandelbrot, in: Nonlinear Dynamics, ed.: R. H. G. Hellemann, Annals of the New York Academy of Sciences 357, 249 (1980)

[9] A. A. Migdal, Sov. Phys. JFTP 42, 743 (1976)

[10] L. P. Kadanoff, Ann. Phys. N. Y. 100, 359 (1976)

[11] A. N. Berker, S. Ostlund, J. Phys. C: Solid State. Phys. 12, 496 (1979)

[12] R. B. Griffiths, M. Kaufmann, Phys. Rev. B 26, 5022 (1982)

[13] J. R. Melrose, J. Phys. A: Math. Gen. 16, 1041 (1983)

[14] D. Sullivan, preprint IHES (1982)

[15] G. Julia, J. Math. Pures et Appl., sér. 8.1, 47 (1918)

[16] P. Fatou, Bulletin Soc. Math. Fr. 47, 161 (1919), 48, 33 (1920)

[17] A. Douady, J. H. Hubbard, C. R. Acad. Sc. Paris, t. 294, série I, 123, and preprint, Cornell University 1984

[18] H.-O. Peitgen, P. H. Richter, GEO June 1984, Gruner + Jahr, Hamburg

[19] H. Brolin, Ark. Math. 6, 103 (1965)

[20] C. L. Siegel, Ann. Math. 43, 607 (1942)

[21] J. Moser, Nachr. Akad. Wiss., Göttingen, Math. Phys. 1, 1962

[22] V. I. Arnold, Usp. Mat. Nauk SSSR 18, 13 (1963)

[23] M. Herman, manuscript 1982

[24] P. Cvitanović, J. Myrheim, Phys. Letters 94 A 8, 329 (1983)

[25] J. H. Curry, L. Garnett, D. Sullivan, Commun. Math. Phys. 91, 267 (1983)

GEOMETRY OF ATTRACTORS: SOME RIGOROUS RESULTS

H.G.Bothe [1])

Our aim is to give an account of some mathematically rigorous results concerning the geometric structure of attractors. Though there are essential differences between attractors of general mappings which are not necessarily one-to-one (i.e. operations of $\mathbb{Z}^+$), diffeomorphisms (i.e. operations of $\mathbb{Z}$), and flows (i.e. operations of $\mathbb{R}$), we shall restrict ourselves to the second case and consider attractors of (invertible) diffeomorphisms only. Two new results are mentioned in this paper: the brief description of a construction in the second part which modifies examples given in [5] and leads to a new type of transversely wild expanding attractors and the criterium for transverse tameness.

Let $f:M \to M$ be a diffeomorphism of a smooth manifold M onto itself. A compact subset Λ of M will be called attracting if it is invariant (i.e. $f(\Lambda) = \Lambda$) and has a neighborhood U in M satisfying

$$f(U) \subset U, \qquad \bigcap_{i=0}^{\infty} f^i(U) = \Lambda$$

(i.e. under iterated application of f the neighborhood U is contracted to Λ, or, equivalently, each point p of U is attracted by the set Λ in the sense that $\lim_{i\to\infty} \mathrm{dist}(f^i(p), \Lambda) = 0$ for any metric of M). To avoid attracting sets which consist of separate independent pieces or which contain inessential parts we shall consider minimal attracting sets, i.e. attracting sets without proper subsets which are also attracting. These minimal attracting sets will be called attractors of f.

Many attracting sets Λ (e.g. the hyperbolic attracting sets as defined below) contain a finite number of disjoint attractors $\Lambda_1, \dots, \Lambda_r$ such that almost each point in M which belongs to the basin $B(\Lambda)$ of Λ, i.e. which is attracted by Λ, lies in the basin of one of the attractors Λ_i. This fact which more precisely means that $B(\Lambda_1) \cup \dots \cup B(\Lambda_r)$ is an open and dense subset of $B(\Lambda)$ justifies to regard $\Lambda_1, \dots, \Lambda_r$ as the essential parts of Λ. In Fig. 1 we show an attracting set of a diffeomorphism of the plane. This set is a straight segment whose essential parts Λ_1, Λ_2 are its end points. The broken lines indicate how the diffeomorphism f acts; i.e. they are f-invariant, and the arrows on them show the direction to

[1]) Institut für Mathematik der Akademie der Wissenschaften der DDR DDR 1080 Berlin, Mohrenstr. 39

which points are moved under f. We see that only the points on the vertical line through p_0 are not attracted by Λ_1 or Λ_2.

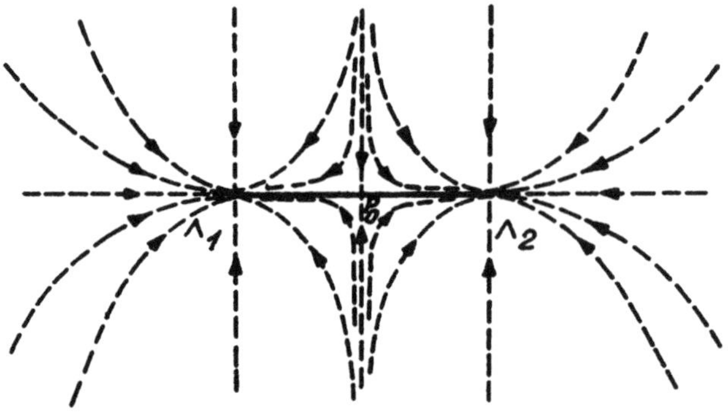

Fig. 1

An attractor Λ will be called stable if small perturbations of f do not destroy its structure, i.e. if each diffeomorphism $f':M \to M$ lying in a sufficiently small neighborhood of f (with respect to the C^1 topology) has an attractor Λ' near Λ which is homeomorphic to Λ. The most important attractors for which stability could be proved are the hyperbolic attractors. To give a precise definition of these attractors we need some technical concepts.

For a point $p \in M$ let $T_p(M)$ be the tangent space of M at p, and denote by $d_p f:T_p(M) \to T_{f(p)}(M)$ the linearization of f at p. for an invariant subset Λ of M $T_\Lambda(M) = \bigcup_{p \in \Lambda} T_p(M)$ denotes the tangent bundle of M restricted to Λ. An invariant set Λ is called hyperbolic if $T_\Lambda(M)$ is the direct sum $T^s \oplus T^u$ of df-invariant subbundles T^s, T^u, and df is contracting on T^s and expanding on T^u in the sense that there are real numbers $a, a' > 0$, $0 < \lambda < 1$ such that the inequalities

$$|df^i(v)| < a\,\lambda^i |v| \qquad (v \in T^s;\ i = 1,2,\dots)$$
$$|df^i(v)| > a'\,\lambda^{-i} |v| \qquad (v \in T^u;\ i = 1,2,\dots)$$

hold, where the length $|\ |$ of tangent vectors is defined with respect to an arbitrarily chosen Riemannian metric of M. More simply this can be expressed as follows: There are defined two complementary directions on Λ, and f is contracting in one of them while it is expanding in the other.

Though hyperbolicity is a strong restriction for attractors which besides stability has several further implications, it was not yet possible to give a satisfactory geometric description for all hyperbolic attractors. But there are classes of hyperbolic attractors for which such a description is possible.

If a hyperbolic attractor Λ is a manifold, then the restriction of f to Λ is an Anosov diffeomorphism, and many results from the theory of these diffeomorphisms are available. Being manifolds these Anosov attractors have a simple local structure. In contrast to this case we are interested in strange attractors whose local structure is rather complicated. We know one class of strange attractors, the so called expanding attractors, whose structure could be described in a quite satisfactory way. The corresponding results are due to R.F. Williams and can be found in [13], [14], [15]. (The special fact from these papers which will be needed here can be found in the introduction to [3].) These expanding attractors are distinguished in the class of all hyperbolic attractors by the fact that their topological dimension coincides with the dimension of the expanding fibres i.e. the fibres in the bundle T^u which appeared in the definition of hyperbolicity. Locally expanding attractors are described by the following theorem [3].

<u>Theorem.</u> Let p be a point in an n-dimensional expanding attractor Λ of a diffeomorphism $f:M \to M$ where M is an m-dimensional manifold. Then there is a homeomorphism h of $\mathbb{D}^n \times \mathbb{D}^{m-n}$ ($\mathbb{D}^k$ the k-dimensional unit ball in the euclidian k-space $\mathbb{R}^k$) onto a neighborhood U of p in M and a Cantor set C in $\mathbb{D}^{m-n}$ such that $U \cap \Lambda = h(\mathbb{D}^n \times C)$. Moreover, for each $c \in C$ the n-cell $h(\mathbb{D}^n \times \{c\})$ is smoothly embedded in M.

This theorem shows that an n-dimensional expanding attractor consists of uncountably many leaves which turn out to be n-dimensional manifolds homeomorphic to $\mathbb{R}^n$. But, since Λ is compact and minimal, these leaves are folded globally so wildly that each of them is dense in Λ. Nevertheless, by Williams' results it is possible to describe rather simply how these leaves behave globally and how they become shuffled if one runs around in Λ. So we can say that the intrinsic structure of expanding attractors is clear.

The name "expanding attractor" is suggested by the following fact: To each expanding attractor Λ there is a Riemannian metric of M and a real number $\lambda > 1$ such that for each point $p \in \Lambda$ all points q in a sufficiently small neighborhood of p in the leaf of p in Λ (i.e. all points q which can be joint with p inside Λ by a sufficiently small curve) satisfy

$$\mathrm{dist}(f(p), f(q)) > \lambda \cdot \mathrm{dist}(p,q).$$

This means that each point of Λ has a neighborhood in its leaf which is expanded under f, i.e. that f locally expands each leaf of Λ.

After this success concerning the intrinsic structure of expanding attractors the question arises how these attractors Λ are embedded in the ambient manifold M, i.e. one asks for the type of the triplet $(M, B(\Lambda), \Lambda)$ where $B(\Lambda)$ denotes the basin of Λ as defined above. In this paper we consider only the pair $(B(\Lambda), \Lambda)$ and ask to what extend it is determined by the interior structure of Λ. (For results about the whole triplet in a special case see [4].)

If the codimension $r = \dim M - \dim \Lambda$ of Λ in M is two, then the leaves of Λ may be globally linked and knotted, and the answer to the last question is negative (see [4]). But even if $r > 2$, where such global linking or knotting is impossible, Λ determines $(B(\Lambda), \Lambda)$ only under a further hypothesis which concerns the Cantor sets which appear if Λ is intersected transversely by an r-dimensional submanifold of M. Indeed, there are Cantor sets in manifolds with properties which are quite surprising for those which are not acquainted with the geometric theory of embeddings. Let us present here a famous example of a Cantor set in $\mathbb{R}^3$ called Antoine's necklace [1].

Start with the solid torus V_0 in $\mathbb{R}^3$ consisting of all point whose distance from the unit circle $\{(x,y,z);\ x^2+y^2 = 1,\ z = 0\}$ is at most $\frac{1}{2}$. Then put into V_0 a chain of four solid tori $V_{1,1},\dots,V_{1,4}$ which are linked as shown in Fig. 2. In the next step in each of the solid tori $V_{1,i}$ we choose a chain of four solid tori $V_{1,i,1},\dots,$ $V_{1,i,4}$ which are linked in $V_{1,i}$ as $V_{1,1},\dots,V_{1,4}$ are in V_0. Continuing in this way we get in the k-th step exactly 4^k solid tori whose union will be denoted by H_k (each link V' of the chain H_{k-1}

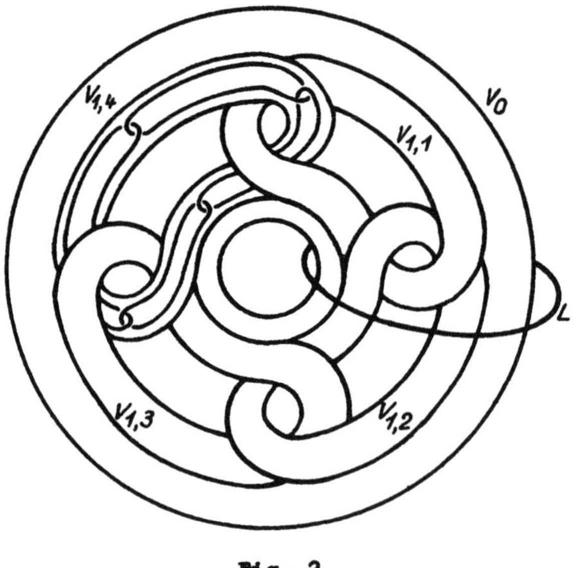

Fig. 2

is replaced in the k-th step by a chain of four linked solid tori in V). We assume that the maximal diameter δ_k of the solid tori in H_k tends to 0 as $k \to \infty$. Then

$$A = \bigcap_{k=1}^{\infty} H_k$$

is a Cantor set in $\mathbb{R}^3$ called Antoine's necklace. One of its surprising properties is the fact that though being 0-dimensional it behaves, with some respects, like a 1-dimensional subset of $\mathbb{R}^3$. So it is impossible to contract the circle L drawn in Fig.2 to a point without intersecting A, i.e. L is linked with A.

Cantor sets in manifolds with properties similar to those of A are called wildly embedded. (A Cantor set C in an m-dimensional manifold M is tame, if for each point $p \in C$ there are arbitrarily small, m-dimensional, topological balls in M which contain p in their interiors and whose boundaries do not intersect C.) While in one- and two-dimensional manifolds all Cantor sets are tame, it was shown by W.A.Blankinship [2] that for $m \geq 3$ each m-dimensional manifold contains wild Cantor sets.

Let us say that an n-dimensional expanding attractor Λ in an m-dimensional manifold M is transversely tame, if for each (m-n)-dimensional submanifold N of M which intersects Λ transversely (i.e. which is transverse to the leaves of Λ) the intersection $N \cap \Lambda$ does not contain Cantor sets which are wild in N.

The following theorem is proved in [7].

Theorem. If Λ is a 1-dimensional transversely tame attractor in a manifold whose dimension is different from three, then the topological type of $(B(\Lambda), \Lambda)$ is determined by the intrinsic structure of Λ.

Wild Cantor sets are rather artificial objects, and the same can be expected for transversely wild expanding attractors. Therefore the question arises whether and to what extend transversely wild expanding attractors do exist at all. First examples of transversely wild expanding attractors can be found in the paper [10]. Later in [5] some general constructions were presented which lead to several types of transversely wild expanding attractors. It turned out that for $n \geq 1$, $n \leq m + 3$ in each m-dimensional manifold n-dimensional transversely wild expanding attractors are possible. Nevertheless, the question whether for each expanding attractor Λ there is a transversely wild attractor with the same intrinsic structure could not be answered even in the 1-dimensional case. In the rest of this paper we shall sketch a new construction for transversely wild expanding attractors which, as we hope, will be general enough to allow a positive answer to the last question.

The following symbols with fixed meaning will be used: $\mathbb{D}^n$ is the n-dimensional closed unit disk in the euclidian n-space $\mathbb{R}^n$ $(n = 1,2,3)$. By S^1 we denote the unit circle i.e. the boundary of $\mathbb{D}^2$, where $\mathbb{R}^2$ is identified with the complex plane so that for each integer r by $t \mapsto t^r$ $(t \in S^1)$ a mapping of S^1 onto itself is defined. An arc is a manifold homeomorphic to $\mathbb{D}^1$. By handlebody we mean a compact 3-dimensional manifold which is homeomorphic to a regular neighborhood of a connected graph in $\mathbb{R}^3$ (see Fig. 3 where a handlebody with three handles is shown).

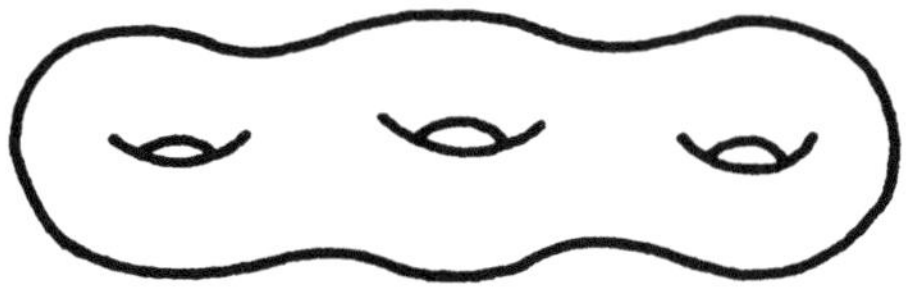

Fig. 3

A topological space Ξ together with a homeomorphism $g: \Xi \to \Xi$ will be called an abstract expanding attractor provided there is a manifold M with a diffeomorphism $f: M \to M$ and a topological embedding $h: \Xi \to M$ such that $\Lambda = h(\Xi)$ is an expanding attractor of f and $hg = fh$ on Ξ. The attractor Λ will be called a realization of the abstract attractor Ξ. These abstract attractors represent just the type of their realizations with respect to the intrinsic structure. Indeed, two expanding attractors have the same intrinsic structure if and only if they are realizations of the same abstract expanding attractors. An expanding attractor of dimension 0 consists of a single point, and we shall exclude this trivial case from our considerations. The simplest (with respect to the intrinsic structure) examples of expanding attractors are the solenoids. Their type is determined by an integer r satisfying $|r| \geq 2$. Let us describe this type by constructing a corresponding abstract expanding attractor (Ξ_r, g_r).

Start with the solid torus $V_0 = S^1 \times \mathbb{D}^2$. Then choose an embedding g of V_0 in its interior whose image $V_1 = g(V_0)$ runs around in V_0 exactly r times in the positive dirction if $r \geq 2$ and $-r$ times in the opposite direction if $r \leq -2$. More precisely, we assume that each disk $g(\{t\} \times \mathbb{D}^2)$ lies in the disk $\{t^r\} \times \mathbb{D}^2$ and that the restriction of g to a disk $\{t\} \times \mathbb{D}^2$ is contracting in the sense that there is a factor $0 < \lambda < 1$ not depending on t such that for $x,y \in \mathbb{D}^2$, $g(t,x) = (t^r,x')$, $g(t,y) = (t^r,y')$ we have $\mathrm{dist}(x',y') = \lambda\, \mathrm{dist}(x,y)$ with respect to the usual metric in $\mathbb{D}^2$. Now the solid tori $V_2, V_3, \ldots$

are defined successively by $V_{i+1} = g(V_i)$. Of course, V_i runs around in V_o exactly r^i times, and the radius of the disks $g^i(\{t\} \times \mathbb{D}^2)$ is λ^i. Therefore the tori V_i become more and more thin as $k \to \infty$. In Fig. 4 the tori V_o and V_1 are drawn for $r = 2$, and the broken line indicates how V_2 runs around in V_o.

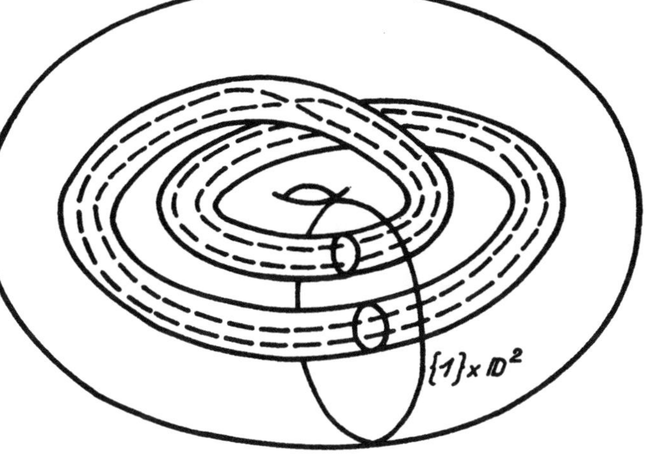

Fig. 4

We define our abstract expanding attractor (Ξ_r, g_r) by

$$\Xi_r = \bigcap_{i=0}^{\infty} V_i , \qquad g_r = g|_{\Xi_r} .$$

The methods of [5] were not general enough to get transversely wild realizations for all solenoids (Ξ_r, g_r). The construction presented there requires $r \geq 6$, and even the announcement in the remark on page 38 concerns only the case $r \geq 3$. As mentioned above, we hope that our new construction will allow to find for each 1-dimensional abstract expanding attractor a transversely wild realization. Let us indicate the idea of this construction by showing that (Ξ_2, g_2) has a transversely wild realization.

The following remark will help us to identify a realization of (Ξ_2, g_2). The proof is simple, if one applies the techniques of [3].

Proposition. Let $f: M \to M$ be a diffeomorphism of an m-dimensional manifold M which is equipped with a Riemannian metric. We assume that we have found a compact m-dimensional submanifold N of M which is mapped by f in its interior (i.e. $f(N) \subset \operatorname{Int} N$) and for which there is a projection $\pi: N \to S^1$ onto S^1 which makes the diagram

$$\begin{array}{ccc} N & \xrightarrow{f} & N \\ \pi \downarrow & & \downarrow \pi \\ S^1 & \xrightarrow{\varphi} & S^1 \end{array}$$

commutative, where $\varphi(t) = t^2$. Moreover we assume that f is contracting on each set $\pi^{-1}(t)$, i.e. that $\pi(p) = \pi(q)$ implies $\operatorname{dist}(f(p),f(q)) < \operatorname{dist}(p,q)$. Then $\Lambda = \bigcap_{i \in \mathbb{Z}} f^i(N)$ is a realization of (Ξ_2, g_2).

We choose two disjoint subarcs A, A' of S^1 for which the restrictions $\varphi|_A$ and $\varphi|_{A'}$ are one-to-one and the following equations and inclusions hold:

$$A \cap A' = \emptyset, \quad A \cup A' \subset \operatorname{Int} \varphi(A),$$

$$A \subset \operatorname{Int} \varphi(A'), \quad A' \cap \varphi(A') = \emptyset.$$

The following example shows that arcs with these properties do exist:

$$A = \{e^{2\pi i t};\ -\tfrac{1}{100} \le t \le \tfrac{3}{8} + \tfrac{1}{100}\}$$

$$A' = \{e^{2\pi i t};\ \tfrac{1}{2} - \tfrac{1}{100} \le t \le \tfrac{3}{4} - \tfrac{\cdot 1}{100}\}.$$

It is not hard to find a compact 4-dimensional submanifold N of the 4-dimensional solid torus $S^1 \times D^3$ such that, if we write D_t for $\{t\} \times D^3$ $(t \in S^1)$, $D_t \cap N$ is a ball if $t \in S^1 \setminus (A \cup A')$ and a handlebody with two handles if $t \in \operatorname{Int} A \cup \operatorname{Int} A'$. If t passes an end point of one of the arcs A, A', then $D_t \cap N$ behaves as shown in Fig. 5 a),b),c) where only one half of the sets $D_t \cap N$ is shown.

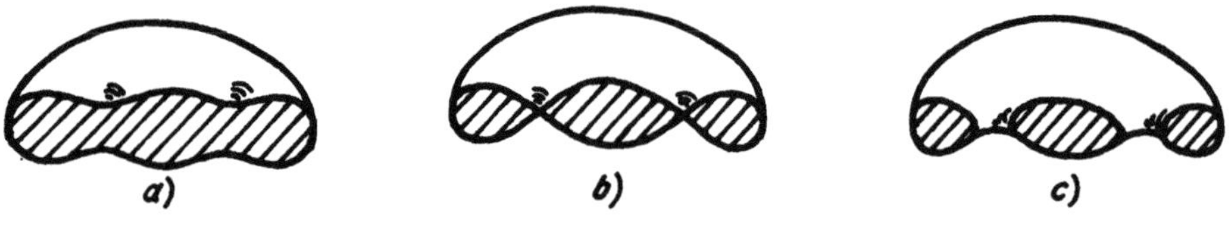

Fig. 5

In the next step of our construction we take an embedding of $S^1 \times D^3$ in $\mathbb{R}^4$. We shall identify $S^1 \times D^3$ with its image under this embedding. Since all smooth simple closed curves in $\mathbb{R}^4$ are unknotted, it is possible to choose a diffeomorphism $f_o : \mathbb{R}^4 \to \mathbb{R}^4$ which maps $S^1 \times D^3$ into itself in such a way that $f_o(D_t) \subset D_{\varphi(t)} = D_{t^2}$, i.e. f_o wraps $S^1 \times D^3$ twice around in itself. Moreover we may assume that f_o has a compact support, i.e. that f_o is the identity outside a compact subset of $\mathbb{R}^4$. now it is not hard to find a diffeomorphism $f_1 : \mathbb{R}^r \to \mathbb{R}^4$

which is the identity outside $S^1 \times D^3$ and has the following properties: (1) $f_1(D_t) = D_t$ for each $t \in S^1$. (2) $f_1 f_o(N) \subset \operatorname{Int} N$. (3) For $t \in A$ or $t \in A'$ the two components of $f_1 f_o(N) \cap D_t$ lie in $N \cap D_t$ as shown in Fig. 6.

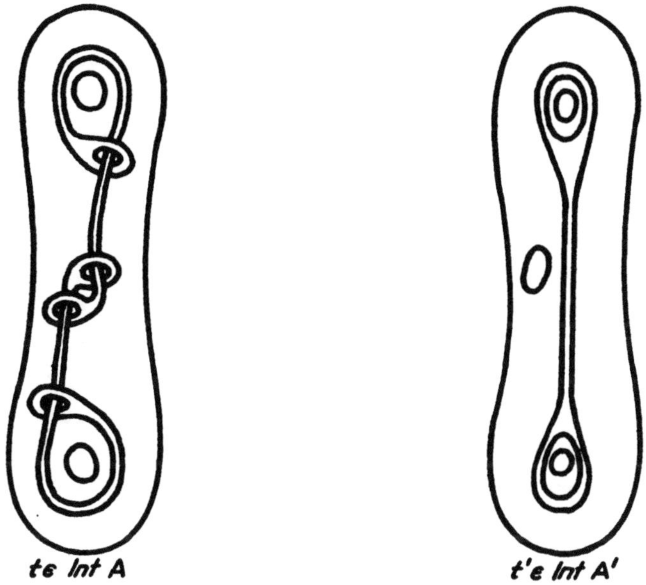

Fig. 6

The diffeomorphism $f_1 f_o : R^4 \to R^4$ will be denoted by f.

An important point which we did not mention above is the simple assumption that for $t \in \operatorname{Int} A$, $t' \in \operatorname{Int} A'$ the handlebody $D_{t'} \cap N$ is smaller than the handlebody $D_t \cap N$. This justifies our further assumption that f is contracting on each set $D_t \cap N$ $(t \in S^1)$. Then the Proposition implies that

$$\Lambda = \bigcap_{i=1}^{\infty} f^i(N)$$

is a realization of (Ξ_2, g_2).

To prove that Λ is transversely wild we look at the intersection $D_1 \cap \Lambda$ (D_1 is a transverse section of Λ). This set contains a Cantor set C which is the intersection of a descending series of 3-dimensional manifolds $Q_o, Q_1, \ldots$ in D_1, where each Q_i has 2^i components each of which is a handlebody with two handles, and each component H of Q_i contains exactly two components of Q_{i+1} which are embedded as shown in Fig. 5 a). This implies that C is a wild Cantor set in D_1 (a proof of this fact can be found in [6]), and we see that Λ is transversely wild.

What we have obtained by our construction is a diffeomorphism $f:\mathbb{R}^4 \to \mathbb{R}^4$ with a transversely wild expanding attractor Λ which is a realization of (Ξ_2, g_2). Since f is the identity outside a compact subset of $\mathbb{R}^4$, we can get a transversely wild realization of (Ξ_2, g_2) in an arbitrary 4-dimensional manifold M by embedding $\mathbb{R}^4$ in M and then extending f by the identity to a diffeomorphism $\bar{f}:M \to M$. To be a transversely wild realization of (Ξ_2, g_2) is a stable property of attractors, and so all diffeomorphisms $\tilde{f}:M \to M$ which are obtained from the special diffeomorphism $\bar{f}$ by sufficiently small perturbations will have attractors of the same kind.

Let us conclude this paper by announcing some sufficient conditions for transverse tameness. Since these conditions are connected with the Hausdorff dimension $\dim_H \Lambda$ of attractors we repeat the definition of this dimension. For a compact subset X of a manifold M $\dim_H X$ is defined to be the infimum of all positive real numbers s for which there are finite coverings $\mathfrak{U}$ of X with $\sum_{U \in \mathfrak{U}} (\operatorname{diam} U)^s$ arbitrarily small. Here the diameters $\operatorname{diam} U$ of the sets $U \in \mathfrak{U}$ have to be taken with respect to a smooth Riemannian metric of M. Then $\dim_H X$ does not depend on this metric but only on the differentiable structure of M. Moreover $\dim_H X$ is bounded from below by the topological dimension of X (which we have used above to define expanding attractors) and from below by the dimension of M.

Combining results of M.A.Štan'ko [11] and J.Väisälä [12] we find that the Hausdorff dimension of wild Cantor sets in a k-dimensional manifold is at least $k - 2$. Therefore expanding attractors in an m-dimensional manifold whose Hausdorff dimensions are less than $m - 2$ must be transversely tame.

We can go a step farther and apply inequalities relating the Lyapounov exponents of an attractor to its Hausdorff dimension (see [8], [9]) to get a sufficient condition for transverse tameness which, very roughly spoken, can be stated as follows: An expanding attractor Λ is transversely tame provided the average of the stretching rate in its leaves is small compared with the rate to which Λ attracts its ambient space. Since all realizations of an abstract expanding attractor (Ξ, g) have the same rate of stretching in their leaves (which is determined by the action of g on Ξ), we see that a realization of (Ξ, g) is transversely tame, provided it is attracting to a sufficiently high degree.

References

[1] Antoine, M.L.: Sur la possibilité d'étendre l'homéomorphie de deux figures à leurs voisinages. C. R. Acad. Sci. Paris 171 (1920), 661-663; Sur l'homéomorphie de deux figures à leurs voisinages. J. Math. Pures Appl. 86 (1921), 221-225

[2] Blankinship, W.A.: Generalization of a construction of Antoine. Ann. of Math. 53(2) (1951), 276-297

[3] Bothe, H.G.: The ambient structure of expanding attractors. I: Local triviality and tubular neighborhoods. Math. Nachr. 107 (1982), 327-348

[4] Bothe, H.G.: The ambient structure of expanding attractors. II: Solenoids in 3-manifolds. Math. Nachr. 112 (1983), 69-102

[5] Bothe, H.G.: Transversely wild attractors. Akad. Wiss. DDR Math. Reports 03/83 (1983)

[6] Bothe, H.G.: The ambient structure of expanding attractors. III: Transversely wild attractors. Math. Nachr. (to appear)

[7] Bothe, H.G.: Transversely tame attractors. Akad. Wiss. DDR Math. Reports (to appear)

[8] Douady, A. and J. Oesterlé: Dimension de Hausdorff des attracteurs. C. R. Acad. Sci. Paris 290 (1980), 1135-1138

[9] Ledrappier, F.: Some relations between dimension and Lyapounov exponents. Comm. Math. Phys. 81 (1981), 229-238

[10] Robinson, C. and R. Williams: Classification of expanding attractors: An example. Topology 15 (1976), 321-323

[11] Štan'ko, M.A.: The embedding of compacta in euclidian space. Mat. Sbornik 83(125) (1970), 234-255

[12] Väisälä, J.: Dimension and measure. Proc. Amer. Math. Soc. 76 (1979), 167-168

[13] Williams, R.F.: One dimensional non wandering sets. Topology 6 (1967), 473-487

[14] Williams, R.F.: Classification of one-dimensional attractors. Global Analysy. Proc. Symp. Pure Math. 14 (1970), 341-361

[15] Williams, R.F.: Expanding attractors. Publ. Math. I.H.E.S. 43 (1974), 169-203

SOME PROBLEMS OF THE PARAMETER IDENTIFICATION OF STRANGE ATTRACTORS

W. Jansen and U. Kriegel [1)]

Since Lorenz [1] originally demonstrated that very simple low dimensional systems could show chaotic behaviour there has been a large development in the investigation of strange or chaotic phenomena. Much of the current interest in chaotic dynamics arises from the possibility that systems of even many degrees of freedom may possess an attractor of only a few dimensions. Therefore only a few variables might be needed to model the asymptotic behaviour once the system's orbit lies on the attractor. This might be important especially for practical purposes. But how can an experimentalist know of what his system really is. Typically he has at his disposal a few time series records, i. e. a discretized part of the system's trajectory or at least a few projections of this. From that he is able to reconstruct the geometrical structure of an attractor [2] and if the time series has a sufficient length more structural constants like Hausdorff dimension [3] or information dimension [4] could be estimated. The aim of this paper is to discuss wether or not it is possible to identify from time series system parameters too.

In the following it is supposed that the measured time series

$$x^{obj} = \{x_1^{obj} , \dots , x_n^{obj}\} \in R^n$$

is generated by a mechanism called the object system which is represented by a map

$$x_{i+1}^{obj} = f^{obj}(x_i^{obj}) . \tag{1}$$

This mechanism then should be approximated by a parametric model of the following form

$$x_{i+1} = f(x_i, p) , \tag{2}$$

where $x \in R^n$ is the model time series and $p \in R^m$ is a parameter vector. The function f is assumed to be twice differentiable in respect to p, and f^{obj} and f should be continous in respect to x. The parameter vector p is determined in such a way that the model behaviour according to a chosen power function gives the best approximation of the

---

1) Central Institute for Cybernetics and Information Processes, Academy of Sciences of the DDR 1086 Berlin, Kurstraße 33, PSF 1298

object system. Usually this will be done by a minimization of the distance between the time series x_i^{obj} and the corresponding model trajectory $x_i(p)$, i. e. a power function of type

$$d_1(p) = \left(\frac{1}{N}\sum_{i=1}^{N}\|x_i^{obj} - x_i(p)\|^2\right)^{1/2} \tag{3}$$

is minimized. Here N denotes the number of points in the time series and $\|\cdot\|$ is an arbitrary norm in R^n. As known this method works well in cases where the system to be modeled has no strange attractor [5] but fails completely in other cases. This is obvious since due to the exponential divergence of trajectories on strange attractors smallest deviations in the initial conditions of the orbit generating maps will increase fastly and reach for sufficient long time series values in the order of the diameter of the attractor.

As a demonstration we generated time series with N=100 points on the attractor using the Henon map [6]

$$\begin{aligned} x_{t+1} &= y_t + 1 - ax_t^2 \\ y_{t+1} &= bx_t \\ a &= 1.4 \\ b &= 0.3 \end{aligned} \tag{4}$$

and the Lorenz system [1] 1)

$$\begin{aligned} \frac{dx}{dt} &= -\sigma x - \sigma y \\ \frac{dy}{dt} &= -xz + rx - y \\ \frac{dz}{dt} &= xy - bz \\ \sigma &= 10.0 \\ r &= 28.0 \\ b &= 8/3 \end{aligned} \tag{5}$$

as object systems, respectively. In fig. 1a,b the power function d_1 of systems (4) and (5) is given as function of the parameter a and r, respectively. In both cases there is no sign of a parameter valley known from conventional identification problems. The $d_1(a)$ distribution for the Henon system shows an irregular behaviour with a very steep and narrow valley around a = 1.4. For values $a > 1.433$ $d_1(a)$ was not calculable because of arithmetical overflow. In the case of the Lorenz system $d_1(r)$ exhibits a nearly uniform shape with no hint of a

---

1) This initial value problem was solved by a Runge-Kutta-Merson method and then the trajectory was discretized with a step size of 1.0 to get a time-one-map.

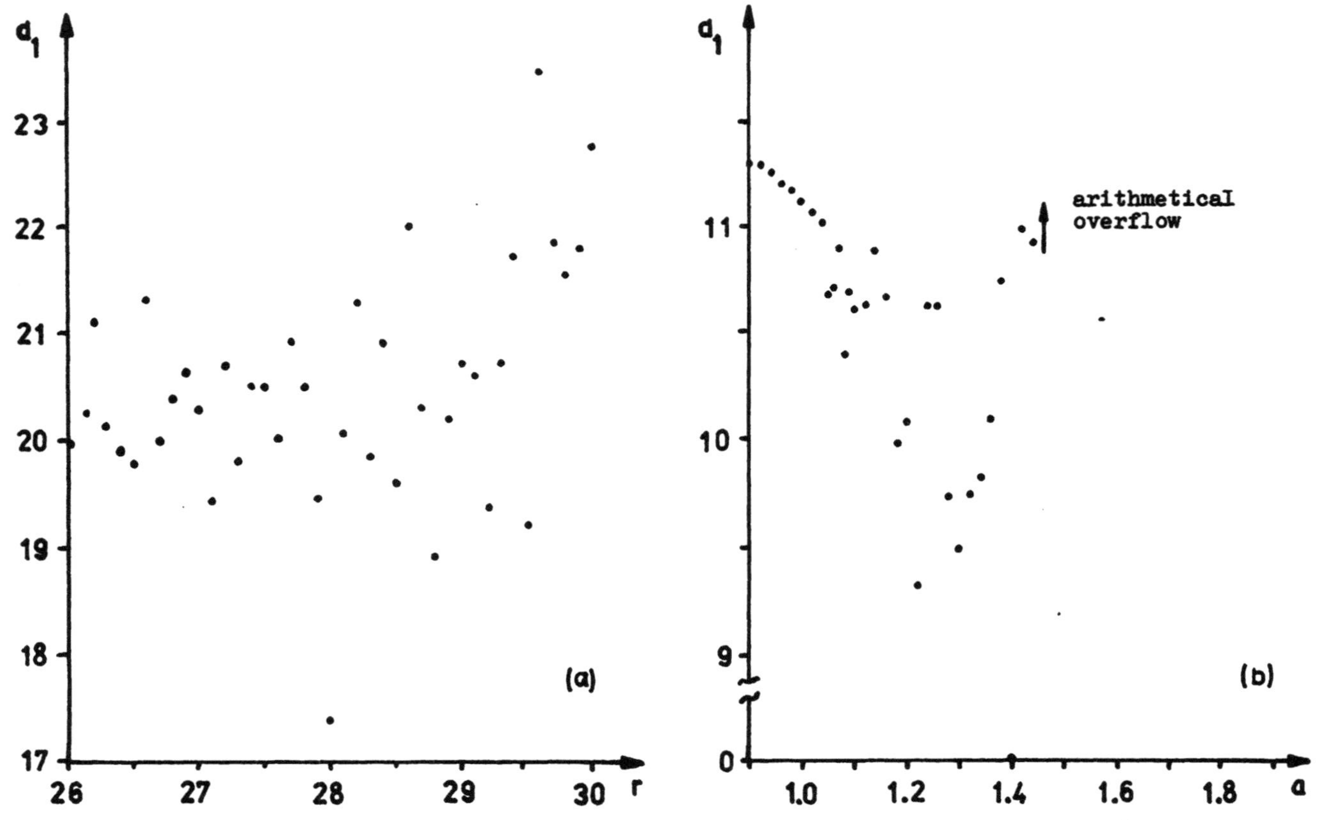

Fig. 1: Power function d_1 for the Lorenz system (a) and the Henon system (b)

parameter valley around the original value r = 28.0. Generalizing this result it seems understandable that the minimum value of a power function of type d_1 could only accidently be found.

Therefore one has to look for another type of power function to enable a successful parameter identification. The basic ingredient of the following considerations is the so called shadowing lemma by R. Bowen [7] : For every sufficient small ε there is a δ such that every δ-pseudo orbit can be ε-approximated - or shadowed - by an ordinary orbit. Here a sequence of points $\tilde{x}_i$ is called a δ-pseudo-orbit if

$\|\tilde{x}_i - f(\tilde{x}_{i-1})\| < \delta$ holds.

That means, if one finds a parameter interval for the model such that the orbit of the object system (1) is also a δ-pseudo-orbit for the model then there exists an ordinary model orbit such that both model and object trajectories approximate each other within a distance of ε. From the proof of the lemma follows immediately that the connection between ε and δ is monotonous. Hence it is sufficient to find a pseudo-orbit with a minimal δ.

For this reason a power function of type d_2 (p)

$$d_2(p) = \max_{i=1,\dots,N} \| x_i^{obj} - f(x_{i-1}^{obj}, p) \| \qquad (6)$$

is investigated next. In fig. 2 we show the power function $d_2(p)$ for different values of parameter a and r, respectively. In the case of the Henon system (4) there is a broad symmetric parameter valley with nearly linear flanks. Contrarily, for the Lorenz system (5) there is only a narrow steep valley around r = 28.0. Near the minimum of $d_2(r)$ a lot of fluctuations are visible making the identification problem difficult. One reason for the occurence of these fluctuations migth be numerical difficulties solving stiff differential equations. On the other hand, looking at the Lyapunov exponents [8,9] of both systems -

$0.42 > \lambda_i > -1.62$ (i=1,2) for system (4),

$0.90 > \lambda_i > -14.6$ (i=1,2,3) for system (5) -

it is obvious that the mean contraction and expansion of a volume element along a trajectory is in the case of the Lorenz system much stronger than for the Henon system. Both arguments suggest that for certain practical purposes the power functions of type d_2 are not applicable.

For that reason it was tried to reduce the strong requirement of the power function d_2 by averageing the distances of all N points consi-

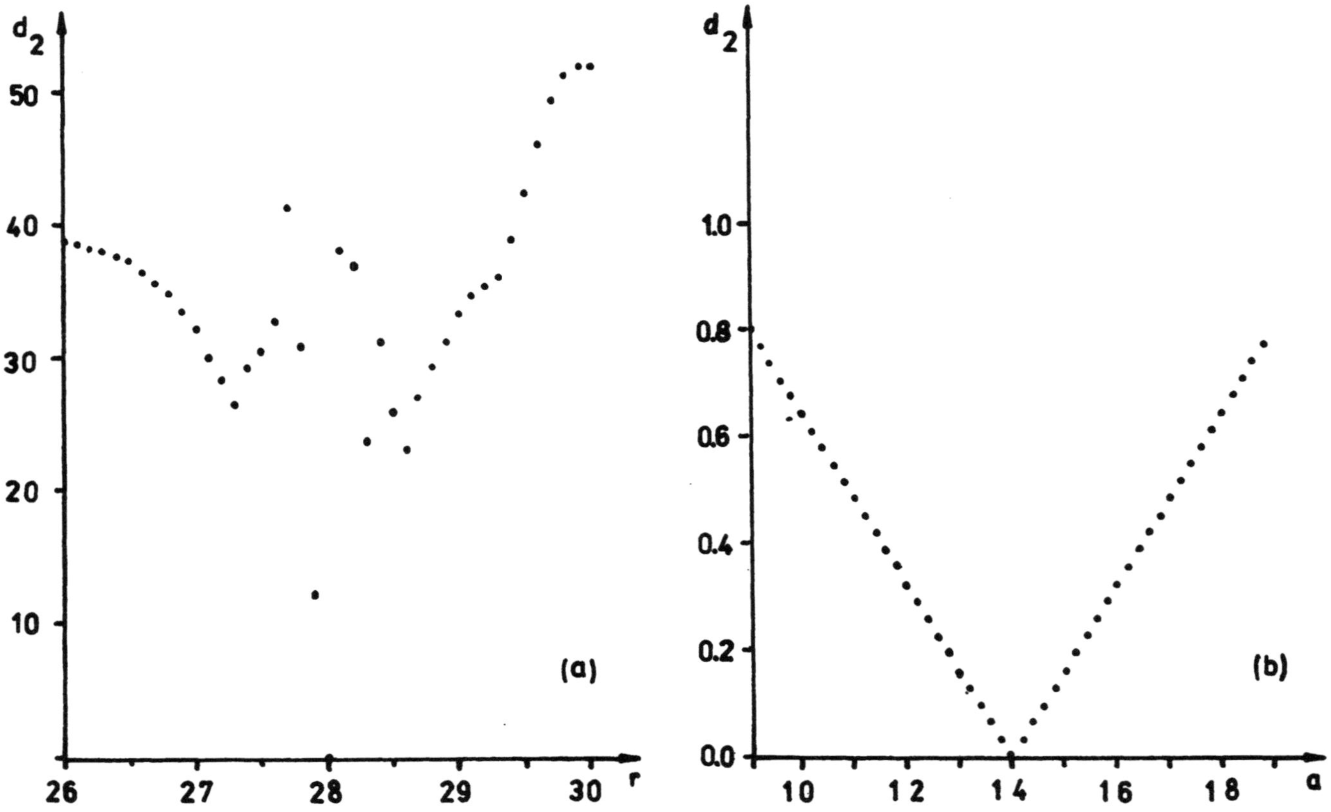

Fig. 2: Power function d_2 for the Lorenz system (a) and the Henon system (b)

dered. This results in a power function

$$d_3(p) = \left(\frac{1}{N}\sum_{i=1}^{N} \| x_i^{obj} - f(x_{i-1}^{obj}, p) \|^2\right)^{1/2} \tag{7}$$

which is similar to d_1 except that the model orbit is replaced by its pseudo-orbit.
As visible in fig. 3 for both cases investigated there exists a broad valley which easyly might be handled by minimization algorithms.

By our assumption about $f(x,p)$ this function is uniform Lipschitz continous

$$\| f(x,p_2) - f(x,p_1) \| < K \| p_2 - p_1 \| \,. \tag{8}$$

$(1/N \sum_i \| x_i \|^2)^{1/2}$ is a norm in the n*N-dimensional space and it follows that

$$\begin{aligned} |d_3(p_2) - d_3(p_1)| &= | (1/N \sum_i \| x_i^{obj} - f(x_{i-1}^{obj}, p_2) \|^2)^{1/2} - \\ &\quad (1/N \sum \| x_i^{obj} - f(x_{i-1}^{obj}, p_1) \|^2)^{1/2} | \\ &\leq (1/N \sum \| f(x_{i-1}^{obj}, p_1) - f(x_{i-1}^{obj}, p_2) \|^2)^{1/2} \\ &< K \| p_2 - p_1 \| \,. \end{aligned} \tag{9}$$

Therefore $d_3(p)$ is also Lipschitz continous with the same Lipschitz constant K, i.e. $d_3(p)$ is not steeper than $f(x,p)$ itself. Furthermore, if we chose

$$\| x \| = \left(\sum_{j=1}^{n} | x^{(j)} |^2 \right)^{1/2}$$

as R^n-norm then $d_3(p)^2$ is twice differentiable and therefore the Taylor series near the minimum exists only up to the quadratic terms. Generally that means that near the optimal parameter values $d_3(p)$ behaves like a positive definite paraboloid in multidimensional identification problems too.

For practical purposes it is important to have an idea about the failure which would arise if one uses the identified model instead of the object system.
In the following, object and model systems are assumed to be axiom-A-systems [10]. Then the entropy h of these systems is given by

$$h = \frac{1}{\log 2} \cdot \sum_{\lambda_i > 0} \lambda_i \tag{10}$$

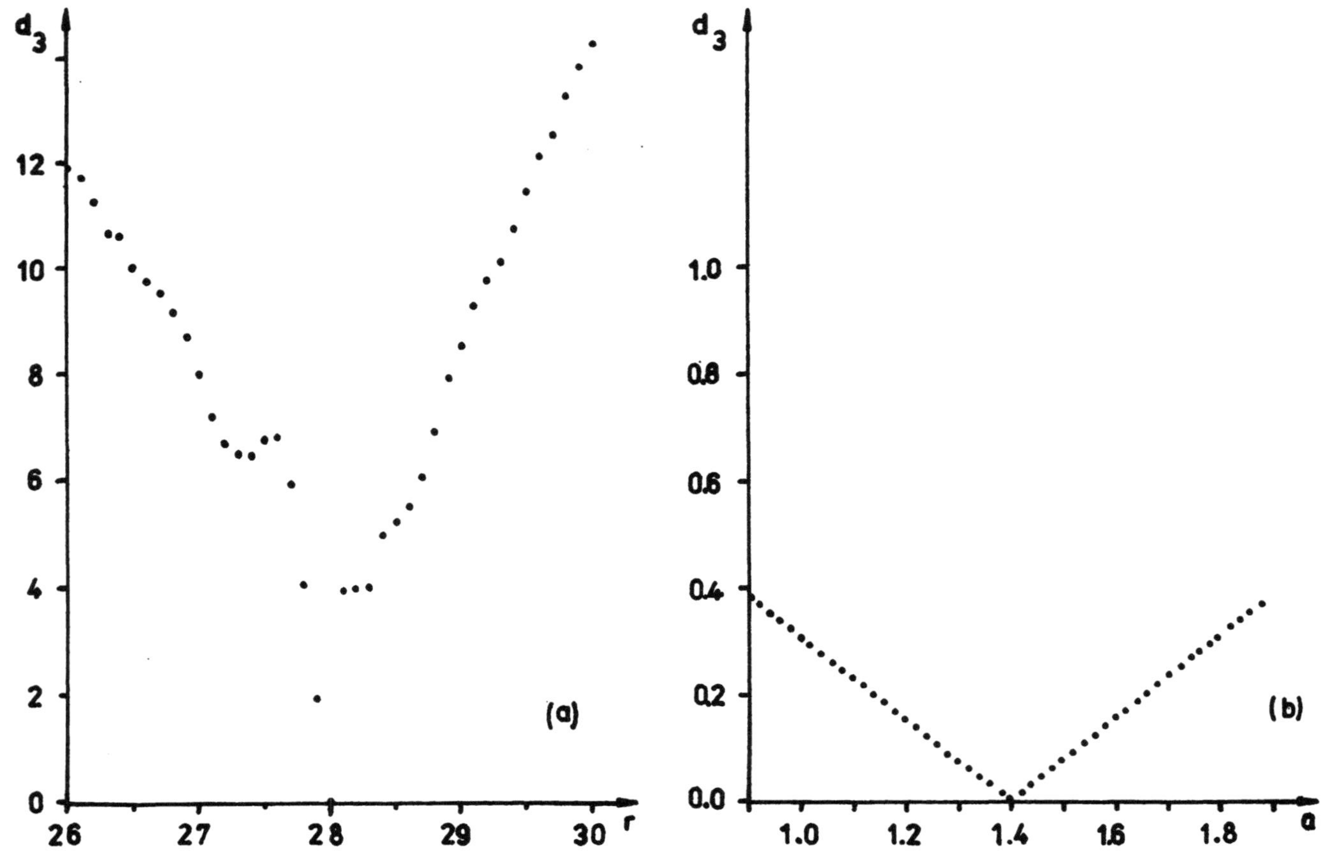

Fig. 3: Power function d_3 for the Lorenz system (a) and the Henon system (b)

where the λ_i are the one-dimensional Lyapunov-exponents [11] .Since the Lyapunov-exponents are average expansion and compression rates of co-moving volume elements along a trajectory equation (10) might be justified phenomenologically in analogy to the information definition given by Shannon [12,13] , i.e. h correspondes to the mean number of bits of initial information lost during every time step.
For the system (4) and (5) we estimated values of h=0.6 bits per iteration and h=1.3 bits per second,respectively. Assuming an initial information of about 25 bits which is a typical value for a 32-bit-computer-word and neglecting numerical and systematical errors this information will be lost after 42 iterations and 19 seconds,respectively. If we incorporate an additive error r per time step which includes all numerical and systematical uncertainties a first estimation of an error propagation law might be

$$e_i = 2^h \, e_{i-1} + r \tag{11}$$

which gives

$$e_i = r \, \frac{2^{ih}-1}{2^h-1} + 2^{ih} \cdot e_o \quad , \tag{12}$$

where e_i is the error of the i'th time step.
Here we have to emphasize, that the occurence of increasing differences e_i between model and object trajectories is not in contradiction to the shadowing lemma [7] which guarantees only the existence of one or a few model trajectories near the object trajectory whereas equation (12) was derived using the measure theoretic aspect of entropy and Lyapunov-exponents and therefore holds for almost all orbits starting near the object trajectory. However, eqn. (11) is a linear approximation and the finite diameter of attractors was not taken into consideration. Therefore these error formulas will hold for a few time steps only.
In order to get an estimation of some kind of a projection horizon the minimal value of the power function $d_3(p)$ might be interpreted as a measure r of systematical and numerical errors and then the number of time steps i for which the distance between model and object orbit is smaller than e_i might be calculated via equation (12).
Otherwise, having an impression of the values of the positive Lyapunov-exponents eqn. (11) can be used in experiment planning to determine a better discretization step size in cases where the object is realized as a continous system. If the error e_1 per step estimated with eqn. (11) is larger than the required accuracy the identification problem cannot be solved. Here the step size has to be divided into a few steps and new time series have to be recorded.

References:

[1] Lorenz, E.N.: "A Deterministic Nonperiodic Flow", J.Atmos. Sci. 20 (1963), 130.

[2] Packard, N.H., J.P. Crutchfield, J.D. Farmer and R.S. Shaw: "Geometry from Time Series", Phys.Rev.Lett. 15 (1980),No. 9, 712.

[3] Termonia,Y. and Z.Alexandrowicz:"Fractal Dimension of Strange Attractors from Radius versus Size Arbitrary Clusters",Phys. Rev.Lett.51 (1983), No. 14,1265.

[4] Grassberger,P. and I. Procaccia:"Characterization of Strange Attractors",Phys.Rev.Lett. 50(1983), No.5,346.

[5] Strobel,H.: "Experimentelle Systemanalyse",Akademie-Verlag, Berlin, 1975.

[6] Henon,M.:"A Two-Dimensional Mapping with a Strange Attractor", Commun.Math.Phys.50(1976),69.

[7] Bowen,R.:"Equilibrium States and the Ergodic Theory of Anosov-Diffeomorphisms",Springer Lecture Notes in Math.470(1975).

[8] Shimada,I. and T.Nagashima:"A Numerical Approach to Ergodic Problems of Dissipative Dynamical Systems",Progr.Theor.Phys. 61 (1979), No.6,1605

[9] Jansen,W. and U.Kriegel:"DEAN - A Differential Equation Analyzer", ZKI-preprint, in preparation.

[10] Bowen,R. and D. Ruelle:"The Ergodic Theory of Axiom A Flows", Invent.Math.29(1975),181.

[11] Oseledec, V.I.:"Mul'tiplikativnaja Ergodiceskaja Teorema", Trudy Mosk. Mat. Obscestva T19(1968),197.

[12] Shannon,C.E. and W.Weaver:"The Mathematical Theory of Communication", University of Illiois Press,1962.

[13] Shaw,R.:"Strange Attractors, Chaotic Behaviour and Information Flow", Z.Naturforsch. 36a(1981),80.

CHAOTIC DYNAMICS OF INFORMATION PROCESSING: THE "MAGIC NUMBER SEVEN PLUS-MINUS TWO" REVISITED.

By

John S. Nicolis, Dpt of Electrical Engineering
University of Patras, Greece

and

Ichiro Tsuda
"Bioholonics", Nissho Blds. 5F, Koishikawa 4-14-24
Bunkyo-Ku Tokyo 112 Japan

I. Introduction.

Central, amongst the aims of Cognitive Psychology, is the evaluation of human performance during the execution of mental tasks. In order to measure this performance in terms of bits of information it is instrumental to emulate the subject as an information channel. To this end a well defined, finite, set, of alternative stimuli (input) is provided by the experimenter. Such stimuli may be strings of digits, letters, words, geometrical symbols, tones, pictures etc. To each of these stimuli a definite apriori probability of occurrence is assigned (although in practice, that is as far as the subject is concerned, all stimuli are presented with equal probability). Then, a well-defined finite set of alternative responses (output) is selected (motor, vocal, reading, calculating, writing etc). Finally, the stimuli are presented to the subject in a randomized order at a steady rate in time and the subject's response to each stimulus is recorded.
From a long, statistically significant, sample of recorded data the measure of the transmitted information by the human channel can be evaluated as the input-output (stimulus-response) crosscorrelation.
The "channel capacity" of the subject is the upper limit of the extent to which the subject can match his responses to the given set of stimuli. Instead of increasing the rate of stimulus presentation the experimenter -giving the subject all the time he wants-may choose to increase the number of alternative stimuli among which the subject has to discriminate and determine the critical set N_c beyond which errors begin to occur. The $\log N_c$ bits defines under these circumstances the channel capacity.

A tremendous number of experiments using the above methodology have been performed in the past 40 years and a very lucid summing up of some most representative cases has been offered by George A. Miller (1974) in a by now classic collection of seven essays. In essence these essays refer to three distinct attributes of a human subject-seen as an information processor: The determination of the span of immediate attention or short-term memory, the span of absolute judgement and the limit of self-embedding in natural language.
Let us review quickly Miller's results:

a) The span of immediate attention: This attribute refers to the maximum number of digits (the longest string) a normal adult can repeat without error.
This number is about seven to eight digits. This "memory span" is limited by the number of symbols and <u>not</u> by the amount of information that these symbols may represent.

b) <u>The span of absolute judgement:</u>

This has to do with the maximum number of alternative stimuli beyond which confusion and errors are reported. The alternative stimuli may differ

mutually in one or more parameters (dimensions). No matter how many alternative stimuli (differing along one dimension only-say pitch) the subject is asked to evaluate, the best we expect him to do is to assign them to about 6-7 different classes without error.
Or, if we know that there are N alternative stimuli, then the subject's judgement enables us to narrow down (or "compress") the particular stimulus to one out of $\sim \frac{N}{7}$
The addition of dimensions to the stimulus (say pitch and loudness)increases the channel capacity but at a decreasing rate(This happens even when the dimensions along which the stimuli differ are not independent).
In other words we increase the total capacity at the expense of accuracy for any particular dimensions (parameter); So we can make relatively crude judgements of several things simultaneously.
It is plausible to assume that in the course of evolution the success of the organism has to do with the ability to respond to the widest possible range of stimuli.
In order to survive in a fluctuating environment it is better to have little information about a lot of things than to have a lot of information about few things (e.g. a restricted part of the environment).
At any rate it seems that there is a clear and definite limit in the accuracy with which we can identify absolutely the magnitude of a unidimensional stimulus variable: we call this limit, after Miller, the "span of absolute judgement". This is usually somewhere in the neighbourhood of seven alternative stimuli (or $\sim$ 2.5 bits).
So there is a span of absolute judgement that can distinguish about seven categories and there is a span of attention that can encompass about seven objects or symbols at a glance. Miller is quick to appreciate however that in spite of the appearance of the "magical number seven plus minus two" in both human attributes, the span of absolute judgement and the span of immediate memory are quite different kinds of limitations that are imposed on our ability to process information: Absolute judgement is limited by the amount of information while Immediate memory is limited by the number of items.
Using hierarchical coding we can in principle squeeze on a given item more and more bits. So, the span of immediate memory seems to be almost independent of the number of bits per symbol; with proper coding our ability to process information in terms of channel capacity may very well increase from 2,5 to 10 or 25 bits [Miller, (1974)].
In the words of Miller the "magical numbers seven" may very well be only "a permicious Pythagorean coincidence".
In spite of the above cautionary remarks we intent-referring to a given hierarchical level of cognition-to pursue the above "coincidence" as a symptom of a more deep and fundamental underlying dynamics.
The fact is that our pitifully small capacity for processing information has made necessary for us to develop ways of "recoding" or abstracting the essential features of the perceived stimuli in a hierarchical fashion thereby expressing our experiences in the form of simple "laws" or with brief words: It is this demonstrable ability to dynamically store and subsequently "compress" variety or to map information from higher to lower dimensionality state space, which gives us the impetus to propose (next section) a dynamical model which in principle takes care, in one stroke, of several seemingly unrelated attributes of human information processing.

c). <u>The recurrent (self-referential) character of natural (and artificial) languages, and the maximum number of discriminable self-embeddings.</u>

As almost everybody knows, clichés aside, learning a language does not involve memorization of particular strings of symbols but rather knowledge of rules for generating admissible strings of symbols.
These rules are given in terms of highly non-linear recursive algorithms. In artificial languages we do possess also Recursive Subroutines whose use

aim at increasing resolution via cascading iterations (see next-section, also).
Occasionaly the main routine is interrupted to perform a subroutine and then we return to main routine when the subroutine is finished.
It is possible also for one subroutine to refer to another subroutine.
Let us see however what happens when a subroutine is referring to itself. The "computer" interrupts the main routine to go into subroutine S. In the middle of execution S however, before S is completed the "computer" is instructed to stop what it is doing and execute subroutine S.
No difficulty arises until subroutine S is completed and the "computer" must decide where to resume its work. Unless special attention is taken in writing the programme the "computer" will not be able to remember that the first time it finishes subroutine S it must reenter subroutine S and the second time it finishes subroutine S it must return to its main routine. The situation gets more difficult with subroutine S calling on itself repeatedly; for each time it does so a new reentry address has to be remembered.
It is also a feature of natural language that sentences can be inserted inside sentences:
Miller gives amongst other the following example:
"The King who said" My Kingdom for a horse" is dead" (one embedding).
"The person who cited" The King who said"
"My Kingdom for a horse" is dead" as an example is a psychologist".
(two embeddings).
"The audience who just heard" The person who cited"
"The King who said" My Kingdom for a horse" is dead" as an example is psychologist" is very patient" (three embeddings)
As the recursive process goes on beyond say the fourth or the fifth self-embedding "chaos" sets in literally and actually; human subjects are very poor also in dealing intelligiblv with more than a few recursive interruptions.
We may ask now. Is there any common feature in the "Linguistic attributes, a) b) and c?
In this paper we intent to propose a dynamical model of an information processing scheme which ensures compressibility of (environmental) strings of digits and also fixes an optimum resolution (that is an optimum code word-length) which maximizes storage capacity.
We are going to see in specific examples that the optimum (average) length of such code is of the order of four to eigth bits. Furthermore the proposed scheme ensures that increasing resolution beyond a certain limit amplifies just microscopic noise thereby destroying compressibility.

II. Possible role of dissipative chaotic dynamics in information processing-Generalities.

Reliable information processing rests upon the existence of a "good" code (or map) or language: namely a set of non-linear recursive rules which generate and store variety (strings of symbols or "templates") at a given hierarchical level and subsequently compress it thereby revealing information at a higher level.
To accomplish this task a language-like good music-should strike at every moment an optimum ratio of variety (stochasticity) versus the ability to detect and correct errors (memory). Is there any dynamics available today which might model this dual objective in state space? The answer is: in principle,Yes. (see for example Nicolis 1982 , a, b, 1983 a, b,c,)
we have been investigating recently dynamical systems described by at least three coupled first order ordinary non-linear differential equations whose repertoire include (for different sets of values of the control parameters) multiple steady states, stable periodic orbits (limit cycles) tori and strange chaotic attractors. We may consider (subscribing to the scientist's credo that events are deduced only by observation and measurement) that variety is generated when the volume in state space expands

along certain directions through the dynamical evolution of our system (thereby decreasing resolution) and gets compressed (dissipated) when the volume in state space occupied by the flow contracts (increases resolution) along other directions towards a "compact" ergodic set-the attractor.
A three-dimensional strange attractor for example creates variety along the direction of his positive Lyapounov exponent λ_+ and constrains variety (thereby revealing information) along the direction of his negative Lyapounov exponent λ_- . Since $|\lambda_-| > \lambda_+$ the attractor is acting literally as a "compressor" -on the average that is* (see next section, below).
For values of the control parameters outside those associated with chaos variety (and complexity) is generated by a dynamical system via cascading bifurcations giving rise to broken symmetry;** within the values of the control parameters which trigger aperiodic trajectories variety is generated (or dissipated) via cascading iterations of the (e.g one-dimensional) map on the interval which is constructed as a "poincare return map" of the attractor. This type of "analog to -digital conversion" is accomplished by parametrizing the attractor along a one-dimensional cut and plotting the position a trajectory crosses the cut versus the position it crosses the next time around the attractor. This stands for a "stroboscopic" pursuit of the (three dimesional) state space flow and gives rise to a **Markov** chain whose number of states **depends** on the partition of the interval
The change in observable information is generally given by the $\log_2$ of the ratio of states $\Sigma(t)$ distinguishable before and after some time interval

$\Delta I = \log_2 \frac{\Sigma_f}{\Sigma_i}$ or $\sim \log_2 \frac{V_f}{V_i}$ where V_f, V_i are the final and initial volumes in state space. The rate of information creation or dissipation is given as $\frac{dI}{dt} = \frac{1}{V}\frac{dV}{dt} \sim \frac{1}{\Sigma}\frac{d\Sigma}{dt}$

In non-chaotic systems the sensitivity of the flow on the intial conditions grows with time at most polynomially say as $\Sigma(t) \sim t^n$; then $\frac{dI}{dt} \simeq \frac{n}{t}$. The variety creation rate from such a system converges to zero as time passes so the systems behavior is predictable; the set of generated strings form rational numbers.
In chaotic systems the sensitivity of the flow on the initial conditions grows with time **exponentially** say as $\Sigma(t) \sim e^{nt}$ and $\frac{dI}{dt} \sim n$; such a system is a continuous information source; this information is not implicit in the whatever initial conditions but is generated by the flow itself in state space: A chaotic dynamics creates strings of digits which constitute aperiodic oscillations or segments of irrational, non-computable numbers. If such a chaotic system is observed by another cognitive system at a sampling rate less than n, no prediction of the system is possible.
We are interested here in cognitive systems characterised by non conservative compact flows where the phenomenon of attraction that is of dimensionality compression (see next section) is possible.
Now for any given typical one-dimensional map in the interval $y=F(x)$ the probability density $P(x)$ for finding the orbit at x can be estimated via successive iterations, from the recursive relation

*The Lyapounov exponents determine the average amount of information production ($\lambda>0$) or dissipation ($\lambda<0$) per iteration.

** For example this is the case of a Hopf bifurcation leading from a destabilized steady state to a stable limit cycle.

$$P_n(x) = \sum_{i=1}^{2^{n-1}} \frac{P_1(x_i)}{\left|\frac{dF^{(n-1)}}{dx}\right|_{x_i}} \qquad \text{where } x_i = (F^{n-1})^{-1}(y)$$

$$x(n+1) = P(x(n)), \ (n,= 2,3,.....)$$

where the first p.d.f $P_1(x)$ can be taken arbitrarily and the resulting P(y)-corresponding to the initial $P_1(x)$ transformed by the action of the map-is the first approximation. Successive iterations of the map will produce closer and closer approximations to the correct asymptotic function of P(x). If the map possesses a stable steady state or a number of stable periodic orbits P(x) will converge to a sharp "δ" function spike or a series of "δ" -like functions on the periodic points. In the chaotic regime of the control parameters the map will possess a continuous P(x) with interspersed spikes.

The information change ΔI per iteration of the particular initial point x on the interval will be determined as

$\Delta I = \log_2\left|\frac{dF}{dx}\right|$ and will amount to variety creation for slopes >1 and information (dissipation) for slopes <1.

The average information change over the whole interval $0 \leq x \leq 1$ will be given by $\langle I\rangle = \int_0^1 P(x)\log_2\left|\frac{dF(x)}{dx}\right| dx$ bits; If t(x) stands for the time between successive passes through the return map, the average information production (or dissipation)rate will be

$$\left\langle\frac{dI}{dt}\right\rangle = \int_0^1 \frac{P(x)}{t(x)} \log \left|\frac{dF(x)}{dx}\right| dx \ \frac{\text{bits}}{\text{sec}}$$

Finally if the initial condition from which the process starts is not known exactly (which is always the case)* i.e it is determined by a apriori probability density distribution $P_0(x)$, one gets for the informational value of this initial condition the expression

$S = \int_0^1 P_0(x)\log_2\left\{\frac{P_0(x)}{P(x)}\right\} dx$ bits within the interval; then the "memory" of the processor measured as the time elapsing until the flow gets completely disconnected from the initial conditions is given as

$$T = \frac{S}{\left\langle\frac{dI}{dt}\right\rangle} = \frac{\int_0^1 P_0(x)\log_2\left\{\frac{P_0(x)}{P(x)}\right\} dx}{\int_0^1 \frac{P(x)}{t(x)} \log_2\left|\frac{dF(x)}{dx}\right| dx} \ \text{sec}$$

Beyond this limit, the iterative process becomes so fine that the processor amplifies essentially intrinsic microscopic noise; due to the great resolution created by the cascade of iterations this noise cannot be "smeared out" any longer and declares itself on the cognitive level thereby rendering any further measurement or observation unreliable. Computing simulation(Shaw,1981)shows that in specific cases the number of iterations under which T is achieved may be rather small. (~20÷30)

III. Information dimension, compressibility and non-uniform processing.

The information dimension D_1 of a strange attractor (see for example Farmer 1982, Farmer et al. 1983) is the number of bits one requires for the unambiguous determination of any point on the attractor; in turn it represents the maximum number of bits we can reliably store in such a self-sustained dynamics-under a given resolution.

The "average degree of compressibility" ensured by the attractor is defined Nicolis et al 1983c) as $\eta = \frac{N-D_1}{N}$ % where N is the dimensionality of

*This initial condition is a point x_0 on the unit interval. It can be expressed as a digital number .011001110011001..; with overwhelming probability this number is incomresible.

state space in which the attractor is embedded as a compact subset. Formally in order to calculate D_1 one proceeds as follows:
One partitions the N-dimensional state space in $n(\varepsilon)$ hypercubes of size (degree of resolution) ε from each of which the flow (or "the trace of the scanning path" of the processor) passes at least once.
Let $P_i(\varepsilon)$ be a number proportional to the relative frequency with which the scanner visits the specific square i. Then the "entropy" of the digitized pattern can be calculated as:

$-\sum_{i=1}^{n(\varepsilon)} P_i(\varepsilon)\log_2 P_i(\varepsilon)$ bits, and as the resolution becomes finer and finer one may be interested in calculating how the distribution $P_i(\varepsilon)$ scales with resolution. One could obtain then the "information dimension" or the bits required for the determination of any point on the "scanning curve" of the processor as:

$$D_I = \lim_{\varepsilon \to 0} \frac{-\sum_{i=0}^{n(\varepsilon)} P_i(\varepsilon)\log_2 P_i(\varepsilon)}{|\log_2 \frac{1}{\varepsilon}|}$$

i.e the asymptotic value of the slope of entropy versus resolution.
To make calculations feasible one should not start from the above expression however, but rather from a relationship between the information dimensionality and the spectrum of the Lyapounov-exponents of the flow (or the discrete map) and calculate not only D_I but also the standard deviation ΔD_I from the second moments of the distributions of local divergence rate of trajectories whose first moments are the Lyapounov exponents; ΔD_I would indicate how much fuzziness enters in the average degree of compressibility.

$\eta = \frac{N-D_I}{N}$ of information realized by the attractor in N-dimensional space.

Since compressibility of a string of digits is the necessary prerequisite for the subsequent simulation of the time series involved, the importance of strange attractors as information processing models cannot be overlooked.
Indeed for an attractor simulating some aspects of a cognitive system one has two basic requirements: Big storage capacity and good compressibility. Let us turn first to the simplest attractors (in any N-dimensional state space) namely stable steady states and stable limit cycles: They have information dimensionality of zero and one respectively.*Consequently they are very poor as information storage units but since they possess only non-positive Lyapounov exponents are ideal as information compressing devices. Strange attractors on the other hand, via an harmonious combination of positive and negative Lyapounov exponents λ_+, λ_- can in principle comply somewhat with both requirements:
Strange attractors may possess a considerable information dimension and this idiom makes them suitable for information storage. On the other hand, being "attractors" i.e possessing also negative Lyapounov exponents λ_- such that $|\lambda_-| > \lambda_+$ they can serve at the same time as information compressors.
The type of the strange attractor however we intend to employ as a model here must be rather "loose" in order to comply with the preceding experimental results.
It must, in other words, possess a large information dimensionality in order to allow for maximum dynamical storage of variety.
We are going to investigate the optimum resolution ε under which this information dimensionality becomes maximum;
the ensuing compressibility factor $C_T = N-D_T$ will be also investigated.
So, in case where our primary aim is **compressibility** the best attractors around are steady states and limit cycles.

*In the limit $\varepsilon \to 0$; For ε finite, the dimensionality of a limit cycle $D_I \ll 1$.

In case however our data (this case) ask for a model which can ensure first of all large storage ability-without jeopardizing compressibility a "loose" strange attractor is the preferred model. In cognition usually one needs both types of attractors-working in "tandem"perhaps. But in same cases(Nicolis and Tsuda 1984)we can do better. Namely we can have just one attractor which under the action of external noise switches from the one model to the other, thereby complying with both requirements for maximum compressibility and maximum storage capacity.

VI. Specific examples and Relevance of the Model to Cognitive psychology-experimental results:

What is the relevance (if any)of the dynamical theory given in sections II and III vis-a-vis the experimental results reported in the introduction-which results essentially refer to empirical limitations of human attributes in information processing? To answer this question let us try to see if there exists in various specific cases of dynamical models of processors an optimum resolution ε (different for each model) or an optimum codeword length for which the cardinal parameter in our theory, the information dimension of the attractor becomes maximum-without jeopardizing the average "compressibility factor"

$$\eta = \frac{N-D_I}{N} \%$$

Let us consider the expression $C_I = N-D_I$ for the degree of average compression realized by the particular attractor and try to calculate the critical resolution ε^* for which D_I becomes maximum or,

$\frac{\partial C_I}{\partial \varepsilon} = 0$ To take into account the best case let us perform the above calculation for the maximum value of

$$D_I = \frac{-\sum_{i=1}^{n(\varepsilon)} P_j(\varepsilon)\log_2(\varepsilon)}{\log 1/\varepsilon}, (\varepsilon \text{ finite})$$

The maximum value of the information dimension under given finite resolution ε is the "fractal dimension" of the attractor and is realized

for $P_i(\varepsilon) = \text{const} = \frac{1}{n(\varepsilon)}$ so,

$$D_I = -\frac{\ln[n(\varepsilon)]}{\ln\varepsilon} \text{ and}$$

$$C_I = N + \frac{\ln[n(\varepsilon)]}{\ln\varepsilon} = \frac{N\ln\varepsilon + \ln n(\varepsilon)}{\ln\varepsilon} = \frac{\ln[\varepsilon^N n(\varepsilon)]}{\ln\varepsilon}$$

We intent now to express $n(\varepsilon)$ in terms of the Lyapounov exponents of the attractor and the resolution length ε.
To this end, we proceed as follows:
Let the number of points on a (one-dimesional)poincare return map of the attractor concerned be M; this means that we determine the orbit with strings of M digits long. The number of points-"boxes" representing the attractor, on the other hand, with a degree of coarsness ε is just $n(\varepsilon)$. The number of points representing the attractor per cycle is therefore

$\frac{n(\varepsilon)}{M} \sim t_c$ where t_c is the time between successive strings of M bits long

The degree of resolution ε has to do with the precision with which we can determine any one of 2^M strings of 0's and 1's that is the Mth refinement of a Markov partition; we have then,

$\varepsilon \sim \frac{1}{2^M}$ and $M = \frac{\ln \frac{1}{\varepsilon}}{\ln 2}$; So,

$$n(\varepsilon) = t_c \frac{\ln(\frac{1}{\varepsilon})}{\ln 2}$$

During the time t_c, $\frac{n(\varepsilon)}{M}$ sampling have occurred with say sampling time period $\Delta\tau \sim 1$.

But after $\frac{n(\varepsilon)}{M}$ iterations in the map the system is completely disengaged from the initial conditions which means that the time t_c can be estimated from the relation:

$\varepsilon e^{\lambda_+ . t_c} \sim 1$ where λ_+ is the positive Lyapounov exponent. So

$t_c \sim \frac{\ln(\frac{1}{\varepsilon})}{\lambda_+}$ and finally $n(\varepsilon) = \frac{\{\ln(\frac{1}{\varepsilon})\}^2}{\lambda_+ . \ln 2}$; substituting in the expression of C_I we get:

$$C_I = \frac{N \ln\varepsilon + 2\ln\left(\ln\frac{1}{\varepsilon}\right) + \ln\left(\frac{1}{\lambda_+ . \ln 2}\right)}{\ln\varepsilon} \quad \text{or,}$$

$$C_I = N + \frac{\gamma + 2\ln\left(\ln \frac{1}{\varepsilon}\right)}{\ln\varepsilon} \quad \text{where} \quad \gamma = \ln\left(\frac{1}{\lambda_+ . \ln 2}\right)$$

The requirement for maximum fractal dimension $\frac{\partial C_I}{\partial \varepsilon} = 0$ gives for the optimum resolution $\varepsilon^* = \exp\left[-e^{\frac{2-\gamma}{2}}\right]$ or the optimum code length

$$M^* = \frac{\ln(\frac{1}{\varepsilon^*})}{\ln 2} \quad (1)$$ from which,

$$C_I(\varepsilon^*) = N - 2e^{\frac{2-\gamma}{2}} \text{ bits} \quad (2)$$

These are our final formulas, for the minimum average compressibility i.e the maximum fractal dimensionality of the attractor concerned and the corresponding optimum code length M^*. It remains to make sure that the extremum of the function.

$$f(\varepsilon) = -D_i(\varepsilon) = \frac{\gamma + 2\ln(\ln \frac{1}{\varepsilon})}{\ln\varepsilon}, \ \varepsilon \neq 0$$

is indeed a minimum or <u>$D_I(\varepsilon)$ is maximum</u>

In fact we have:

$$\frac{\partial f(\varepsilon)}{\partial \varepsilon} = \frac{2 - \left[\gamma + 2\ln(\ln\frac{1}{\varepsilon})\right]}{\varepsilon(\ln\varepsilon)^2} \quad \text{and}$$

$$\frac{\partial^2 f(\varepsilon)}{\partial\varepsilon^2} = \frac{-6+2\gamma+4\ln(\ln\frac{1}{\varepsilon})+(\ln\varepsilon)\left[\gamma-2+2\ln(\ln\frac{1}{\varepsilon})\right]}{\varepsilon^2(\ln\varepsilon)^3};$$

for $\varepsilon = \varepsilon^* = \exp\left[-\exp(\frac{2-\gamma}{2})\right]$ we get

$$\left(\frac{\partial^2 f(\varepsilon)}{\partial\varepsilon^2}\right)_{\varepsilon=\varepsilon^*} = \frac{2}{\exp\left[-2\exp(\frac{2-\gamma}{2})\right]\exp(\frac{3}{2}(2-\gamma))}$$ which

is always positive. So $f(\varepsilon)$ possesses a minimum at ε^* and consequently $D_I(\varepsilon) = -f(\varepsilon)$ becomes maximum.

$D_I(\varepsilon)$ becomes zero for $\varepsilon_0 = \exp\left\{-e^{-\gamma/2}\right\}$. Negative values of D_I do not possess of course any physical meaning.

Let us now apply formulas (1) and (2) to a number of simple case:

a) <u>The Lorenz attractor</u> $\dot{x} = \sigma(y-x)$, $\dot{y} = -y+rx-xz$, $\dot{z} = xy-bz$ (for $b = 4$, $\sigma = 16$ and $r \simeq 45.92$)

we calculate with the computer $\lambda_+ \simeq 1,5$ $\gamma \sim -0.03895$,

$C_I(\varepsilon^*) \sim 3-0,72157 \sim 2,28$ $\varepsilon^* \sim 0.062$ and $M^* \sim 4$.

That is, the optimum code which maximizes storing capacity by the Lorenz attractor is $\sim$4 bits long; Nevertheless the resulting compressibility is impressive too-much larger of what one calculates from the information dimensionality with $\varepsilon \to 0$. $(C_I(0) \simeq 3-2.06=0.94)$

b) The Bernoulli map: $x' = \mathrm{Mod}_1(2x)$. In this case of one-dimensional map-assumed here to be uniform-the number of covering segments of resolution ε is $n(\varepsilon) \sim \frac{\ln(1/\varepsilon)}{\lambda_+}$ from which we get,

$C_I \sim N + \frac{\ln(\ln\frac{1}{\varepsilon})-\ln\lambda_+}{\ln\varepsilon}$ The requirement $\frac{\partial C_I}{\partial\varepsilon} = 0$ gives for the optimum resolution

$\varepsilon^* \sim \exp\left\{-e^{1+\ln\lambda_+}\right\}$ from which $C_I(\varepsilon^*) = N - e^{-(1+\ln\lambda_+)}$

The numerical calculation for $\lambda_+ = 2$ in this case, give

$C_I(\varepsilon^*) = 0,81606$, $\varepsilon^* = 0,004354$.

and $M^* \sim 7 \div 8$ bits.

The second case above corresponds to a map with uniform p.d.f $P(x)$.

So the NUF[1] here is zero. In other cases however (the assymetric tent map the logistic map, the Henon map, the Rössler attractor, Nicolis et al 1983c) the NUF may, for certain intervals of the control parameters greatly exceed the Lyapounov exponent.

In such cases the maximization of the average storing capacity does not provide the optimum code length: One expects a great variance ΔM^*-which is in conformity with most of Miller s experiments about the limiting length of errorless recall of strings of items or digits.

(1) Non-uniformity-factor. Describes the variance of the local divergence rate of a chaotic trajectory.

References

1) G.A. Miller "The psychology of communication",Penquin (1974)

2) J.S. Nicolis "The role of chaos in Reliable information processing" J.of the Franklin Inst. In press. Also in "Synergetics" (Ed. H. Haken)Proceeding of the Elmau Conference 1-8 May (1983)a

3) J.S. Nicolis "Sketch for a dynamical theory of Language" Kybernetes (1982a) April, 11 pp 123-132.

4) J.S. Nicolis "Should a reliable information processor be chaotic"? Kybernetes (1982b) Oct. 11 pp. 269-274

5) J.S. Nicolis "Dynamics of Hierarchical Systems. An evolutionary approach" To appear in Springer series of Synergetics (H.Haken Ed) (1983)b.

9) J.Doyne Farmer E.Ott,J.A.Yorke "The Dimension of chaotic attractors" Physica 7D 1983 p153.

10) J.S.Nicolis G Mayer-Kress G Haubs "Non uniform chaotic dynamics with implications to information processing" Z. Naturforsch (1983c) 38a pp1157-1169

11) Robert Shaw "Strange Attractors, Chaotic Behavior, and Information Flow" Z. Naturforsch. 36a, pp 80-112 (1981)

12) J.Doyne Farmer "Information Dimension a-d the probabilistic structure of chaos" Z.Zaturforsch 37a 1304 (1982)

13) J.S.Nicolis I.Tsuda Submitted to: Zeit. für Naturforsch.(1984)

INFORMATION - THEORETICAL METHODS IN THE STOCHASTIC THEORY OF OPEN SYSTEMS

G. Czajkowski[1]

1. The information-theoretical scheme

To understand the phenomena in open systems such as e.g. the transition between multiple stationary states it is necessary to go over to a stochastic formulation via master equations or related stochastic equations. Mostly we deal with a master equation ME of the type

$$P(x,t) = \sum_{x'} w(x,x')\, P(x',t) \tag{1}$$

where x denotes the possible states of the system and $P(x,t)$ the probability at the time t see, for example, [1] . The next two steps in the stochastic theory are :

1. the determination of a stationary solution of (1) if it exists,
2. the determination of the general time-dependent solution of the ME (1) .

* In particular, we are interested in the time dependence of fluctuations, correlation functions and other statistical characteristics of the relevant variables.

Unfortunately, the steady state is known explicitly only for step-by-step processes, and in some particular cases; otherwise one has to make use of some approximations. Even if the steady state is known the determination of the time dependent solutions is very complicated. The problem of a reasonable approximation appears. In recent papers [2-6] the use of information-theoretical methods was proposed. The scheme is as follows

1. Let $X \ni x$ be the set of the possible states and $Q^j(x)$ $j=1,\dots,n$ a set of linearly independent functions.
2. The mean values of $Q^j(x)$ at the time t are given by

$$M_j(t) = \sum_{x} Q^j(x)\; P(x,t) \tag{2}$$

[1]Address: Institute of Mathematics and Physics, Academy of Technology and Agriculture, Al. Prof.S.Kaliskiego 7, 85-790 Bydgoszcz, Poland.

3. Given a set $M(t) = (M_j(t))$ one can construct an "accompanying ensemble" a macrostate

$$K_M = \{ P(x,t) \; : \; P \geq 0, \sum_x P(x,t)\, Q^j(x) = M_j(t), \sum_x P(x,t) = 1 \} \quad (3)$$

4. The information $S[K]$ of the macrostate K is defined as

$$S[K] = \sup_{P \in K} S[P] \qquad , \; S[P] = -\sum_x P \ln P \qquad (4)$$

5. It was proved that the maximum, if it exists, is attained for a distribution $P_K(x,t)$ which is called the representative distribution for the macrostate K_M. It has the form

$$P_K(x,t) = \exp\{ -\lambda_0(t) - \sum_j \lambda_j(t)\, Q^j(x) \} \qquad (5)$$

with

$$\lambda_0 = \ln \sum_x \exp\,(-\sum \lambda_j(t)\, Q^j(x)) \qquad (6)$$

cf., e.g., [7-9] . The functions $\lambda_j(t)$ are related to the M_j's by the equations

$$\partial \lambda_0 / \partial \lambda_j = M_j \; , \; j = 1,\dots,n \; . \qquad (7)$$

6. Sometimes we will look for a distribution which minimizes the information-gain $I\,[P,P^0]$,

$$I\,[P,P^0] = \sum_x P \ln P/P^0 \qquad (8)$$

over a macrostate K ,P^0 is a certain reference state. It has the form

$$P_K' = \exp\,(-\lambda_0 - \sum_j \lambda_j Q^j)\; P^0 \qquad (9)$$

where

$$\lambda_0 = \ln \sum_x P^0 \exp(-\sum \lambda_j Q^j) \qquad (10)$$

7. The main point of the proposed approximation is : the exact solution $P(x,t)$ of the equation (1) can be approximated either by a representative distribution (5) or, if a certain distribution P^0 is distinguished (e.g., the stationary solution is explicitly known) by the distribution P_K' (9).

The unknown functions $\lambda_j(t)$ are determined by the equations

$$\sum_{\underline{x}} Q^j(\underline{x}) \dot{P}_K(\underline{x},t) = \sum_{\underline{x}} \sum_{\underline{x}'} Q_j(\underline{x}) \; w(\underline{x},\underline{x}') \; P_K(\underline{x}',t) \tag{11}$$

j=1,...,.n, due to some initial conditions $\lambda_j(0)$.

8. In particular, the stationary solution P^s of the ME (1)can be assumed in the form (8) or (9) . The coefficients λ_j will be found by the equations (11)where we put $\dot{P}^s = 0$.

2. An approximated stationary solution of a master equation

The above scheme will be, for demonstration, applied for the well-known Schlögl's model of the 2nd order nonequilibrium phase transition [10] . The appropriate master equation has the form

$$\dot{P}(x,t) = w(x,x-1) \; P(x-1,t) + w(x,x+1) \; P(x+1,t)$$

$$- \left[w(x+1,x) + w(x-1,x) \right] \; P(x,t) \tag{12}$$

with

$$w(x-1,x) = ax + x(x-1) \quad ; \quad w(x+1,x) = b(x+1) \tag{13}$$

a,b are external parameters [11] , and x is assumed to be a discrete variable x = 0,1,...,N. Here the exact stationary solution of [12] is known and has the form

$$P^s(x) = P^s(0) \prod_{m=1}^{x} b\,(a+m-1)^{-1} \tag{14}$$

We can compare it with the approximation obtained by the information-theoretical scheme.Let us take, for simplicity, $Q^1 = x$ and

$$P_K = \exp(-\lambda_0 - \lambda x) \tag{15}$$

For $N \longrightarrow \infty$ we have

$$\lambda_0 = \ln(1 - e^{-\lambda})^{-1} \quad , \quad M_1 = e^{-\lambda}(1 - e^{-\lambda})^{-1} \tag{16}$$

where M_1 denotes the corresponding average of x. On the other hand we obtain for the stationary values of the statistical moments M_1^s, M_2^s from (12)

$$\dot{M}_1^s = \langle w(x+1,x)\rangle^s - \langle w(x-1,x)\rangle^s = 0 \qquad (17)$$

$$0 = b(M_1^s + 1) - a M_1^s - (M_2^s - M_1^s) \qquad (18)$$

Due to the information-theoretical scheme we put M_1 from (16) into (18) and set M_2 in the form

$$M_2 = M_1 + 2 M_1^2 \qquad (19)$$

following from (15). So we obtain

$$2 M_1^2 + M_1 (a-b) - b = 0 \qquad (20)$$

The positive root of the equation is

$$M_1 = (1/4) (- (a-b) + \sqrt{(a-b)^2 + 8b}) \qquad (21)$$

One usually starts from the situation that $a \gg b$ (below the threshold) Here we obtain the value

$$M_1 \simeq b / (a-b) \qquad (22)$$

which agrees well with the exact stationary value [11]. The test for the approximation may be done by considering the equations for higher statistical moments. We have

$$\dot{M}_2^s = 0 = b + (a + 3b - 1) M_1^s + (2b - 2a + 3) M_2^s - 2 M_3^s \qquad (23)$$

For our approximation (15)

$$M_3 = 6 M_1^3 + 6 M_1^2 + M_1 \qquad (24)$$

On the other hand it follows from (23) for $b \ll a$

$$M_3^s = M_1^s \qquad (25)$$

which is compatible with (24) only for $M_1^s \to 0$. The other extremum $b \gg a$, for $b \gg 1$, leads to

$$M_1 = b/2 \gg 1$$

which does not agree with the stationary value $M_1 \sim b$. We may also prove that the relation (23) is not satisfied now. Substituting (25) and $M_2 \approx 2 M_1^2$ we obtain

$$M_3 \simeq 4 M_1^3 \tag{26}$$

which disagrees with (24) . So we obtained the well-known result that the stationary distribution above the threshold is not of the type (15) . One deduces that it has at least the form

$$P_K(x) = \exp(-\lambda_0 - \lambda_1 x - \lambda_2 x^2) \tag{27}$$

Let us assume, for simplicity, that λ_1, λ_2 are those of the Gaussian distribution. Then

$$M_3 = 3 M_1 M_2 - 2 M_1^3 \tag{28}$$

Substituting M_2 from (18) to the equation (23) we obtain for M_1 :

$$b(2b-2a+4) + M_1[2b^2 + 2b(1-2a) + 4a^2 - 2a - 1] - 6 M_1^2 (b-a+1) + 4 M_1^3 = 0 \tag{29}$$

Note that for $b \gg a > 1$ we obtain approximately $M_1 \approx b$, in agreement with the exact stationary results [11] . Otherwise we have to solve the eqs. (18) , (23) together with

$$M_j = \sum_{K=1}^{N} x^j \exp(-\lambda_0 - \lambda_1 x - \lambda_2 x^2) \quad , j = 1,2,3 \tag{30}$$

which is possible only numerically.

There is a non-negligible region where the approximations (15) and (27) are not valid. It is, in particular, the critical region where $a \approx b$. At the critical point we have $a_c = b + 1$. We proved previously [2 , 12] that for $a,b \gg 1$ the critical distribution behaves like $\exp(-\lambda_0 - \lambda x^2)$. From the eqs. (18) we get

$$M_2^{cr} = b$$

$$M_3^{cr} = b(1 + 2 M_1^{cr}) = M_2^{cr} + 2 M_1^{cr} M_2^{cr} \tag{31}$$

By means of the distribution $\exp(-\lambda_0 - \lambda x^2)$ we have

$$M_1 = (2M_2/\pi)^{1/2}, \quad M_3 = 2M_1M_2, \quad \lambda = 1/2M_2 \tag{32}$$

the relation for M_3 in full analogy to that obtained by the dynamical equations (8). Since for $b \gg 1$ the distribution $\exp(-\lambda_o - x^2/2b)$ becomes very smooth we may test a "linear approximation" of (8) in the form

$$P(x) = \begin{cases} z^{-1}\left(1 - \sum_j \lambda_j x^j\right) & x \leqslant N_0 \\ 0 & N_0 < x < N \end{cases} \tag{33}$$

with,

$$Z = N_0\left[1 - \sum_i \lambda_i A_i\right], \quad A_i = (1/N_0)\sum_{x=1}^{N_0} x^i \tag{34}$$

see also [13], p. 96. The relations (6) take the form

$$M_j = A_j - \sum_i c_{ij}\lambda_i \quad ; \quad c_{ij} = A_{i+j} - A_iA_j, \tag{35}$$

N_0 is a new variable which may be considered as a "fitting parameter". We considered; for illustration, the following numerical example. Let $b = 38.5$. The approximation $P \sim \exp(-\lambda x^2)$ yields

$$M_1^{app} = 4.95, \quad M_3^{app} = 381.2 \tag{36}$$

On the other hand by the relation (31) we obtain

$$M_3^{cr} = 419.7, \quad M_3^{app} / M_3^{cr} = 90\ \%$$

The linear approximation (33-35) in a very crude approximation of an equally likely distribution

$$P\,x = \begin{cases} 1/N_0 & 0 \leqslant x \leqslant N_0 \\ 0 & \text{otherwise} \end{cases}$$

gives

$$N_0 = 10 \quad , \quad M_1^{app} = 5.5 \quad , \quad M_3^{app} = 302.5 \quad .$$

The relation (31) gives here

$$M_3^{cr} = 462 \quad , \qquad M_3^{app} / M_3^{cr} = 65\,\% \quad .$$

Comparing the two distributions one obtains a physical meaning of N_0. We have

$$P(x > N_0) \simeq .1 \, ,$$

$$P(x \leqslant N_0) \simeq e^{-\lambda_0} \int_0^{N_0} e^{-\lambda_* x^2} \, dx = .9$$

so N_0 is a certain "correlation length".

3. Time dependent solutions of Markovian Master Equations

Let us assume that the stationary solution $P^s(\underline{x})$ of the ME (1) is known. Due to the proposed information-theoretical scheme a time-dependent solution of (1) can be taken in the form (9) where we take $P^s(\underline{x})$ as the reference state. The coefficients $\lambda_j(t)$ will be determined by the equations (11) due to some initial conditions. We tested the method for the Schlögl's models of nonequilibrium phase transitions by solving numerically the equations (11) with P in the form (9) and $j=1,\dots,7$ [5]. Note that if the initial distribution $P(\underline{x},0)$ is not very different from the steady state P^s (e.g. if we consider relaxation phenomena near the steady state) a linear approximation may be used where the eqs. (11) take the form

$$\dot{\lambda}_i = \sum_j b_{ij} \lambda_j \quad , \; i,j = 1,\dots,n \tag{37}$$

with

$$\underline{B} = (b_{ij}) \, , \quad \underline{B} = \underline{C}^{-1} \underline{A} \quad , \quad \underline{A} = (a_{ij}) \, , \quad \underline{C} = (c_{ij})$$

$$c_{ij} = \langle \alpha^i \alpha^j \rangle^s \quad , \quad \alpha^i = Q^i - \langle Q^i \rangle^s \tag{38}$$

$$a_{ij} = \sum_{\underline{x}} \sum_{\underline{x}'} Q^i(\underline{x}) \; w(\underline{x},\underline{x}') \, Q^j(\underline{x}') \; P^s(\underline{x}') \quad .$$

4. Time-dependent solution of a Fokker-Planck equation

The information-theoretical scheme can be also applied to a Fokker-Planck equation FPE

$$\dot{P}(\underline{x},t) = \mathcal{L}\left[P(\underline{x},t)\right] = -\sum_k \frac{\partial}{\partial x_k}\left(F_k - 1/2 \sum_m D_{km} \frac{\partial}{\partial x_m}\right) P \qquad (39)$$

where $F_k(\underline{x})$ are drift coefficients and D_{km} are diffusion coefficients. We look for a solution in the form (8) or (9) . The equations (11) will take the form

$$\int d\underline{x}^n \; Q^j(\underline{x}) \; \dot{P}_K(\underline{x},t) = \int d\underline{x}^n \, Q^j(\underline{x}) \, \mathcal{L}\left[P_K \, \underline{x},t\right] \qquad (40)$$

$j = 1,\dots,r$, where $Q^j \, \underline{x}$ are linearly independent functions. In previous papers we discussed examples of one- [3-4] and multidimensional [6] FPE. For one-dimensional FPE the use of the form (9) seems to be natural since the steady state, if exists, is quite easy to obtain. The same conclusion holds true for the multidimensional FPE in the case where the drift coefficients fulfil the so-called potential condition

$$F_k = -(\partial / \partial x_k) \; V(\underline{x}) \qquad (41)$$

If, furthermore, the diffusion coefficients fulfil the condition

$$D_{km} = \delta_{km} \, D \qquad (42)$$

then the stationary solution of (39) has the form

$$P^s(\underline{x}) = \mathcal{N} \exp\left\{ - 2 \, V(\underline{x}) \, / \, D \right\} \qquad (43)$$

($\mathcal{N}$: normalization factor) . To illustrate the information-theoretical method we treat the "potential case" in two dimensions. Let

$$\begin{aligned} F_1(\underline{x}) &= -\alpha x_1 , \qquad F_2(\underline{x}) = -\alpha x_2 \\ V(\underline{x}) &= (\alpha/2)(x_1^2 + x_2^2) \qquad \text{potential} \end{aligned} \qquad (44)$$

The stationary solution of (39) has the form

$$P^s(\mathbf{x}) = \mathcal{N} \exp\{(-\alpha/D)(x_1^2 + x_2^2)\} \quad (45)$$

We assume that at the time 0 the distribution $P(\mathbf{x},0)$ has the form

$$P(\mathbf{x},0) = \exp(-\lambda_{00} - \lambda_{10}x_1 - \lambda_{20}x_2)P^s(\mathbf{x}) \quad (46)$$

and consider its evolution due to the equation (39) which takes here the form

$$\dot{P} = 2\alpha P + \alpha x_1(\partial P/\partial x_1) + \alpha x_2(\partial P/\partial x_2) +$$
$$+(D/2)(\partial^2/\partial x_1^2 + \partial^2/\partial x_2^2) P \quad (47)$$

We look for a solution in the following form

$$P(\mathbf{x},t) = \exp(-\lambda_0(t) - \lambda_1(t)x_1 - \lambda_2(t)x_2)P^s(\mathbf{x}) \quad (48)$$

In the case considered it is possible to find the exact solution of (47) Substituting (48) to (47) and comparing the corresponding terms by the powers of x_k we may check that (48) is the exact solution when λ_i $i = 1,2$ satisfy the equations

$$\dot{\lambda}_i = -\alpha\lambda_i, \quad \dot{\lambda}_0 = -D/2(\lambda_1^2 + \lambda_2^2) \quad (49)$$
$$\lambda_i(0) = \lambda_{i0}, \quad \lambda_0(0) = \lambda_{00}$$

For other two-dimensional examples see [6] .

5. Conclusion

We hope that the information-theoretical scheme may be an effective method to solve stochastic evolution equations, both discrete master equations and continuous Fokker-Planck equations.

This research was supported by the Polish Ministry of Science, Higher Education and Technology, Projest M.R.I.7.

References

[1] Haken,H.: Synergetics, An Introduction. Springer-Verlag, Berlin-Heidelberg-New York, 1977 .

[2] Czajkowski, G., in : D. Wolf edit. : Noise in Physical Systems. Proc. 5th Int.Conf.on Noise, Bad Nauheim, March 13-16, 1978. Springer-Verlag, Berlin - Heidelberg - New York 1978, p. 327.

[3] Czajkowski, G.: Z.Phys. B 43 1981 ,87 .

[4] Kociszewski, A.: Acta Phys.Polon. A 60 1981 , 291.

[5] Czajkowski, G,. and A. Kociszewski : Z.Phys. B 53 1983 ,307.

[6] Owedyk, J.: Acta Phys.Polon. A 63 1983 , 317 .

[7] Ingarden, R.S.: Fortschr.Phys. 13 1965 , 755 .

[8] Ingarden, R.S., and A. Kossakowski : Ann.Phys. NY 89 1975 , 451.

[9] Schlögl, F.: Phys.Rep. 62 1980 , 267.

[10] Schlögl, F.: Z.Phys. 253 1972 , 147 .

[11] McNeil, K.J., and D.F. Walls : J.Stat.Phys. 10 1974 , 439.

[12] Czajkowski, G., in : H.Haken edit. : Synergetics. Proc. Int. Workshop on Synergetics at Schloss Elmau, May 2-7 , 1977. Springer-Verlag, Berlin - Heidelberg- New York 1977, p. 155.

[13] Ingarden, R.S. : Statistical Thermodynamics, Lecture Notes by G. Czajkowski. Nicolaus Copernicus University Press, Toruń 1979, in Polish.

LOTKA-VOLTERRA STATISTICAL MECHANICS IN CELL-MEMBRANE/CELL-METABOLISM SYSTEMS

Charles J. Lumsden[1]

Introduction

The question of a statistical mechanics for biological systems has received some consideration in recent years (Goel et al., 1971; Lumsden and Trainor, 1979a, b; Lumsden, 1982). The influential early work is that of Kerner (1957, 1959, 1971, 1972), in which an attempt was made to erect a theory of Gibbs equilibrium ensembles on partially Hamiltonized versions of biologically interesting dynamical models. Applications of this approach to the determination of collective or whole-system properties have ranged across genetic and biochemical networks, morphogenesis, behavioral processes, demography, and ecology. While the resulting "bioensemble theory" is largely an ad hoc construct, the approach is intuitively appealing and can be said to have enjoyed some early successes (Goodwin, 1963; Leigh, Jr., 1969; Cowan, 1970; Kerner, 1971). In my own work I have tried to develop a more comprehensive approach designed to replace the ad hoc arguments with firmer foundations and expand the modeling effort into a statistical mechanics suited to both equilibrium and nonequilibrium processes. Lotka-Volterra models have proven to be a valuable adjunct to this effort, and have been of particular interest to me in modeling cell-membrane/cell-metabolism systems.

Linear Langrangians

The formulation of Langrangian and Hamiltonian equations for the very large class of biomathematical models characterized by rate laws of the form

$$\frac{dx_k}{dt} \equiv \dot{x}_k = X_k(\underset{\sim}{x},t) \quad , \quad k = 1, \ldots , N \tag{1}$$

has become a subject of continuing research (e.g., Lotka, 1956; Kerner, 1957, 1959; Goodwin, 1963; Leigh, Jr., 1969; Lumsden and Trainor, 1979a, b; Lumsden, 1982; Paine, 1982; Steeb and Kunick, 1982). In these ordinary differential equations, the X_k are continuous and suitably smooth functions of the configuration space vector $\underset{\sim}{x} = (x_1,\ldots,x_N)$ and possibly the time, t. Whereas physical systems are specified by generalized coordinates and momenta which usually correspond to particle motions, biological configuration spaces may be coordinatized by such variables as biochemical concentrations, organism population densities, rates of growth, gene frequencies, shape parameters, and so on, depending upon the phenomena represented by the model.

If the Euler-Lagrange equations

$$\frac{d}{dt} \frac{\partial L}{\partial \dot{x}_k} - \frac{\partial L}{\partial x_k} = 0 \tag{2}$$

corresponding to the variational principle (fixed end points)

---

(1) Membrane Biology Group, Clinical Science Division, Room 7313, Medical Sciences Building, University of Toronto, Toronto, Ontario, Canada M5S 1A8.

$$\delta \int_{t_1}^{t_2} L\, dt = 0 \tag{3}$$

for the Lagrangian $L(\underset{\sim}{x}, \dot{\underset{\sim}{x}}, t)$ are to directly reduce to the rate laws (1), then $\partial L/\partial \dot{x}_k$ must be independent of $\dot{x}_i$ for all $\underline{i}$. The Langrangian itself must therefore be at most linear in the rates:

$$L(\underset{\sim}{x}, \dot{\underset{\sim}{x}}, t) = U_k(\underset{\sim}{x}, t)\, \dot{x}_k - U_o(\underset{\sim}{x}, t) \quad \text{(summation convention)} \tag{4}$$

The Euler-Langrange equations (2) for the Langrangian (4) can be written as

$$\Gamma_{km}\, \dot{x}_m = \Lambda_k \tag{5a}$$

where the quantities Γ_{km} are given by

$$\Gamma_{km} = -\Gamma_{mk} = \frac{\partial U_k}{\partial x_m} - \frac{\partial U_m}{\partial x_k} \tag{5b}$$

anu the quantity

$$\Lambda_k = -\frac{\partial U_k}{\partial t} - \frac{\partial U_o}{\partial x_k} \tag{5c}$$

plays the role of a generalized force for the dynamics. Granted invertibility of the matrix $||\Gamma_{km}||$, equation (5a) can be transformed to read

$$\dot{x}_k = \Gamma^{-1}_{km}\, \Lambda_m \tag{6}$$

Comparison of equations (2) through (6) with the rate laws (1) shows that the existence problem for L can be reformulated as the existence of a set of functions

$$\{U_o(\underset{\sim}{x},t),\ U_k(\underset{\sim}{x},t),\ k = 1,\ldots,N\}$$

such that the relations

$$\Lambda_k = \Gamma_{km}\, X_m \tag{7}$$

are identities over the domain of interest. The Γ_{km} themselves obey the coupled linear partial differential equations

$$\frac{\partial \Gamma_{kn}}{\partial t} = \frac{\partial}{\partial x_k}(\Gamma_{nm} X_m) - \frac{\partial}{\partial x_n}(\Gamma_{km} X_n) \quad . \tag{8}$$

When the dynamics (1) is autonomous (i.e., the X_k are time independent) it is natural to seek linear Langrangians that are also explicitly independent of time. Equation (6) for the x_k can then be expressed as

$$\dot{x}_k = -\Gamma^{-1}_{km} \frac{\partial U_o}{\partial x_m} \tag{9}$$

from which we see that $U_o(\underset{\sim}{x})$ is a constant of the motion:

$$\frac{dU_0}{dt} = \frac{\partial U_0}{\partial x_k}\dot{x}_k = \frac{\partial U_0}{\partial x_k}(-\Gamma^{-1}_{km})\frac{\partial U_0}{\partial x_m} \equiv 0 \tag{10}$$

by Γ_{km} antisymmetry. The system dynamics is therefore constrained to hypersurfaces of constant U_0.

If $\rho(\underset{\sim}{x},t)$ is a statistical mechanical ensemble of system replicas such that $\rho(\underset{\sim}{x},t)d\underset{\sim}{x}$ is the likelihood that the system is in hypervolume element $d\underset{\sim}{x}$ at $(\underset{\sim}{x},t)$, then the dynamics for ρ obeys a continuity equation

$$\frac{\partial}{\partial t}\rho + \frac{\partial(\rho\dot{x}_j)}{\partial x_j} = 0 \tag{11}$$

where the velocities x_j are given by the equations (1) or (9). For a dynamics in the form of ordinary differential equations (1) describing biological subsystems (such as cells or animals or societies, depending on the application), their interactions, and their couplings to the environment, in a statistical mechanics for biological systems one seeks a related set of equations expressing the behavior of the subsystem assemblage <u>as a whole</u>. If $A_m(\underset{\sim}{x})$, $m = 1, \ldots, M << N$ are the system properties expressed in terms of the subsystem (= microscopic) variables, then the statistical-mechanical constructs representing the system-level of organization are the expectation values

$$\langle A_m\rangle(t) = \int d\underset{\sim}{x}\, A_m(\underset{\sim}{x})\rho(\underset{\sim}{x},t) \tag{12}$$

If, for example, it can be shown that

$$\frac{d\langle A_m\rangle}{dt}(t) = F_m(\langle A_1\rangle, \ldots, \langle A_m\rangle) \quad , \tag{13}$$

where we have in general that

$$\frac{d}{dt}\langle A_m\rangle(t) = \int d\underset{\sim}{x}\, A_m(x)\frac{\partial\rho}{\partial t}(\underset{\sim}{x},t) \quad , \tag{14}$$

then the subsystem interactions sustain a domain of whole-system behavior described by its own closed set of autonomous dynamical laws. The identification of biologically meaningful models (1) that give rise to such remarkable forms of collective behavior as (13) is one of the central problems to be addressed by a statistical mechanics for biological systems.

Comparatively little is known about maps of the form

$$\int d\underset{\sim}{x}\, A_m(\underset{\sim}{x})\frac{\partial\rho}{\partial t}(\underset{\sim}{x},t) \to F_m(\langle\underset{\sim}{A}\rangle) \tag{15}$$

for nonequilibrium ensembles ρ given only that $\underset{\sim}{x}$ follows rate laws of the form (1). In contrast, the transition is relatively well understood when $\underset{\sim}{x}$ obeys a Hamiltonian mechanics and has been studied extensively (e.g., Zwanzig, 1960; Mori, 1965; Balescu and Wallenborn, 1971; Haake, 1973; Kenkre, 1977; Prigogine, 1980; Grabert, 1982). This circumstance greatly increases the interest and relevance of Hamiltonian formulations in biomathematical systems theory.

In physical statistical mechanics, the continuity equation for ρ is the Liouville equation, which expresses the phase flow $\dot{\underset{\sim}{x}}$ in terms of the system Hamiltonian, H. Unfortunately, for linear Langrangians (4) the determinant $||\partial^2 L/\partial \dot{x}_k \partial \dot{x}_n||$ for the Legendre transformation from Langrangian to Hamiltonian form is totally singular, blocking the standard route to expressing (11) as a Liouville equation. We have, however, found that for systems of the Lotka-Volterra type the time development of ρ is given by the generalized Liouville equation

$$\frac{\partial}{\partial t} \rho \approx \mathcal{L} \rho \quad , \tag{16}$$

where the generalized Liouville operator is given by

$$\mathcal{L} = - [\cdot, H] \ , \tag{17}$$

$\approx$ is the Dirac weak equality, $[\cdot, \cdot]$ is the Poisson bracket, and H is the Dirac total Hamiltonian of the system (Dirac, 1950, 1964), not the conventional Hamiltonian (Lumsden and Trainor, 1979a,b). For the linear Langrangians (4)

$$N = U_0 + \phi_k x_k \quad , \tag{18}$$

in which the ϕ_k are primary constraints in the sense of Dirac (1950, 1964), with $\phi_k = p_k - U_k$ and p_k the standard canonical momenta $\partial L/\partial \dot{x}_k$. The constraints ϕ_k vanish on the N-dimensional hypersurface of allowed motion in the 2N-dimensional phase space of coordinates $(\underset{\sim}{x},\underset{\sim}{p})$. The generalized Liouville equation (16) constitutes a starting point for a rigorous statistical mechanics of Lotka-Volterra systems equivalent to the basis for physical statistical mechanics provided by the conventional Louiville equation. Extensions to more general systems that include dissipations have also been developed (e.g., Lumsden and Trainor, 1979a,b; Lumsden, 1982).

In a nonequilibrium statistical mechanics, one of the most interesting questions concerns the asymptotic properties of the ensembles as $t \to \infty$ in equation (16). Attention generally focusses on the existence of stationary ensembles and the extent to which time-dependent ensembles approach the stationary solutions as $t \to \infty$. If a stationary attractor $\rho_0(\underset{\sim}{x})$ exists for $\rho(\underset{\sim}{x},t)$ then the expectation values $\langle A_m \rangle(t)$ of macrovariables $A(\underset{\sim}{x})$ approach steady-state values $\langle A_m \rangle_0$ and the system can be said to possess self-regulating or homeostatic tendencies on the $\underset{\sim}{A}$-level of macroscopic order. From the Liouville equation (16) it is apparent that the stationary-ensemble condition is $-[\rho,H] \approx 0$ with $[H,H] \approx 0$. The constant of the motion $H \approx U_0$ appears to be isolating in the conservative models studied to date (Kerner, 1957; Goodwin, 1963; Cowan, 1970; Kerner, 1971). Equal initial a priori probabilities are motivated in the symplectic geometries of the Liouville dynamics (Hobson, 1971). The ensemble of optimal information entropy is then (Jaynes, 1963; Hobson, 1971) proportional to $e^{\lambda H}$ where λ is a constant analogous to the inverse temperature. Such an ensemble is stationary, and the properties of the Dirac weak equalities imply that (up to normalization factors)

$$\rho \approx e^{-\lambda H} \approx e^{-\lambda U_0} \tag{19}$$

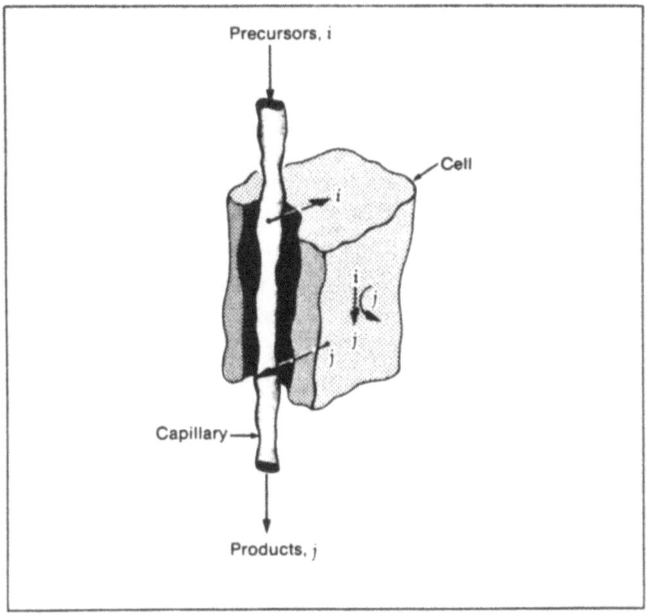

Figure 1. Physical arrangement of the capillary and cell for the membrane transport/metabolism example discussed in the text. The microcirculatory flow delivered in the capillary supplies precursor molecules i which are autocatalytically converted within the cell into products j. Exchange between the vascular compartment and intracellular space is mediated by membrane transport mechanisms. Both precursors i and products j act as intracellular messengers modulating steps in membrane transport and in the biosynthesis of the product molecules.

Lotka-Volterra Transport-Metabolism in a Cell

Among other applications (Lumsden and Trainor, 1979a, b; Lumsden and Wilson, 1981; Lumsden, 1982) I have begun to use these general developments to model the collective properties inherent in the process by which a cell utilizes precursor materials to synthesize product molecules such as hormones or metabolites, which are then transported across the cell membrane and passed into the microcirculation for shipment to target sites within the organism. The anatomical arrangement is illustrated in Fig. 1. Blood flow carries the precursors to the vicinity of the cell, where they move from a nearby capillary into the extravascular space and are taken across the plasma membrane of the cell by a transport process. Lotka-Volterra properties for the biochemical dynamics greatly facilitate the initial analysis, and are motivated in the following way. The precursor substances act as intracellular messengers in the transport mechanism, stimulating their own uptake: if $N_i(t)$ is the number of precursor molecules of type i within the cell at time t, then during the interval [t, t + dt] a total of $dN_i = \varepsilon_i N_i(t)dt$, $\varepsilon_i > 0$ additional molecules of this type are extracted from the extracellular pool by the membrane transport machinery and injected into the cytoplasm. Conversely, if $N_j(t)$

is the number of product molecules of type j within the cell at time t then during the same time interval a total of $dN_j = -\epsilon_j N_j(t)dt$, $\epsilon_j < 0$ move out of the cell and into the capillary, where they are convected away from the synthesis site by the flowing blood.

Within the cell, biochemical reactions transform the precursors into product molecules. For this class of systems models we consider steps in biochemical synthesis that are autocatalytic in the sense that the conversion of precursors i to products j will not occur unless a product molecule is also active in the process. In this way we model the action of molecules that function as messengers to switch on their own synthesis. In the simplest case one product molecule and one precursor must be present in order for the transformation of the precursor to occur, and the supply of catalysts is not limiting. If σ_{ji} is the total cross section for binary encounters between precursor i and product j within the cell and ω_{ji} is the fraction that occurs in the catalytic sites of the $i \to j$ enzyme molecule, then during $[t, t + dt]$ the quantity N_i decreases by the amount $\omega_{ji}\sigma_{ji}N_iN_j dt$ while N_j increases by $(\beta_i/\beta_j)\,[\omega_{ji}\sigma_{ji}N_iN_j dt]$, where β_i/β_j models the change in j per precursor i consumed. Factors $\beta_i/\beta_j < 1$ allow for wastage effects due to errors in the transformation process, which result in molecules i being converted to some other combination of products than j. If α_{ij} is defined to be the product $\beta_i\sigma_{ji}\omega_{ji}$, then the intracellular changes in i and j are

$$dN_j = \frac{\alpha_{ij}}{\beta_j} N_j dt \quad \text{and} \quad dN_i = -\frac{\alpha_{ij}}{\beta_i} N_j N_i dt = \frac{\alpha_{ji}}{\beta_j} N_i N_j dt \quad ,$$

where $\alpha_{ij} = -\alpha_{ji}$ and $\alpha_{ii} = 0$. Combining the membrane transport terms with the cytoplasmic reaction terms then summing over the possible end products of each precursor, one finds that the abundance N_k of each molecular species follows a kinetic equation of the Lotka-Volterra form

$$\dot{N}_k = \epsilon_k N_k + \beta_k^{-1} \alpha_{nk} N_k N_n \tag{20}$$

where $\epsilon_k > 0$ if k is a precursor and $\epsilon_k < 0$ if k is a product molecule.

Let N_k^0 denote the steady-state solutions of equations (20), which require that $||\alpha_{nk}||$ be nonsingular, i.e., that the total number of interacting molecular species is even. Then in the scaled coordinates $v_k \equiv \ell n(N_k/N_k^0)$ the kinetics (20) can be written as the incompressible flow (div $\dot{\underset{\sim}{v}} = 0$)

$$\dot{v}_k = \frac{\alpha_{nk}}{\beta_k} q_n \left(e^{v_n} - 1\right) \approx [v_k, H] \tag{21}$$

where H is the Dirac total Hamiltonian

$$H = U_o(\underset{\sim}{v}) + \dot{v}_k \phi_k \tag{22}$$

with the primary constraints $\phi_k = p_k - U_k(\underset{\sim}{v})$ vanishing for the allowed motion. The $U_k(\underset{\sim}{v})$ functions for the Lotka-Volterra dynamics (20) are the linear combinations (Kerner, 1971)

$$U_k(\underset{\sim}{v}) = \tfrac{1}{2} \Lambda_{ik} v_i \tag{23}$$

where the matrix $||\Lambda_{ik}||$ is the inverse $||\alpha_{ik}/\beta_i\beta_k||^{-1}$ and $U_o(\underset{\sim}{v})$ is the conserved additive function $U_o(\underset{\sim}{v}) = N_\ell^o \beta_\ell(e^{v_\ell} - v_\ell)$ such that $\dot{U}_o \equiv 0$. The linear Langrangian for the transport-metabolism network is the sum of the terms involving the $U_k(\underset{\sim}{v})$ and $U_o(\underset{\sim}{v})$:

$$L = U_k(\underset{\sim}{v})\dot{v}_k - U_o(\underset{\sim}{v}) \tag{24}$$

The generalized Liouville equation for the time development of the statistical ensemble $\rho(\underset{\sim}{v},t)$ is thus

$$\frac{\partial\rho}{\partial t} \approx - [\rho, H] \ . \tag{25}$$

The Gibbs ensemble

$$\rho \propto e^{\lambda U_o} \tag{26}$$

is by the previous section the stationary, information-entropy optimal solution to this Liouville equation in the sense defined in the preceding section. Since $U_o(\underset{\sim}{v})$ is simple-additive in the molecular species $\underline{\ell}$, the ensemble ρ can be expressed as a product of one-species densities ρ_ℓ:

$$\rho = \pi_\ell \rho_\ell \tag{27}$$

where up to normalization constants the ρ are given by

$$\rho_\ell \equiv \rho(v_\ell) = \exp[\lambda N_\ell^o \beta_\ell (e^{v_\ell} - v_\ell)] \tag{28}$$

in the v-coordinates.

The exponential factor λ in equation (26) is analogous to the inverse temperature factor $(k_B T)^{-1}$, k_B = Boltzmann's constant, in a physical system. It is therefore a macroscopic parameter characteristic of the behavior of the transport-metabolism network as a whole. Just as the cytoplasm has a physical temperature, T, the network (20) has a "transport/metabolism temperature" $T^* \propto \lambda^{-1}$ that expresses the mean value of U_o driving the reaction system. Temperature-analogues for kinetics of the Lotka-Volterra type have previously been considered in some detail for ecological systems, and a number of relationships between them and the fluctuation behavior and steady-state properties of the kinetics (20) have been reported (e.g., Kerner, 1972). Their study and generalization in membrane-transport/cell-metabolism models is continuing.

References

Balescu, R., and J. Wallenborn. 1971. On the structure of the time-evolution process in many-body systems. Physica 54: 477-503.

Cowan, J.D. 1970. A statistical mechanics of nervous activity. In Some Mathematical Questions in Biology: Lectures on Mathematics in the Life Sciences, Vol. 2., pp. 3-57. Amer. Math. Soc., Providence, RI.

Goel, N.S., S.C. Maitra, and E.W. Montroll. 1971. On the Volterra and other nonlinear models of interacting populations. Review of Modern Physics 43: 231-276.

Goodwin, B.C. 1963. Temporal Organization in Cells. Academic Press, New York.

Grabert, H. 1982. Projection Operator Techniques in Nonequilibrium Statistical Mechanics. Springer-Verlag, New York.

Hobson, A. 1971. Concepts in Statistical Mechanics. Gordon and Breach, New York.

Haake, F. 1973. Statistical treatment of open systems by generalized master equations. Springer Tracts in Modern Physics 66: 98-168.

Jaynes, E.T. 1963. Information Theory and Statistical Mechanics. In G. Uhlenbeck et al., eds., Statistical Physics 3: Brandeis 1962, pp. 181-218, W.A. Benjamin, New York.

Kenkre, V.M. 1977. The generalized master equation and its applications. In U. Landman, ed., Statistical Mechanics and Statistical Methods in Theory and Applications, pp. 441-461. Plenum Press, New York.

Kerner, E.H. 1957. A statistical mechanics of interacting biological species. Bulletin of Mathematical Biophysics 19: 121-146.

Kerner, E.H. 1959. Further considerations on the statistical mechanics of biological associations. Bulletin of Mathematical Biophysics 21: 217-255.

Kerner, E.H. 1971. Statistical-mechanical theories in biology. In I. Prigogine and S.A. Rice, eds., Advances in Chemical Physics, Vol. 19, pp. 325-352. John Wiley and Sons, New York.

Kerner, E.H. 1972. Gibbs Ensemble: Biological Ensemble. Gordon and Breach, New York.

Leigh, Jr., E.G. 1969. The ecological role of Volterra's equations. In Some Mathematical Problems in Biology: Lectures on Mathematics in the Life Sciences, Vol. 1, pp. 1-61. American Mathematical Society, Providence, RI, U.S.A.

Lotka, A.J. 1956. Elements of Mathematical Biology. Dover, Toronto, Canada.

Lumsden, C.J. 1982. The social regulation of physical caste: the superorganism revived. Journal of theoretical Biology 59: 749-781.

Lumsden, C.J., and L.E.H. Trainor. 1979a. On the statistical mechanics of constrained biophysical systems. Journal of Statistical Physics 20: 657-669.

Lumsden, C.J., and L.E.H. Trainor. 1979b. Nonequilibrium ensembles of self-organizing systems. Canadian Journal of Physics 57: 23-38.

Lumsden, C.J., and E.O. Wilson. 1981. Genes, Mind, and Culture: The Coevolutionary Process. Harvard University Press, Cambridge, MA, U.S.A.

Mori, H. 1965. Transport, collective motion, and Brownian motion. Progress of Theoretical Physics 33: 423-455.

Paine, G.H. 1982. The development of Langrangians for biological models. Bulletin of Mathematical Biology 44: 749-760.

Prigogine, I. 1980. From Being to Becoming: Time and Complexity in the Physical Sciences. W.H. Freeman and Company, San Francisco, U.S.A.

Steeb, W.-H, and A. Kunick. 1982. Lagrange functions of a class of dynamical systems with limit-cycle and chaotic behavior. Physical Review A 25: 2889-2892.

Zwanzig, R. 1960. Ensemble method in the theory of irreversibility. Journal of Chemical Physics 33: 1338-1341.

THE INFLUENCE OF NOISE ON AN BIOCHEMICAL OSCILLATOR

H. Herzel, W. Ebeling, L. Schimansky-Geier,[1] E.E. Sel'kov[2]

1. Introduction

A Volterra system admits stable limit cycles if the dimension is at least three [6] . The transformation of a standard model of biochemical oscillations [11] leads to such a three-dimensional system of Volterra type. We investigate this model including stochastic sources into the deterministic equations. The noise could be caused by a fluctuating supply, thermal noise or by the finite number of reacting particles. We assume that a description by additive, short-correlated noise is possible. The stationary Fokker-Planck equation is solved locally by Taylor expansions. The global form of the probability distribution and correlation function are derived by numerical simulation. Noise induced oscillations are discussed and compared with "deterministic chaos", which occurs in an extended model.

2. The model

The following model describes the regulating mechanisms of a living cell (e.g. glycolytic oscillations) [11].

$$\dot{x} = 1 - Bx - xy^2$$
$$\dot{y} = A(xy^2 - y) \qquad (1)$$

Here x and y are the concentrations of certain chemical substances and A,B are regulatory parameters.
Because x is strictly positive the following transformation to Volterra equations is possible:

$$X_1 = \frac{1}{x} \quad ; \quad \dot{X}_1 = X_1\ (B-X_1+X_2)$$
$$X_2 = y^2 \quad ; \quad \dot{X}_2 = X_2\ (-2A+2AX_3) \qquad (2)$$
$$X_3 = xy \quad ; \quad \dot{X}_3 = X_3\ (-A-B+X_1-X_2+AX_3)$$

1) Sektion Physik, Humboldt-Universität Berlin, 1040 Berlin, Invalidenstr. 42, GDR
2) Institute of Biophysics "Academy of Sciences, Puscino, USSR

Fig. 1 gives an impression of the deterministic behaviour, which is in detail discussed elsewhere [11] .

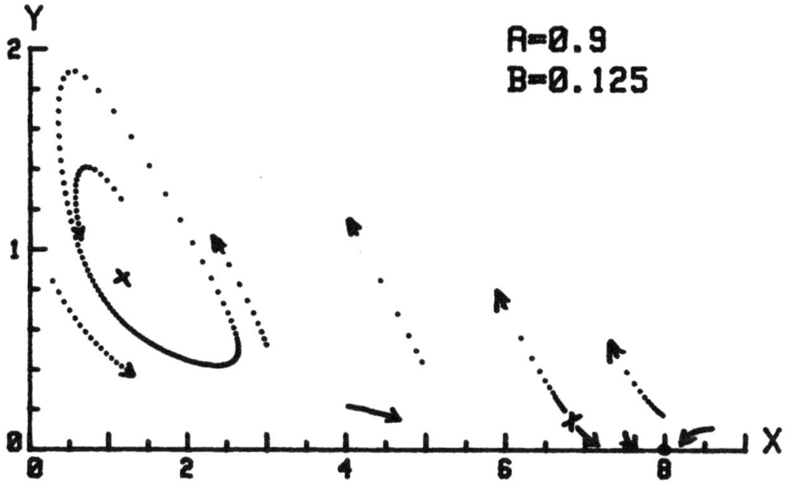

Fig. 1. Representative stroboscopic phase portrait with stable node (•) unstable stationary points (x) and a stable limit cycle. Every 0.1 time units a point was plotted. So we can distinguish fast and slow "motion".

The following bifurcation diagram shows the Hopf bifurcation. In the case B = 0 at A = 1 an elliptic limit cycle with an angular frequency of one is generated. By increasing A further the oscillations become saw-tooth-like and at A = 1.234 the limit cycle disappears.

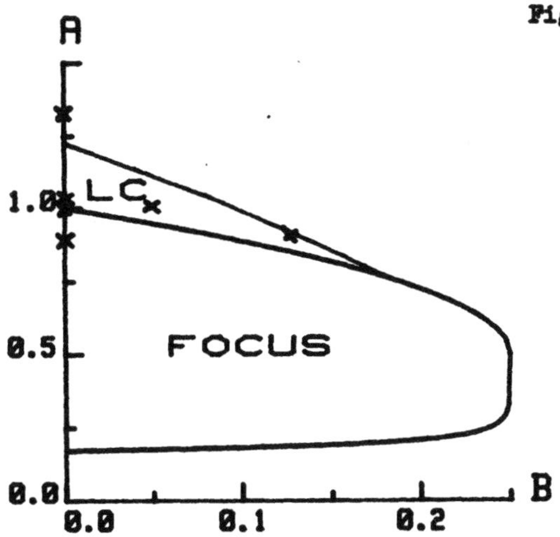

Fig. 2. Parameter region of a stable limit cycle and a stable focus (the crosses denote the parameter values examined in this paper)

3. A local solution of the Fokker-Planck equation

We assume additive Gaussian white noise.

$$\dot{x} = 1 - Bx - xy^2 + G_1(t) \quad ; \quad \langle G_i(t) \rangle = 0$$
$$\dot{y} = A(xy^2 - y) + G_2(t) \quad ; \langle G_i(t)\, G_j(t + \tau) \rangle = 2K_i\, \delta_{ij}\, \delta(\tau) \tag{3}$$

The ansatz $P^o = Z^{-1}\exp(-F(x,y))$ for the stationary probability leads to the following Fokker-Planck equation for the stochastic potential F [4,9].

$$\frac{\partial F}{\partial x}\left(1-Bx-xy^2+K_1\frac{\partial F}{\partial x}\right) + \frac{\partial F}{\partial y}\left(A(xy^2-y) + K_2\frac{\partial F}{\partial y}\right)$$
$$- K_1\frac{\partial^2 F}{\partial x^2} - K_2\frac{\partial^2 F}{\partial y^2} - B - y^2 - A(1-2xy) = 0 \tag{4}$$

We use a polynomial expansion of F to the 4th order around stationary points:

$$F = A_{10}(x-x_o) + A_{01}(y-y_o) + \dots + A_{04}(y-y_o)^4 \tag{5}$$

By inserting this expansion into (4) a nonlinear algebraic system for the expansion coefficients A_{ij} is derived. To overcome the difficulties of nonlinear systems we combine this method with the approximation for small noise due to Stratonovich [14,1]. By that method the polynomial coefficient A_{ij} can be computed for small K_i-values by solving linear algebraic systems. We take these results as a starting point of a Gauss-Newton iteration for the original nonlinear system. In this way the Fokker-Planck equation was solved locally.

4. Numerical simulation

For simulation we have chosen the following procedure [8]. First we solve eq. (1) using a Runge-Kutta scheme. During the calculation a series of random numbers of Gaussian type is introduced as fluctuation terms. K denotes the variance of the random numbers in a unit time interval. By integrating (1) over sufficiently long time we find the autocorrelation function $\langle x(o)\, x(t) \rangle$ (averaging over some hundred intervals) and a stationary probability distribution as a histogram.

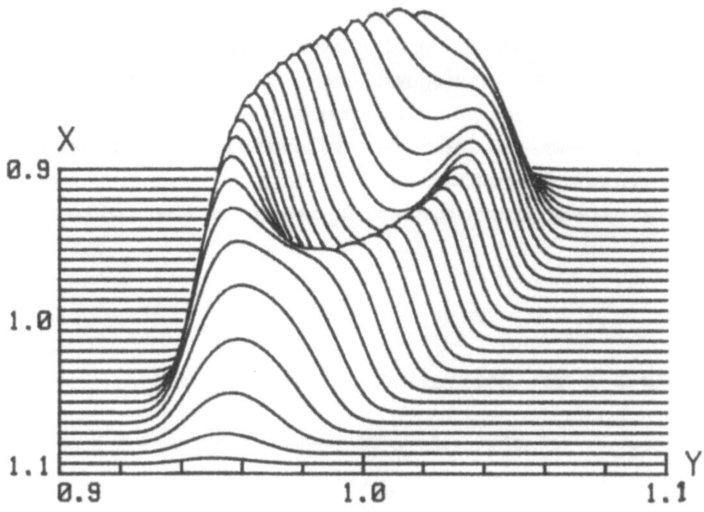

Fig. 3. Probability distribution computed by the local method ($A = 1.0004$, $K_i = 10^{-7}$)

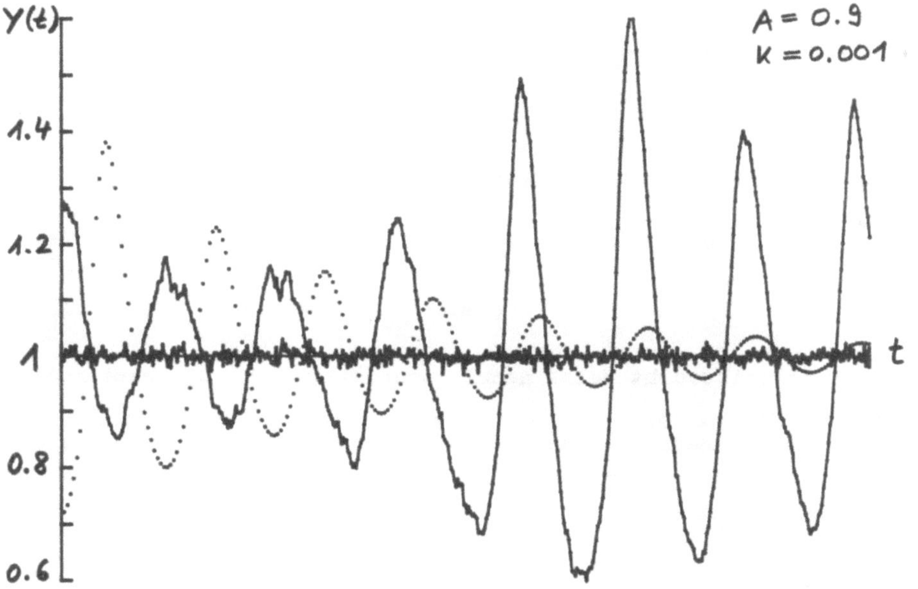

Fig. 4. Typical realization of trajectory and noise in the (deterministically) damped region (the dotted line denotes a trajectory without noise)

5. The influence of noise on the Hopf bifurcation

Roughly seen, a stable focus in the deterministic case corresponds to an elliptic one-peak distribution with the maximum near the stationary point. A stable limit cycle corresponds to a distribution shaped like a crater (Fig. 3). Since (1) is not a system with "detailed balance" even additive noise not only "smears out" the deterministic attractors but also effects deterministic predictions [7, 15].
In the case B = 0 a linear transformation leads to the following equations:

$$q = 1 - x + \frac{1-y}{A} \quad ; \quad p = y-1 \tag{6}$$

$$\dot{q} = p \quad ; \quad \dot{p} = -Aq-p\,(-\alpha + 2Aq + (2-A)\,p + Apq) + G(t)$$

$$\langle G(t) \rangle = 0 \quad ; \quad \langle G(t)\, G(t+\tau) \rangle = 2\,\varepsilon\, \delta(\tau) \quad ;$$

$$\alpha = A - 1 \qquad \text{(bifurcation parameter)}$$

The close analogy to mechanical systems examined in [1] is to be seen. In the close vicinity of the bifurcation the probability distribution is mostly determined by the noise and the nonlinearities. The coefficients computed by the approximation due to Stratonovich diverge at the bifurcation. For this reason here a method due to Tolstopjatenko and Schimansky-Geier [15] is used. In the parameter region $\alpha^2 \ll \varepsilon$ the following approximation is valid:

$$P^0 = Z^{-1} \exp\left[2q-2p + \frac{\alpha+4\varepsilon}{\varepsilon}(q^2+p^2) - \frac{1}{4}\,qp - \frac{1}{16\varepsilon}\,(q^2+p^2)^2\right] \tag{7}$$

The bifurcation can be characterized by the extrema of the distribution:

$$\frac{\partial P^0}{\partial q} = \frac{\partial P^0}{\partial p} = 0 \quad ; \quad q = -p$$

$$q^3 - (2\alpha + \tfrac{15}{2}\varepsilon)\, q - 4\,\varepsilon = 0 \tag{8}$$

The maximum is shifted by noise nearly proportional to the cube root of the noise value.
In addition to the maximum a minimum and a saddle point arise at a critical positive value α_c. In this way the bifurcation of forming a crater is shifted by noise.

$$\alpha_c = \frac{3}{\sqrt[3]{2}}\,\varepsilon^{\frac{2}{3}} - \frac{15}{4}\,\varepsilon \quad ; \quad \varepsilon \ll 1 \Rightarrow \alpha_c \sim \varepsilon^{\frac{2}{3}} \tag{9}$$

These analytical results are in agreement with numerical calculations using the method described in section3.

6. Noise-induced oscillations

Remarkable effects of noise are the noise-induced oscillations which take place in the parameter region of a stable focus [3,8]. In that region the deterministic system is trapped into the stable steady state with damped oscillations (Fig. 4). Since the amplitude is not constant the shape of the probability distribution remains a peak. To examine these oscillations we evaluated the auto-correlation function:

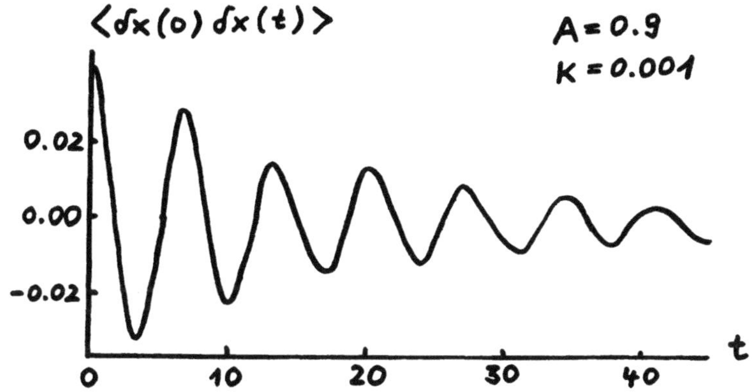

Fig. 5. Correlation function $\langle \delta x(o)\ \delta x(t) \rangle$ of noise-induced oscillations

Though the excitation by the noise is irregular the oscillations themselves are relatively regular. In agreement with experiments [5] the system acts as a narrow-band filter. In this way the parameter region of oscillations is largely increased.

7. Fully developed oscillations

The following representative probability density is computed by numerical simulation.

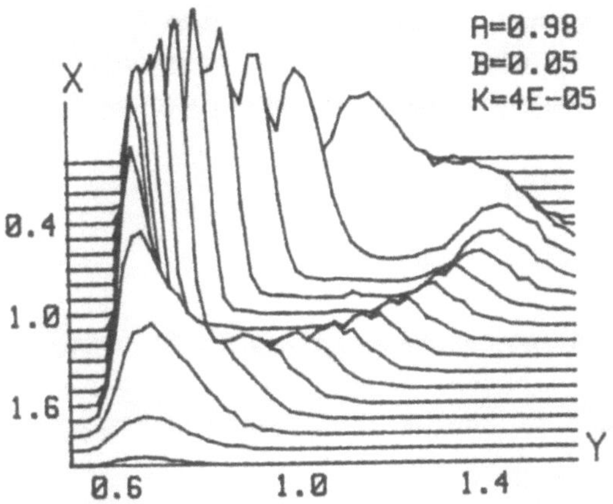

Fig. 6. Perspective plot of a probability distribution

The asymmetric form is a result of the different "speed" along the limit cycle (Fig. 1) and of the rather different values of the fluctuations.
The sensitivity of the orbit is connected with the Lyapunov exponents without noise. One of them is zero, i.e. the fluctuations along the orbit are neither damped nor increased. That leads to the "phase diffusion" under the influence of noise. The other Lyapunov exponent (Floquet exponent) is negative if the limit cycle is stable. For computation we solve the linearized system simultaneous to (1):

$$\dot{\bar{x}}(t) = \bar{F}(\bar{x}(t)) \quad ; \quad \dot{\overline{\delta x}}(t) = \frac{\partial \bar{F}}{\partial \bar{x}}(\bar{x}(t))\, \overline{\delta x}(t) \tag{10}$$

$$\lambda_1(t) = \frac{1}{\tau} \ln \frac{\| \overline{\delta x}(t+\tau) \|}{\| \overline{\delta x}(t) \|} \qquad (\tau \text{ sufficiently small})$$

The time-average of $\lambda_1(t)$ gives the maximal Lyapunov exponent [13]. More information contains the "local" Lyapunov exponent $\lambda_1(t)$ [10]. It describes the sensitivity of the orbit during the time.

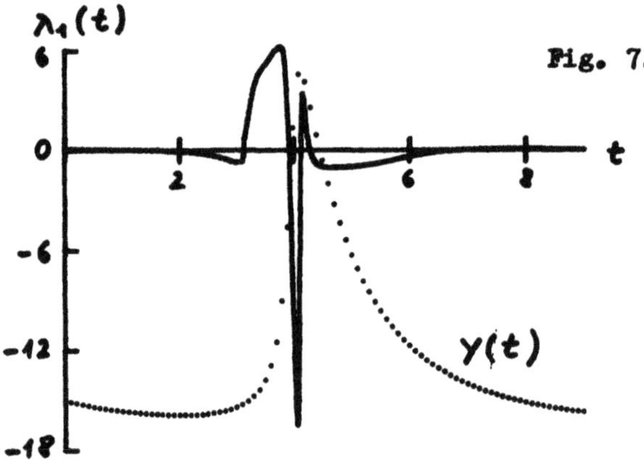

Fig. 7. The "local" maximal Lyapunov exponent during a period
$\langle \lambda_1 \rangle = 0$

The time dependence of the second Lyapunov exponent and of the divergence is similar. In agreement with the simulation it can be seen, that fluctuations increase if the nonlinearity becomes dominant (around the minimum of y (t)).

8. Irregular oscillations

It is of considerable interest to distinguish between aperiodic behaviour caused by noise and deterministic chaos. In the following section two methods are presented: Computation of the maximal Lyapunov exponent and the Lorenz map.
In the following way small noise produces strong irregular behaviour in

our model. We consider the parameter region corresponding to an unstable focus. Here the deterministic trajectory always approaches the stable node on the x-axes (Fig. 1), but even small noise leads to a return of the trajectory to the focus at irregular times.

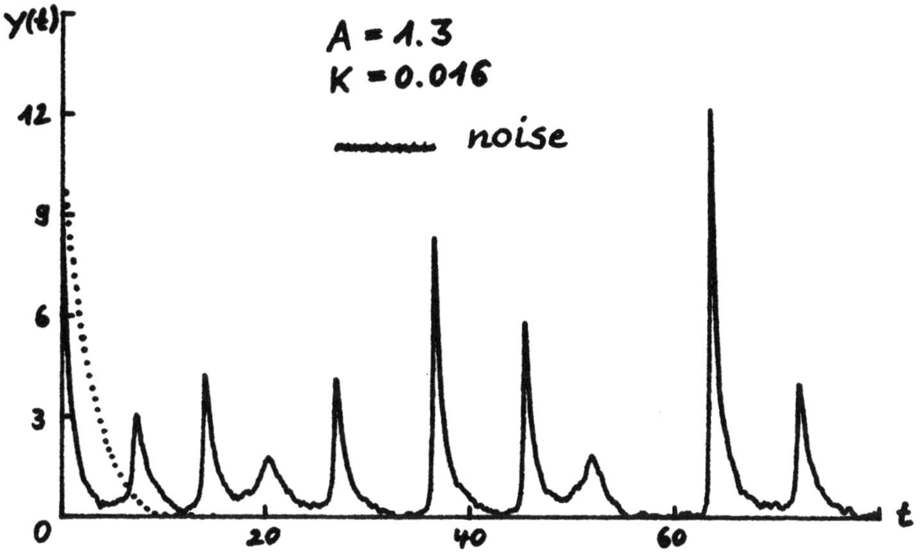

Fig. 8. Irregular behaviour caused by noise

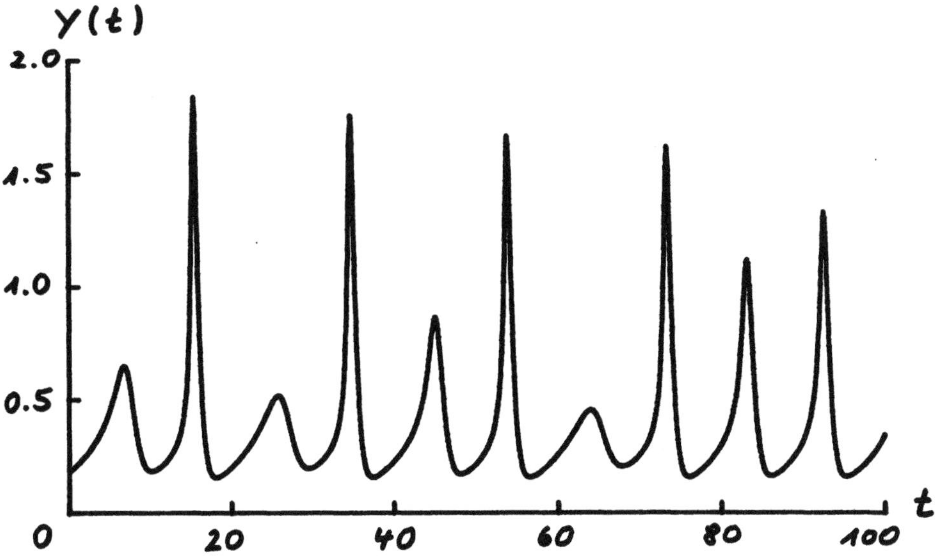

Fig. 9. Chaotic time dependence (A = 4; B = 0.35; D = 0.1; E = 1.5; F = 0.2)

The deterministic chaos was found in an extended model [12]

$$\dot{x} = 1 - Bx - xy^2 - Exy + z$$

$$\dot{y} = A\,(xy^2 - y + D) \tag{11}$$

$$\dot{z} = F\,(Exy - z)$$

In this case the maximal Lyapunov exponent is positiv ($\lambda_1 = 0.027 \pm 0.002$), but in the examined irregular oscillations caused by noise (Fig. 3, Fig. 8) it is negative($\lambda_1 = -0.07 \pm 0.01$). In Fig. 10 the Lorenz maps of both cases are presented, i.e. the maps of the maxima of y(t).

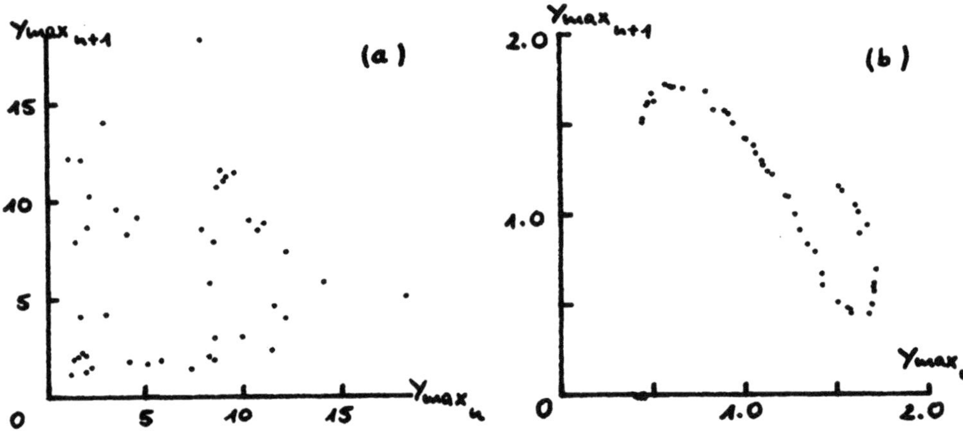

Fig. 10 Lorenz map of noise-induced irregular behaviour (a) and deterministic chaos (b)

In this way the qualitative differences are shown.

9. Conclusion

For discussing the influence of noise it is useful to distinguish different parameter regions. The main effects in the region of developed oscillations are quite different amplitude fluctuations and "phase diffusion". Near the bifurcation threshold even additive noise shifts the maximum and the bifurcation value. In the deterministically damped region relativly regular noise-induced oscillations occur. Finally small noise can lead to irregular behaviour with large amplitudes.
Most of these effects are not restricted to our concrete model.

References

[1] Ebeling, W., H. Engel-Herbert, Adv. Mechanics 5 (1982) 3/4, 41-60
[2] Ebeling, W. H. Herzel, E.E. Sel'kov, studia biophysica 98 (1983) 3
[3] Feistel, R, W. Ebeling, Physica 93A (1978), 114 - 137
[4] Graham, R., in Scattering Techniques Applied to Supramolecular and Nonequilibrium systems, Plenum Press, New York 1981
[5] Hess, H., A. Boiteux, Bunsenges. Phys. Chem. 84 (1980) 346 - 351
[6] Hofbauer, J., Nonl. Anal., Theory, Methods, Applications 5 (1981) 9, 1003 - 1007
[7] Hoffmann, K.H., Z. Phys. B 49 (1982) 245 - 252
[8] Iwamoto, K., M. Seno, J. Chem. Phys. 70 (1979) 12, 5851 - 5857
[9] Krasovskii, A.A., Fasovoe prostranstvo staticheskaya teoriya dinamicheskikh system, Nauka 1974
[10] Nicolis, J.S., G. Mayer-Kress, G. Haubs, Z. Naturforsch. 38a (1983) 1157 - 1169
[11] Schulmeister, Th., E.E. Sel'kov, studia biophysica 65 (1977) 2, 121 - 132
[12] Schulmeister, Th., studia biophysica, 72 (1978) 3, 205 - 206
[13] Shimada, I., T. Nagashima, Prog. Theor. Phys. 61 (1979) 1605-1616
[14] Stratonovich, R.L., Radiofizika 23 (1980), 942 - 949
[15] Tolstopjatenko, A. L. Schimansky-Geier, Acta Phys. Pol. in press

MODELS FOR PARASITIC DISEASES

K. P. Hadeler [1]

The classical models for epidemic diseases classify the host population as susceptibles, exposed, infected, recovered etc. The transmission between these classes is governed by a mass action kinetics. Such models are called prevalence models. Whereas these models are well-suited to describe the spread of bacterial or viral infections they appear inappropriate for most helminthic infections, where the host acquires a small number of parasites at different times, in many cases transmitted by vectors. Models for such diseases, in the form of ordinary differential equations have been proposed by Anderson and May in several papers [1], [12]. In [2] K. Dietz and the author have constructed a hybrid (semi-stochastic) model for helminth diseases which takes account to the age structure of the host population, the discrete structure of the parasite population and the transmission by vectors. This model has been generalized and thoroughly investigated in a series of papers [2] - [9].

These models have three basic components. The first is the classical Lotka model for a population with age structure. Let $u(t,a)$ give the age distribution of the host population at time t, i.e. $u(t,a)da$ is the size of the cohort with age in $[a,a+da)$. Let $b(a)$ be the rate with which individuals of age a produce offspring, and let $\mu(a)$ be the age-depending death rate. Then u satisfies the partial differential equation

$$u_t + u_a + \mu(a)u = 0. \qquad (1)$$

The newborns are given by

$$u(t,0) = \int_0^\infty b(a)u(t,a)da. \qquad (2)$$

In a more general version, the Gurtin-McCamy model, the birth rate $b = b(a,P)$, and the death rate $\mu = \mu(a,P)$ depend on the total size

$$P(t) = \int_0^\infty u(t,a)da. \qquad (3)$$

The second component of the model is a birth and death process with killing which describes the parasite population within the host. The process counts the number of parasites. The states are $r = 0, 1, 2, \ldots$ and the state ϕ, where there is no host. The probabilities of transition during $\Delta t > 0$ are $\varrho r\Delta t$ for $r \to r+1$ (birth), $\sigma r\Delta t$ for $r \to r-1$ (death), and

1) Lehrstuhl für Biomathematik, Universität Tübingen,
Auf der Morgenstelle 28, D7400 Tübingen

$\alpha r\Delta t$ for $r \to \phi$ (death of the host), the probability of immigration of an individual is $\varphi\Delta t$. Except for the trivial transition $r \to r$, $\phi \to \phi$ all other transitions have probability zero.

Such processes have been studied by Puri [13] and Karlin and Tavaré [11]. The main tool in studying these processes is the generating function

$$v(t,z) = \sum_{r=0}^{\infty} v_r(t) z^r \tag{4}$$

for the functions $v_r(t)$, where $v_r(t)$ is the probability for r individuals at time t. The function v satisfies the partial differential equation

$$v_t + g(z)v_z - (z-1)\varphi(t)v = 0 \tag{5}$$

where g is the polynomial

$$g(z) = (\alpha + \sigma + \varrho)z - \sigma - \varrho z^2 . \tag{6}$$

The third component is the transmission law which describes the acquisition of parasites by hosts or, equivalently, the immigration of parasites into the hosts. We assume that the acquisition rate $\varphi(t)$ is a nonlinear function

$$\varphi(t) = \beta f(P(t), \bar{w}(t)) \tag{7}$$

of the total population size and of the average parasite load $\bar{w}(t)$.

In the general model ([2] [7] [9]) these three components are unified. The host population is partitioned according to age and individual parasite load. Let $u_r(t,a)da$ denote the size of the host cohort with age in $[a,a+da)$ carrying r parasites. Then the generating function

$$u(t,a,z) = \sum_{r=0}^{\infty} u_r(t,a) z^r \tag{8}$$

gives the state of the population at time t. The total population size is given by

$$P(t) = \int_0^{\infty} u(t,a,1)da, \tag{9}$$

and the average parasite load is

$$\bar{w}(t) = \frac{\int_0^{\infty} u_z(t,a,1)da}{\int_0^{\infty} u(t,a,1)da} \tag{10}$$

With the above assumptions on the age structure of the host population and the stochastic process for the parasite population one can derive a partial differential equation for the function u which is a generalization of equations (1) and (5)

$$u_t + u_a + g(z)u_z - [(z-1)\varphi(t) - \mu(a,P)]u = 0 \tag{11}$$

The function satisfies an initial condition

$$u(0,a,z) = u_o(a,z) \tag{12}$$

and a boundary condition of Lotka type

$$N(t) \equiv u(t,0) = \int_0^\infty b(a,P(t))\, u(t,a,\omega)da \tag{13}$$

Here it is assumed, as is typical for helminthic diseases, that neonatals are not infected. The parameter $\omega \in [0,1]$ describes a geometric decrease of fertility of hosts with individual parasite load.

In several papers [2] - [4], [6] - [8] the model has been investigated with the simpler boundary condition

$$u(t,0) = N(t) \tag{14}$$

where N(t) is a given function for newborn hosts.

Equation (11) has to be considered together with the initial condition, the boundary condition (13) or (14), respectively, and the transmission law (7), (10). Thus the problem posed is a first order partial differential equation with nonlinearities containing quotients of functionals of the solution.

An approach to the initial value problem is the following. For the moment assume that the three functions φ,N,P of t are known. Then solve the linear initial value problem (11)(12)(14). The solution u depends on the functions φ,N,P. Then introduce the function u into the equations (7), (10)(9)(13). Thus one obtains three coupled integral equations of Volterra type for the three functions φ,N,P.

If the parameter functions b,μ, and f do not depend explicitely on the total population size P, then the equation for P becomes obsolete, and the problem reduces to two equations for the variables φ and N [5]. Finally, in the case of the simpler boundary condition (14) one has only a single integral equation for the parasite acquisition rate φ [2] [7]. Once the functions φ,N,P have been determined, the function u is again obtained from a linear problem (11)(12)(14).

In [2], [3], [7] an existence theorem for the case of the boundary condition (14) has been given, which takes account of the fact, that u is a generating function.

I shall not present these equations here, but restrict the discussion to stationary solutions. For stationary solutions the functions φ, N, P, and also $\bar{w}$ are constants, which can be obtained from certain transcendental equations. In order to formulate these equations the following notions are introduced.

As is well known, the birth and death process can be described in terms of the solution operator of the Riccati equation $\dot{z} = g(z)$. The function $G(t,z_o)$ is the solution of the equation $\dot{z} = g(z)$ with initial datum $z(0) = z_o$, evaluated at t.

$$G(t,z) = \frac{z_1(z-z_2) - z_2(z-z_1)e^{-\varkappa t}}{(z-z_2) - (z-z_1)e^{-\varkappa t}} \qquad \text{for} \quad \rho > 0$$

$$G(t,z) = 1 - (1-z)e^{\varkappa t} - \frac{\alpha}{\varkappa}(1-e^{\varkappa t}) \qquad \text{for} \quad \rho = 0 \tag{15}$$

where $g(z) = -\rho(z-z_1)(z-z_2)$, $z_1 \geq 1 \geq z_2$, and $\varkappa = \rho(z_1-z_2)$ or $\varkappa = \alpha + \sigma$, respectively. From (15) one derives the following functions

$$\Gamma_\omega(s) = G(-s,1) - 1$$

$$\Gamma_z(s) = G_z(-s,1)$$

$$Q_\omega(a) = -\int_0^\infty \Gamma_\omega(s)\,ds$$

$$q(a) = \int_0^a \Gamma_z(s)\,ds$$

$$M(a,P) = \int_0^a \mu(a,P)\,ds$$

In terms of these functions one has the following equations. The average parasite load is given by

$$w(P,\varphi) = \frac{\int_0^\infty e^{-Q(a)\varphi - M(a,P)} q(a)\,da\,\varphi}{\int_0^\infty e^{-Q(a)\varphi - M(a,P)}\,da} \tag{16}$$

The birth law assumes the form.

$$N = N\int_0^\infty b(a,P)e^{-M(a,P)}e^{-Q_\omega(a)\varphi}\,da \tag{17}$$

and the total population size relates to the neonatals as

$$P = N\int_0^\infty e^{-Q(a)\varphi - M(a,P)}\,da \tag{18}$$

The parasite acquisition law is

$$\varphi = \beta f(P,\bar{w}(P,\varphi)) \tag{19}$$

First consider the case where the parameter functions do not depend on P and where the neonatals are prescribed, i.e. where the boundary condition (14) with given N is considered. Then only the equation (19) is relevant. It assumes the simpler form

$$\varphi = \beta f(\bar{w}(\varphi)) \tag{20}$$

Here we take f normalized as $f(0) = 0$, $f'(0) = 1$. Equation (20) has the solution $\varphi = 0$ corresponding to a non-infected population. Also there is a branch of nontrivial solutions $\beta = \varphi/f(\bar{w}(\varphi))$ parametrized by $\varphi > 0$. This branch bifurcates from $\varphi = 0$ at $\beta_o = \int_0^\infty e^{-M(a)}q(a)da/\int_0^\infty e^{-M(a)}da$. It has been shown that the direction $d\beta/d\varphi$ may be positive or negative, i.e. the bifurcation may be forward or backward. Even if the function f is monotone there may be turning points, and hysteresis occurs. In [2] a very natural condition for monotonicity of the nontrivial branch has been given: If the function f is concave in the sense of Krasnoselskij, i.e.if

$$\frac{df}{dw} > 0 \quad , \quad \frac{d}{dw}\left[\frac{f(w)}{w}\right] > 0 \tag{21}$$

then for given β, equation (20) has at most one solution $\varphi > 0$.

In [2], [4] it has been shown for the initial value problem (11)(12)(14) that the solution $\varphi = 0$ is stable for $\beta < \beta_o$ and unstable for $\beta > \beta_o$. In a forthcoming publication [8] it will be shown that the direction of the bifurcation diagram indicates the stability properties of the stationary solution: If $d\beta/d\varphi > 0$ then the stationary solution with $\varphi > 0$ is stable, if $d\beta/d\varphi < 0$ then it is unstable.

The next step is to consider the problem with the Lotka birth law (13) but still with parameter functions independent of P. Then in general there are no stationary solutions, but there are persistent solutions, i.e. solutions with stable age distributions and exponential growth or decay

$$u(t,a,z) = \bar{u}(a,z)\exp(\lambda t). \tag{22}$$

Of course the number of newborns N is arbitrary, the corresponding parameters φ and λ are given by the equations [5]

$$\int_0^\infty b(a)e^{-M(a) - Q_\omega(a)\varphi - \lambda a}da = 1 \tag{23}$$

$$\varphi = \beta f(\tilde{w}(\varphi,\lambda)) \tag{24}$$

$$\tilde{w}(\varphi,\lambda) = \frac{\int_0^\infty e^{-Q(a)\varphi - M(a) - \lambda a} q(a)\,da}{\int_0^\infty e^{-Q(a)\varphi - M(a) - \lambda a}\,da}\,\varphi \tag{25}$$

From (23) one can obtain λ as a function of φ. If this function $\lambda = \Lambda(\varphi)$ is introduced into equation (24)(25) then a nontrivial branch of solutions is obtained which is parametrized by $\varphi > 0$. Again one can show that the concavity property (21) ensures that the nontrivial branch is monotone [9].

Finally consider the full problem, where the parameter functions b,μ,f depend on P. Assume that the birth rate $b(a,P)$ decreases and the death rate $\mu(a,P)$ increases with total population size. Furthermore the transition function $f(P,w)$ is assumed to be increasing in φ.

The constants φ,P satisfy the equations

$$h(P,\varphi) \equiv \int_0^\infty b(a,P)e^{-M(a,P)-Q(a)\varphi}da = 1 \tag{26}$$

$$\varphi = \beta f(P,\bar{w}(P,\varphi)) \tag{27}$$

whereas the constant N can be determined from equation (18). Assume that $h(0,0) > 1$ and that the equation $h(P,0) = 1$ has a solution $P_o < \infty$. Then from (26) one obtains $\varphi = \phi(P)$ as a decreasing function of P, where $0 \le P \le P_o$. If this expression is introduced into (27) then

$$\beta = \frac{\phi(P)}{f(P,w(P,\phi(P)))} \tag{28}$$

Thus there is a branch of solutions for which the host population stabilizes with positive parasite load at a level below its natural density P_o

In addition to the monotonicity properties of b,μ and f required earlier assume that $f(P,\bar{w})$ is concave with respect to the second argument (in the sense of (21)) and increasing in the first argument. Then one can show: For every $\beta > 0$ there is at most one stationary solution (P,φ) with $\varphi > 0$ (see [9]).

The case where the equation $h(P,0) = 1$ has no finite solution P_o has been discussed in [9].

Finally I shall indicate a relation between the present epidemic models and ordinary differential equations. It is well known that certain Lotka age structure models (1) can be reduced to systems of ordinary differential equations. A complete classification of those Gurtin-McCamy systems which can be reduced is given in [10]. For the moment consider the model (11) with

$$b(a,P) = ae^{-\nu a}, \quad \mu(a,P) = \mu \tag{29}$$

where ν and μ are constants.

If one introduces

$$\begin{aligned} N(t,z) &= \int_0^\infty b(a)\, u(t,a,z)\,da \\ H(t,z) &= \int_0^\infty e^{-\nu a}\, u(t,a,z)\,da \\ P(t,z) &= \int_0^\infty u(t,a,z)\,da \end{aligned} \tag{30}$$

then these functions satisfy the equations

$$\frac{\partial}{\partial t}\begin{pmatrix} N \\ H \\ P \end{pmatrix} + g(z)\frac{\partial}{\partial z}\begin{pmatrix} N \\ H \\ P \end{pmatrix} + (\mu-(z-1)\varphi)\begin{pmatrix} N \\ H \\ P \end{pmatrix} + \nu\begin{pmatrix} N \\ H \\ O \end{pmatrix} - N(t,1)\begin{pmatrix} O \\ 1 \\ 1 \end{pmatrix} = 0 \tag{31}$$

This problem can be solved using the method of characteristics.

References

[1] Anderson, R. M., May, R. M., Coevolution of hosts and parasites, Parasitology 85, 411-426 (1982).

[2] Hadeler, K. P., Dietz, K., Nonlinear hyperbolic equations for the dynamics of parasite populations, Computers and Math. with Appl. 9, No. 9, 415-430 (1983) ed. M. Witten .

[3] Hadeler, K. P., Dietz, K., An integral equation for helminthic infections: Global existence of solutions. In: Conf. Proc. Recent Trends in Mathematics, Reinhardsbrunn 1982, Teubner Texte zur Mathematik 50, Teubner Leipzig 1982/83.

[4] Hadeler, K. P., An integral equation for helminthic infections: Stability of the non-infected population. Proc. Vth Int. Conf. on Trends in Theory and Practice of Nonl. Diff. Equ. 1982 (Ed. V. Lakshmikantham), 231-240.

[5] Hadeler, K. P., Integral equations for infections with discrete parasites: Hosts with Lotka birth law. In: Autumn Course on Mathematical ecology, Trieste 1982 (Eds. S. Levin, T. Hallam).

[6] Hadeler, K. P., Hysteresis in a model for parasitic infection. Proc. Numerical methods for bifurcation problems Dortmund 1983 (Eds. H. D. Mittelmann, T. Küpper) Birkhäuser Verlag.

[7] Hadeler, K.P., Dietz, K., Population dynamics of killing parasites which reproduce in the host. J. Math. Biol., submitted.

[8] Hadeler, K. P., Vector models for infectious diseases: stability of the non-trivial stationary states. Proc. Conf. Diff. Equ. Dundee 1984.

[9] Hadeler, K. P., Dietz, K., Models for parasitic diseases: transmission rate depending on total population size. in preparation

[10] Hadeler, K. P., Reduction of Lotka age structure models to ordinary differential equations, in preparation.

[11] Karlin, S., Tavaré, S., Linear birth and death process with killing. J. Appl. Prob. 19, 477-487 (1982).

[12] May, R., Anderson, R. M., Population dynamics of infectious diseases II. Nature 280, 455-461 (1979).

[13] Puri, P. S., A method for studying the integral functions of stochastic processes with applications III., Proc. 6. Berkeley Symp. on Math. Stat. and Prob. Vol. III, 481-500. (Eds. LeCam, Neyman, Scott) Berkeley and Los Angeles 1972.

Acknowledgement: This work was supported by the Deutsche Forschungsgemeinschaft.

SPONTANEOUS BIOLOGICAL PATTERN FORMATION IN THREE DIMENSIONS
THE MORPHOGEN PREPATTERN THEORY OF MITOSIS AND CYTOKINESIS

Axel Hunding*

1. Introduction

Spontaneous pattern formation may arise in (bio)chemical networks coupled to diffusion, i.e. an initial homogeneous distribution of certain chemical species may become unstable by changing variables such as enzyme activity or simply the size (radius) of the sphere. Hereby a new inhomogeneous, yet stable, concentration distribution is set up within the sphere, without any outside control imposing the geometry of the pattern, which is created spontaneously. Such spatial dissipative structures, or Turing structures, give rise to gradientformation, i.e. high concentration at one (spontaneously created) pole and low at the opposite pole, a phenomenon of particular interest in the context of prepattern formation in blastulas. They also give rise to a bipolar concentration pattern, which should be an ideal prepattern for spindleformation and chromosome distribution in the process of mitosis (cell division).

The prepatterns arise through bifurcation of the chemical kinetic reaction-diffusion equations, which are nonlinear partial differential equations subject to boundary conditions. The process is thus mathematical analogous to other well-known phenomena within the area of synergetics. Prepattern formation has been studied theoretically by means of analytical bifurcation theory, and simulated numerically by a fast code, thereby establishing series of stable prepatterns within the sphere, as they appear by changing simple parameters such as the radius. The recorded prepatterns are shown to account for previously experimentally determined chromosome distributions in nuclear division of the radiolarian Aulacantha scolymantha, thus indicating a possible role for these prepatterns in the evolution of mitosis. It is argued here, that prepatterns may play a role even in animal and higher plant cells.

2. Bifurcation analysis, and selection rules

The spherical system under consideration is thermodynamically open. Thus chemical substances are added and removed from the system with uniform rate, i.e. without imposing any symmetry upon the system from the outside. Inside the sphere, an autocatalytic chemical network is assumed to exist

*Adress: Panum Institute, Department of Chemistry 6-2-30, University of Copenhagen, Bleqdamsvej 3, DK-2200 Copenhagen N, Denmark.

the components of which are free to diffuse, but which are confined to the system, i.e. the gradient of the concentrations is zero at the surface of the sphere. In general the set of partial differential equations describing the system is then

$$\frac{\partial c}{\partial t} = F_o(c) + D\Delta c, \tag{1}$$

where c is a vector, the components of which are the concentrations of the chemical network, the vector function F_o contains the nonlinear chemical kinetic terms and the last term arises from Ficks law. Here D will be taken as a constant diagonal matrix. The appearance of spatial dissipative structures is not confined to any specific form of the rate equations F_o, although a common feature of such networks is the appearance of autocatalytic terms in F_o. For the sake of argument,consider a kinetic scheme, which derives from a biochemical system with enzymatic product activation. Here F_o takes the form

$$F_o(1) = \nu - k_1 e c_1 c_2^\gamma/(K + c_2^\gamma) \tag{2}$$

$$F_o(2) = k_1 e c_1 c_2^\gamma/(K + c_2^\gamma) - k_2 c_2. \tag{3}$$

This describes an open two-component system, where component 1 is fed homogeneously by the constant term ν, and reacts to component 2 under catalysis of an enzyme e, which is activated by component 2 through Hill-type kinetics. Component 2 is degraded in turn by first-order kinetics and the degradation products leave the sphere homogeneously. In its most simplified form, one may omit c_2^γ in the denominator, which allows (2-3) to be normalized to the form used by SELKOV

$$F(1) = 1 - c_1 c_2^\gamma \tag{4}$$

$$F(2) = \alpha c_1 c_2^\gamma - \alpha c_2, \tag{5}$$

where c_1 and c_2 are proportional to the original concentrations and α becomes a characteristic rate constant proportional to the enzyme concentration e. Normalizing (1), the homogeneous stationary state becomes $c_o = 1$. This stationary state may be examined by standard linear stability analysis, and conditions (inequalities) established under which the homogeneous solution is unstable with respect to small fluctuations. The result shows that spatial inhomogeneous fluctuations amplify provided the diffusion constants D_1 and D_2 are different by about an order of magnitude.

Writing $z = c - c_o$ and splitting the right hand side of (1) in linear and remaining nonlinear part, stationary solutions to (1) satisfy

$$0 = L(z) + N(z), \tag{6}$$

where $L(z) = D\Delta z + Mz$ with M a constant matrix in general. In particular for SELKOVs scheme (4-5)

$$L = D\Delta + \begin{bmatrix} -1 & -\gamma \\ \alpha & \alpha(\gamma-1) \end{bmatrix} \tag{7}$$

Eq. (6) admits the trivial solution $z = 0$ (the homogeneous solution). Inhomogeneous, stationary solutions to the nonlinear equation (6) may arise, when parameters such as the rate constant α exceeds critical values $\alpha_c(nl)$ where $\alpha_c(nl)$ are given as those values for which the linear problem

$$L_o z = 0 \tag{8}$$

admits nontrivial solutions $z \neq 0$. Here L_o is L with α replaced by $\alpha_c(nl)$. The nontrivial solutions z are proportional to the eigenfunctions ψ_{nlm} of the Laplacian

$$\psi_{nlm}(r,\Omega) = j_n(\kappa_{nl}r)Y_{nm}(\Omega) \tag{9}$$

satisfying $\Delta\psi_{nlm} = -\kappa_{nl}^2\psi_{nlm}$ with $\frac{\partial}{\partial r}\psi_{nlm} = 0$ at $r = R$. $z = d_{nl}\psi_{nlm}$ satisfy (8) provided

$$|-D\kappa_{nl}^2 + M(\alpha_c)| = 0. \tag{10}$$

This determinand equation defines $\alpha_c(nl)$ and vector d_{nl}. In particular, for (4-5),

$$\alpha_c(nl) = \frac{D_2}{D_1} \cdot \frac{D_R^2k_{nl}^4 + D_Rk_{nl}^2}{(\gamma - 1)D_Rk_{nl}^2 - 1} \tag{11}$$

where $D_R = D_1/R^2$ and $k_{nl} = R\kappa_{nl}$. Inhomogeneous stationary solutions z_p (primary bifurcation) to (6) may be obtained by expansion in a small parameter ε. Writing

$$z_p = \sum_{j=1}^{\infty} \varepsilon^j \phi_{j-1} \tag{12}$$

$$\alpha - \alpha_c(nl) = \sum_{j=1}^{\infty} \varepsilon^j \tau_j \tag{13}$$

and collecting terms of the same order in ε, one obtains to first order in ε

$$\phi_o = d_{nl}j_n(\kappa_{nl}r)\sum_{m=-n}^{n} \nu_m Y_{nm}(\Omega). \tag{14}$$

The linear combination expansion coefficients ν_m, and τ_1, may be obtained from the terms to second order in ε, which yield

$$L_o(\phi_1) + \tau_1 L_1(\phi_o) + N_2(\phi_o) = 0. \tag{15}$$

Here, $L_1(\phi_o)$ is a constant matrix multiplying ϕ_o.

Let [u,v] define the inner product of vectors u and v, i.e. the integral of u*·v over the sphere. Consider the adjoint operator L_o^+ which obtains from L_o by transposing $M(\alpha_{nl})$. Then the solution space of L_o^+ is spanned by $\phi_o^+(m) = d_{nl}^+ j_n(\kappa_{nl} r) Y_{nm}(\Omega), m \in [-n,n]$. The Fredholm solvability condition of (15) which equation is of the form $L_o(\phi_1) = f$, is $[f, \phi_o^+(m)] = 0$, $m \in [-n,n]$. That is

$$\tau_1 [L_1(\phi_o), \phi_o^+(m)] = -[N_2(\phi_o), \phi_o^+(m)] \qquad m \in [-n,n] \tag{16}$$

which together with normalisation of ϕ_o, i.e.

$$[\phi_o, \phi_o] = 1, \tag{17}$$

yields a set of 2n + 2 equations in the 2n + 1 ν_m's and τ_1. $N_2(\phi_o)$ is proportional to the square of $j_n \Sigma \nu_m Y_{nm}$, with a proportionality constant dependent of the chemical network in question. This constant is common for all 2n + 1 equations of (16) and the same holds for the therm $[L_2(\phi_o), \phi_o^+]$. Both constants may then be absorbed by scaling τ_1 to τ_1'. Evaluation of $[N_2, \phi_o^+]$, involving integrals over three spherical harmonics is most easily done using tables of 3-j or Clebsch-Gordan coefficients, since

$$\int Y_{nl} Y_{NM} Y_{n'l'} d\Omega = (-1)^l (4\pi)^{-\frac{1}{2}} [(2n+1)(2N+1)(2n'+1)]^{\frac{1}{2}} \begin{pmatrix} n' & n & N \\ 0 & 0 & 0 \end{pmatrix} \begin{pmatrix} n & n' & N \\ -l & l' & M \end{pmatrix}. \tag{18}$$

The integral vanishes unless n' + n + N is even and M + l' = l. Thus (16) - (17) reduce to a set of 2n + 2 simple algebraic equations. To be specific, for $\alpha_c(nl) = \alpha_c(21)$, ϕ_o in (14) is spanned by the five Y_{2m}'s, $m \in [-2,2]$, and (16) - (17) evaluate to

$$\tau_1' \nu_{2\pm2} + 2\nu_{20}\nu_{2\pm2} - \nu_{2\pm1}^2 (3/2)^{\frac{1}{2}} = 0 \tag{19}$$

$$\tau_1' \nu_{2\pm1} + 2\nu_{2\pm2}\nu_{2\mp1} (3/2)^{\frac{1}{2}} - \nu_{20}\nu_{2\pm1} = 0 \tag{20}$$

$$\tau_1' \nu_{20} - \nu_{20}^2 + 2\nu_{22}\nu_{2-2} + \nu_{21}\nu_{2-1} = 0 \tag{21}$$

$$\mu_o^2 - \nu_{20}^2 - 2\nu_{22}\nu_{22}^* - 2\nu_{21}\nu_{21}^* = 0 \tag{22}$$

where μ_o^{-2} is a normalisation integral over the spherical Bessel function. The structure of these equations is *independent* of the chemical network but arises through vanishing or nonvanishing of spatial integrals over appropriate products of eigensolutions to the Laplacian. The explicit solution of (19) - (22) may be simplified further. Introducing a rotation in the Euler angles $(\alpha\beta\gamma)$ and writing

$$Y_{2m}(\Omega) = \sum_{m=-2}^{2} D^{(2)}_{m'm}(\alpha\beta\gamma) Y_{2m'}(\omega), \tag{23}$$

it is possible to find a rotation, which carries the general solution (14) of (19) - (22) into the particular simple form

$$\phi_o = d_{21}\tau_1' j_2(\kappa_2 r) Y_{20}(\omega) \qquad \tau_1' = \pm\mu_o \tag{24}$$

i.e. out of the five spherical harmonics Y_{2m}, the selection rule is that transitions from the homogeneous stationary state to Y_{2m}, $m \neq 0$ are forbidden, whereas only Y_{20} appears. Thus the *geometry* of the bifurcating branch is common for a large class of chemical networks, although the amplitude depends upon the details of the kinetic equations.

Bifurcationtheory thus indicates, that the selection rules and number of secondary branches may be common to a broad class of chemical reaction networks. The analytical construction of such branches is only feasible to low order in the expansion parameters. The above analysis shows, however, that it is meaningful to study bifurcation numerically for a particular reaction network, since the pattern formation, from a geometrical point of view, should be representative of a broad class of such networks.

3. Numerical solution of nonlinear reaction-diffusion systems

The time-evolution of patterns according to (1) may be studied by discretizing the Laplacian (the Method of Lines). The resulting system of nonlinear ordinary differential equations is large, but the Jacobian is sparse. The system is stiff, and is solved accordingly. GEARs method is used with a build-in sparse-matrix package to handle the corrector-step, which requires the solution of a large linear system of algebraic equations. This is accomplished by iterative methods. Solving $Au = v$, matrix A is written as a sum of the diagonal matrix D, lower L and upper U triangular matrices. With the notation $M_D = -D^{-1}M$ and $b = D^{-1}v$ one may solve $Au = v$ iteratively, given u_n, by

$$u_{n+1} = (E - \omega L_D)^{-1}\{[\omega U_D + (1-\omega)E]u_n + \omega b\}. \tag{25}$$

E is the identity matrix. This may be followed by a backwards iteration with L_D and U_D interchanged. Finally, this symmetric iteration is followed by Chebyshev acceleration (or semiiteration) with little additional work (the SSORSI method). The overrelaxation parameter ω is varied from shell to shell in the mesh imposed on the sphere. The result is an impressive 4 order of magnitude gain in integration step length over non-stiff methods. Due to the larger overlay per step of the stiff code, the actual gain on a CPU-time basis is smaller, but still a factor 470 even for a fairly coarse mesh (2000 equations). This large speed-up makes numerical simulation in three spatial coordinates feasible.

Parameter space of (1), (4) and (5), i.e. $\alpha_c(nl)$ in (11) as a function of $D_R = D_1/R^2$ is given in Fig. 1 for the three lowest values of κ_{nl}, which are κ_{11}, κ_{21} and κ_{02}. There results three regions denoted j_1, j_2 and $j_o + j_2$, since the prepatterns which arise in these regions are $j_1(\kappa_{11}r)Y_{10}$, $j_2(\kappa_{21}r)Y_{20}$ and $j_o(\kappa_{02}r) + j_2(\kappa_{21}r)Y_{20}$, respectively. A cell operating on a crude coordination between enzyme activity α and size (R) of the sphere, such as dilution of an enzyme inhibitor by swelling, would move the system in parameter space to the left and upwards simultaneously. Depending on the initial cell size, paths like Ser 1 or Ser 2 in Fig. 1 would result.

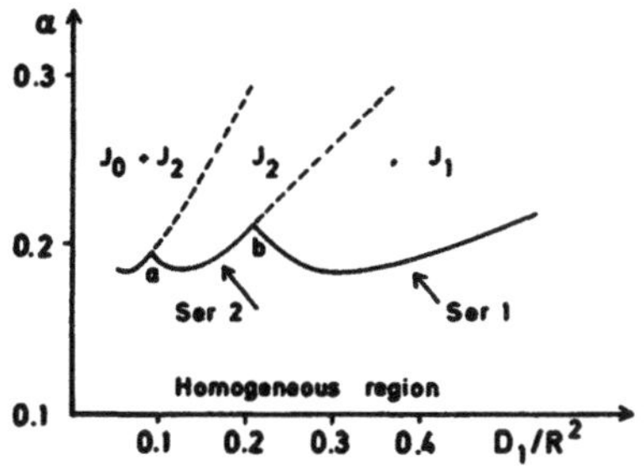

Fig. 1 Parameter space displaying (11), i.e. critical enzyme activity $\alpha_c(nl)$ as a function of D_1/R^2. The spherical reaction-diffusion system is unpatterned for parameters at the bottom of Fig.1. Crossing over the solid curves, spontaneous pattern formation takes place to prepatterns of geometry closely related to Laplacian eigenfunctions j_1Y_{10}, j_2Y_{20} or $j_o + j_2Y_{20}$

Starting in the bottom of Fig. 1 (low enzyme activity α) the system is in the homogeneous state. By crossing a critical curve $\alpha_c(nl)$, the homogeneous state looses its stability and spontaneous pattern formation occurs. If the cells initial size is such, that the critical region is encountered between points a and b in Fig. 1, the j_2-region obtains and primary bifurcation as discussed in Section 2 results, with selection rules arising from (19) - (22), i.e. the homogeneous state vanishes and a bipolar concentration prepattern of form (24) arises. This pattern is most pronounced near the surface, since $j_2(\kappa_{21}r)$, $r \in [0,R]$ rises monotonically from zero in the center of the sphere. The pattern has high concentration at the two spontaneously created poles, and low concentration in a ring close to the surface in the equator plane. This prepattern should be an ideal platform for spatial organisation for the process of mitosis and consequently it is called the "Mitosis-pattern". Moving further into parameter space, eventually the $j_o + j_2$ region is encountered. The result is a new prepattern (secondary bifurcation) which essentially adds a ball, in the center of the sphere, to the bipolar j_2Y_{20} pattern, which is retained as depicted in Fig. 2. This pattern should in turn be an ideal platform for cell-division (cytokinesis), since matter previously located in the equatorial ring of the mitosis-prepattern now becomes attracted to the center of the sphere. Consequently, this prepattern is called the "Cytokinesis-pattern". The quantitative bifurcation diagram for this series of transitions has been recorded numerically.

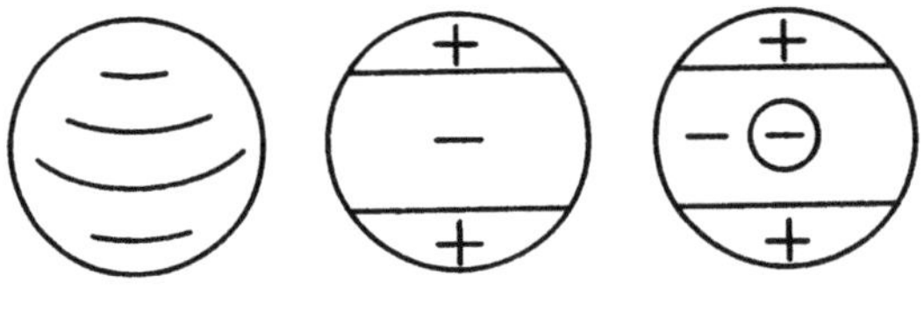

Fig. 2 Prepatterns recorded along the path "Ser 2" in Fig. 1. The homogeneous state bifurcates to the bipolar "Mitosis"-prepattern j_2Y_{20} which further into parameter space undergoes secondary bifurcation to $j_o + j_2Y_{20}$, which is called the "Cytokinesis"-prepattern

When the cell starts swelling from a smaller initial size the region j_1 in Fig. 1 is encountered through pathway Ser 1. The series of prepatterns which arises along this path is describable as a) Homogeneous solution, b) $j_1(\kappa_{11}r)Y_{10}(\Omega)$, c) (transient) $j_1(\kappa_{11}r)Y_{10}(\Omega)+j_2(\kappa_{21}r)Y_{22}(\Omega)$, d) $j_2(\kappa_{21}r)Y_{20}(\Omega)$ and e) $j_2(\kappa_{21}r)Y_{20}(\Omega)+j_o(\kappa_{02}r)$. The first inhomogeneous pattern, (b), has high concentration at one spontaneously generated pole and low concentration at the opposite pole, i.e. spontaneous gradient formation occurs. This pattern changes into the "mitosis-pattern" (d), and "cytokinesis-pattern" (e), as before, if the system moves rapidly along the pathway Ser 1. If, however, the system grows slowly compared to the time it takes to build up the prepatterns, the system bifurcates to a second branch through the long-lived transient (c), which eventually lead to the "mitosis-pattern" (d) again, but this time with the original polar axis tilted 90 degrees toward the original (b) axis.

4. Nuclear division in Aulacantha scolymantha

It is reasonable to assume, that the highly precise, but complex, process of mitosis in higher organisms developed from a less precise, more haphazard, but simple, mechanism. Rudiments of such crude mitotic schemes may be sought in present-day unicellular eukaryotes. The radiolarian Aulacantha scolymantha provides an excellent such system. During nuclear division, the nucleus obtains approximately spherical form. The more than 1000 presumably identical chromosomes are distributed to each hemisphere of the nucleus in a crude segregation, in which the chromosomes occupy certain well defined regions of particular symmetrical shape, inside the spherical nucleus. The governing principle of this pattern formation in A.s. has hitherto been entirely obscure. It is difficult to attribute the confinement of the chromosomes to highly symmetrical space regions to conventional spindle forces.

The present author has argued, that the governing principle in A.s. is the prepattern-sequences recorded in the previous section. If it is assumed, that the chromosomes in A.s. are confined to those regions of space, where the prepatterns have neither high, nor low concentration, i.e. the "null-region" of the functions $j_n(\kappa_{nl}r)Y_{nm}(\Omega)$, then a substantial agreement is found between theoretically predicted series of prepatterns and actual experimentally recorded chromosome distributions. Two of these are of special interest here.

Chromosomal regions equivalent to the null-regions of the mitosis-prepattern are experimentally observed. This may be taken as evidence that this prepattern actually exists in nature, and contributes to spatial governing of chromosomes to the two half-spheres of the cell. However, this chromosomal distribution is closely related to what is usually

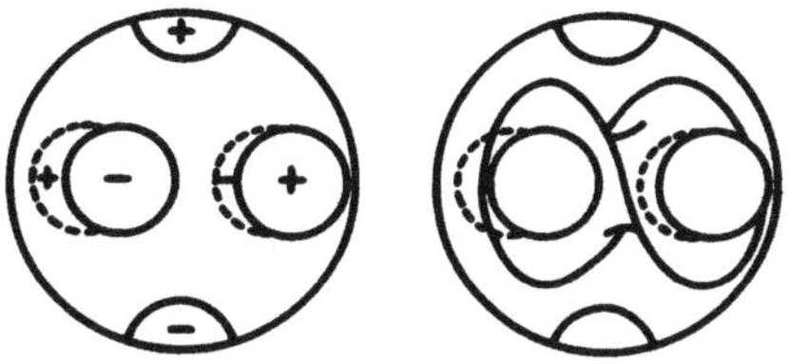

Fig. 3 The null-region of the prepattern $j_1Y_{10} + j_2Y_{22}$ (left) creates a saddle-shaped chromosomal distribution (right), photographed in Aulacantha scolymantha. Such highly symmetrical distributions are difficult to explain by conventional spindle forces and may be experimental evidence of Laplacian prepatterns

encountered during standard anaphase in cells, a process which is controlled by a bipolar spindle.

A more decisive argument for the presence of prepattern-governing in A.S. is provided by the null regions of $j_1Y_{10} + j_2Y_{22}$. At first, the chromosomes are confined to the equatorial region (observed experimentally) as this is the null-region of j_1Y_{10}. When j_2Y_{22} appears, effectively four balls of alternating sign are squeezed into this equatorial region, with the result, that the equatorial plate is folded into a highly symmetrical twisted, saddle-shaped object, Fig. 3. A highly structured chromosomal distribution with this particular symmetry is very difficult to explain by conventional spindle forces. Yet, this structure is found experimentally in A.s. where it has been photographed repeatedly (the "windschief mutterplatte", recorded by K. GRELL). According to the prepattern theory, this prepattern changes into the bipolar mitosis prepattern, with its axis perpendicular to the original (j_1Y_{10}) axis. During this process, the saddle-shaped object turns into two disconnected regions, of the form of two parallel horseshoes with their openings in the same direction, which eventually close to the two symmetrical null regions of the mitosis-prepattern. Such horseshoe chromosomal regions are reported experimentally as well. The experimentally established appearance of the highly symmetrical saddle-shaped object and the transition involving horseshoeregions strongly suggest that the chromosomal mass is under spatial control of the prepatterns recorded numerically.

Bifurcations in prolate spheroids (elongated spheres) and oblate spheroids (compressed spheres) have been obtained numerically, and the main result is, that the sequences of prepatterns found in the sphere and discussed above, are stable toward spatial deformations of the sphere. In particular, the creation of the saddle-shaped chromosonal distribution, Fig. 3, has been shown to occur even in spatially deformed cells. Experimentally recorded peculiarities of the saddle-shaped object in A.s. may be better understood on this basis. Also, an alternative route to this structure was found: If the cell penetrates parameter space along the path designated Ser 2 in Fig. 1, in part of this region the saddle-shaped structure emerges, if the cell is not a perfect sphere but slightly compressed. Secondly, an alternative route to the "cytokinesis" prepattern, Fig. 2, was found in prolate spheroids: If the cell is started in the "mitosis" prepattern in a spherical cell, and the cell then elongates (as many cells do in anaphase), the "cytokinesis" prepattern builds in.

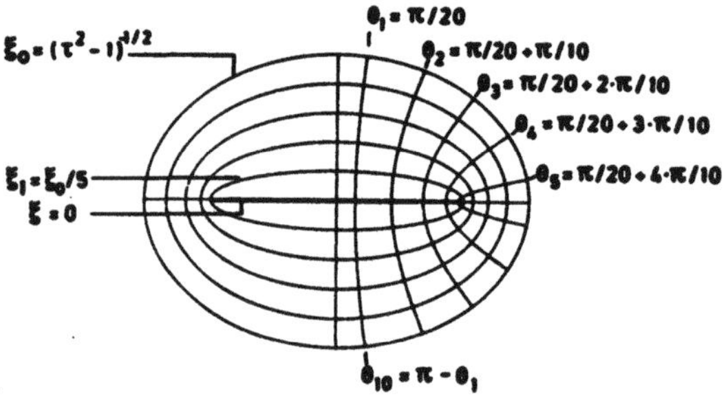

<u>Fig. 4</u> Oblate spheroidal coordinates. A set of confocal ellipses are constructed, and intersected by hyperbolas. Bifurcations obtained in these and prolate spheroidal coordinates confirm the results of the sphere and yield alternative pathways to the "mitosis"- and "cytokinesis"-prepatterns, Fig. 2

5. Prepatterns in mitosis and cytokinesis

The mechanism of mitosis is to most people linked to the centrioles, which are assumed to give rise to bipolarity. However, noncentriolar mitosis occur in a number of species, most notably in higher plants. Also, it has been demonstrated experimentally, that the centrioles are not needed. When the centriolar region is displaced off the spindle pole of the cranefly, the cells were still perfectly capable of undergoing anaphase and cytokinesis (Dietz, 1964). Recently, laser microbeam destruction of prophase centrioles was accompanied by normal anaphase movements of chromosomes (Berns and Richardson, 1977). Modern theories on mitosis and cytokinesis are thus centriolar free models, based on intrinsic properties of the spindle (Zimmerman and Forer, 1981). All these models predict a force proportional to the length of the kinetochore fiber. Such theories have been tested quantitatively on asymmetrically resting trivalents (Luykx, 1970), Fig. 5. If force balance is achieved, the length $b + c$ should match a, but the data show, that the equatorcrossing fibre a is much too long, even if the cross over part of fiber a is omitted $(a - \Delta)$. It was shown (Hunding, 1981), however, that the prepattern theory of mitosis yields a much better force balance. The idea is, that the spindle fibers interact with the mitosis prepattern along the entire fibre with a force given by the local gradient of the prepattern. In one mechanism to achieve such a force one may assume that the microtubules carry an electric charge per subunit and the concentration gradient of the prepattern give rise to an electrical diffusion potential. This force is easy to quantify, and a very good force balance obtains. This may indicate that prepatterns exist even in centriolar containing cells, and contribute to pole formation and force determination in the spindle. It is thus suggested, that prepatterns should be added to existing theories on the mechanism of mitosis.

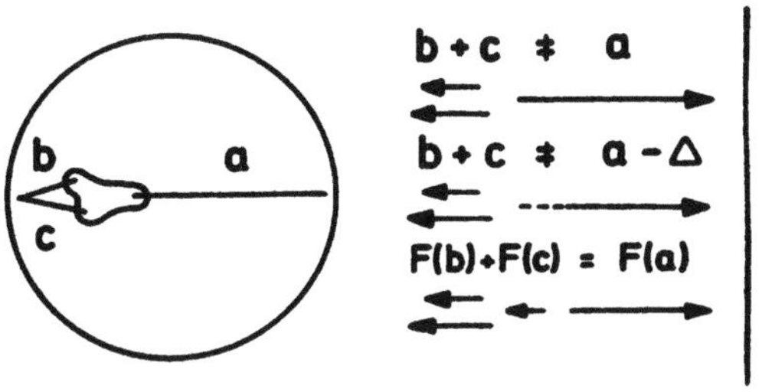

Fig. 5 Spindle forces on trivalents. Force balance is not achieved by two theories assuming the force proportional to the length of the fibers. The prepattern theory, however, yields force balance because the equator-crossing part Δ of a is attracted to the left pole by the symmetrical bipolar prepattern.

The process of cytokinesis (cell cleavage) seems to be as complex to understand as that of mitosis (chromosome distribution to the two cell halves). Cytokinesis as it takes place in most asters containing cells has been reviewed (Rappaport, 1971). This work indicated a mechanism based on overlapping astral rays. However, recently it was found, that a needle may be inserted in the zone where such overlap might occur. If the needle was rotated around equator parallel to the mitotic axis, it should prevent any such overlap formation. However, it was found that furrow formation took place despite this drastic procedure. It was concluded, that furrow establishment does not seem to depend upon a precise stable arrangements of astral rays (Rappaport, 1978).

It has been argued (Hunding, 1982), that the prepatterns under discussion may account not only for normal cytokinesis, but for a number of experiments, like the abovementioned, which are difficult to explain by other current models. The stirring process reported is likely to be slow (2 min pr cycle) relative to the regeneration time of the reaction diffusion prepatterns and consequently the prepatterns may be basically unaffected.

The prepattern theory of mitosis and cytokinesis is in agreement with a number of experiments which are not explainable by other current models. Such experiments are quadripartition in flattened cells (Scott, 1960), and in torus deformed cells (Rappaport, 1971), both discussed in (Hunding, 1982). It has been shown that anaphase movements of chromosomes are not affected by disruption of interzonal microtubules (Snyder and Vogt, 1982), a result also found in plant spindels (Mole-Bajer and Bajer, 1982). Finally, tripartition is observed in elongated spermatids and quite remarkably, no contractile rings form. Rather the division is accompanied by vesicle formation, thus resembling plant cell division (Ward, Argon and Nelson, 1981).

In conclusion a number of experimental facts are in support of the assumption, that prepatterns exist in cells and play a governing role both in mitosis and cytokinesis. Thus the interesting possibility exists, that a number of diverse phenomena, like centriolar and noncentriolar mitosis, and plant and animal cytokinesis, are different exploitations by nature of a unifying principle, based on spatial control by prepatterns.

Acknowledgement

Calculations were performed at RECKU, with support from the Danish National Research Council.

Literature

[1] Berns, M.W. and Richardson, S.M.: Continuation of Mitosis after selective laser microbeam destruction of the centriolar region. J. Cell Biol. 75 (1977) 977.

[2] Billing, G.D. and Hunding, A.: Bifurcation analysis of nonlinear reaction-diffusion systems: Dissipative structures in a sphere. J. Chem. Phys. 69 (1978) 3603.

[3] Dietz, R.: The dispensability of the centrioles in the spermatocytes divisions of *Pales Ferruginea (Nematocera)*. Chromosomes Today 1 (1964) 161.

[4] Hunding, A. and Billing, G.D.: Secondary bifurcations in spherical reaction-diffusion systems. Chem. Phys. 45 (1980) 359.

[5] Hunding, A.: Dissipative structures in reaction-diffusion systems. Numerical determination of bifurcations in the sphere. J. Chem. Phys. 72 (1980) 5241.

[6] Hunding, A.: Possible prepatterns governing mitosis: The mechanism of spindle-free chromosome movement in *Aulacantha Scolymantha*. J. Theor. Biol. 89 (1981) 353.

[7] Hunding, A. and Billing, G.D.: Spontaneous pattern formation in spherical nonlinear reaction-diffusion systems: Selection rules favor the bipolar "mitosis" pattern. J. Chem. Phys. 75 (1981) 486.

[8] Hunding, A.: Spontaneous biological pattern formation in the three dimensional sphere. Prepatterns in mitosis and cytokinesis. In Haken, H., Ed.: Evolution of Order and Chaos in Physics, Chemistry and Biology. Springer Series in Synergetics, vol XVII, Springer 1982.

[9] Hunding, A.: Bifurcations in nonlinear reaction diffusion systems in prolate spheroids. J. Math. Biol. 17 (1983) 223.

[10] Luykx, P.: Cellular mechanisms of chromosome distribution. Int. Rev. Cyt. Suppl. 2 (1970).

[11] Molé-Bajer, J. and Bajer, A.S.: Action of Taxol on Mitosis: Modification of Microtubule Arrangements and Function of the Mitosis Spindle in *Haemanthus Endosperm*. J. Cell Biol. 96 (1983) 527.

[12] Rappaport, R.: Cytokinesis in Animal Cells. Int. Rev. Cyt. 31 (1971) 169.

[13] Rappaport, R.: Effects of Continual Mechanical Agitation prior to Cleavage in Echinoderm Eggs.
J. Exp. Zool. 206 (1978) 1.

[14] Scott, A.C.: Furrowing in flattened sea urchin eggs.
Biol. Bull. 119 (1960) 246.

[15] Snyder, J.A. and Vogt, S.I.: Anaphase motions in the presence of disrupted astral and interzonal microtubules.
J. Cell Biol. 95 (1982) 305a.

[16] Ward, S., Argon, Y. and Nelson, G.A.: Sperm morphogenesis in wild-type and fertilization-defective mutants of *Caenorhabditis elegans*.
J. Cell. Biol. 91 (1981) 26.

[17] Zimmerman, A.M. and Forer, A., Eds.: Mitosis/Cytokinesis.
Academic Press, New York 1981.

EXISTENCE OF WAVE TRAINS IN REACTION-DIFFUSION SYSTEMS

Klaus R. Schneider[1)]

1. INTRODUCTION

The problem of formation of self-organized patterns is a fundamental problem not only of the natural sciences. Much experimental work has been done in different fields [4,5,6,1o,17,21], there are also theoretical investigations to find out general principles governing pattern formation in irreversible processes [2-9,2o,21]. In this connection the concept "dissipative structure" has been introduced by I. Prigogine [8]. Examples of dissipative structures are spatially inhomogeneous structures (Turing structures), temporally periodic structures (selfsustained oscillations), spatio-temporally periodic structures (periodic travelling waves).

In what follows we are concerned with irreversible processes the characteristic subprocesses of which are diffusion and kinetics. We will study pattern formation basing only on the interaction of these processes. As a well-known example we consider the Belousov-Zhabotinskii reaction [17,2o]: When four or five chemical compounds (for example sodium bromate, malonic acid, sulfuric acid and ferroin) are mixed in the appropriate concentration ranges and at the appropriate temperature, the B-Z-system organizes itself into temporal or spatio-temporal dissipative structures of macroscopic dimension: Synchronous oscillations, i.e. spatially homogeneous time periodic oscillations with the same phase can be observed in a well-stirred reactor. When the stirring is interrupted wave-like structures appear: wave trains, wave fronts, spiral waves [9,17,2o]. In the sequel we restrict ourselves to the formation of wave trains, i.e. concentric waves propagating from a leading center in reaction-diffusion systems (see fig. 1). Our main purpose is to investigate the interaction of diffusion and kinetics generating periodic travelling waves.

Fig. 1. Target patterns in the Belousov-Zhabotinskii-reaction.

1) Institut für Mathematik der AdW der DDR, 1086 Berlin, Mohrenstr. 39

2. BASIC ASSUMPTIONS

We assume that the process under consideration may be described by a reaction-diffusion equation of the form

$$\frac{\partial u}{\partial t} = D\,\Delta u + f(u), \tag{2.1}$$

where $u = (u_1,\dots,u_n)^T$ is the vector of state variables depending on space $x = (x_1,\dots,x_m)^T$ and time t, D is the matrix of diffusivities, Δ is the Laplace operator in the spatial coordinates. The vector function f describes all reactions and their interactions. The corresponding system

$$\frac{du}{dt} = f(u) \tag{2.2}$$

is called the kinetic system.

Our mathematical description of propagating concentric waves bases on the following idealization [7]: We focus on a small segment about a radial line excluding the leading center (here the wave length is no longer defined) and the outside ring of the target pattern which meets and interacts with an other expanding target pattern. In this approximation the pattern appears to be almost exactly a plane wave solution of (2.1), that means a solution of the form

$$u(x,t) = \tilde{u}(x\cdot\nu - ct) \tag{2.3}$$

where ν is the unit vector in the direction of propagation, $c > 0$ is the velocity of propagation. The travelling wave solution $\tilde{u}(x\cdot\nu - ct)$ is called a wave train if $\tilde{u}(z)$, $z := x\cdot\nu - ct$, is periodic in z with period $\omega > 0$.

Substituting (2.3) into (2.1) and taking into account $\nu\cdot\nu = 1$ we obtain

$$D\tilde{u}'' + c\tilde{u}' + f(\tilde{u}) = 0 \tag{2.4}$$

where $' := d/dz$. Thus, the existence of a wave train to (2.1) is equivalent to the existence of a nontrivial periodic solution of the system (2.4). Assuming the appearence of target patterns to be a consequence of the interaction of diffusion and kinetics we ask the following question: Which properties of the kinetics and of the matrix D of diffusivities imply the existence of wave trains? The important problem of the stability of wave trains must be excluded from wants of space.

3. EXISTENCE OF WAVE TRAINS

Plane wave solutions of (2.1) satisfy the basic equation (2.4). Since this equation does not depend on ν we have

LEMMA 3.1. If (2.1) has a plane wave solution belonging to the direction ν_0, then (2.1) possesses to any direction ν a travelling wave.

Using the negative criterion of I. Bendixson we obtain

LEMMA 3.2. If $n = 1$ (one species!), $f \in C^1(R,R)$, and $c \neq 0$, then (2.1)

has no wave train.

The basic problem is to prove the existence of periodic solutions of the differential system (2.4). To this end we embed (2.4) into the family of equations

$$\varrho D\tilde{u}'' + c\tilde{u}' + f(\tilde{u}) = 0 , \quad \varrho \text{ small real number.} \tag{3.1}$$

This embedding can be interpreted as follows: We assume that to given c the process of diffusion is small compared with the process of reaction. Setting $z = c\zeta$ and using the abbreviations

$$\varepsilon := \varrho/c^2 , \ \bar{u}(\zeta) := \tilde{u}(c\zeta), \ \dot{} := d/d\zeta \tag{3.2}$$

we get from (3.1)

$$\varepsilon D\ddot{\bar{u}} + \dot{\bar{u}} + f(\bar{u}) = 0 . \tag{3.3}$$

Another imbedding of (2.4) is the following

$$D\tilde{u}'' + c\tilde{u}' + \varrho f(u) = 0 , \quad \varrho \text{ small,} \tag{3.4}$$

that means we suppose the diffusivities to be large compared with the velocity of reaction to given c. Setting $\varrho z = c\zeta$ and using (3.2) we get the equation (3.3) again.

Under the assumption

(D_1). det $D \neq 0$

the equation (3.3) is equivalent to the system

$$\begin{aligned} \dot{\bar{u}} &= \bar{v}, \\ \varepsilon \dot{\bar{v}} &= - D^{-1} [\bar{v} + f(\bar{u})] . \end{aligned} \tag{3.5}$$

The system (3.5) is a singularly perturbed system of autonomous differential equations. Some basic properties of such systems are noted in the appendix.

In what follows we assume

(F_1). $f \in C^1(G,R^n)$.

Thus, the system (3.5) satisfies the hypotheses (H_1) and (H_2) in the appendix where the function φ coincides with $-f$. The corresponding degenerated system reads

$$\dot{\bar{u}} = - f(\bar{u}) . \tag{3.6}$$

Disregarded stability properties this system has the same topological structure of the trajectories as the kinetic system (2.2). Especially, the system (3.6) has a non-degenerated periodic solution if we assume

(F_2). The kinetic system (2.2) has a non-degenerated periodic solution $p_o(t)$ with period τ_o.

To be able to apply Theorem 5.4 we still need the hypothesis (H_3). It is easy to verify that this assumption holds if we assume

(D_2). D has no eigenvalues on the imaginary axis.

Applying Theorem 5.4 to the system (3.5) we get

PROPOSITION 3.3. Assume the assumptions (D_2), (F_1), (F_2) hold. Then there is a positive number ε_1 such that for each ε with $|\varepsilon| < \varepsilon_1$ the system (3.5) has a $\tau(\varepsilon)$ periodic solution with $\tau(\varepsilon) \to \tau_0$ as $\varepsilon \to 0$.

Taking into account the transformation $z = c\zeta$ in case of embedding (3.1) and $\varrho z = c\zeta$ in case of embedding (3.4) we obtain from Proposition 3.3

COROLLARY 3.4. Assume the assumptions of Proposition 3.3 hold. Then we have

(i) The equation (3.1) has to any $c > 0$ for sufficiently small ϱ an $\omega(\varepsilon)$-periodic solution such that $\omega(\varrho) \to c\,\tau_0$ as $\varrho \to 0$.

(ii) The equation (3.4) has to any $c > 0$ for sufficiently small ϱ an $\omega(\varepsilon)$-periodic solution such that $\varrho\,\omega(\varrho) \to c\,\tau_0$ as $\varrho \to 0$, that is $\omega(\varrho) \to \infty$ as $\varrho \to 0$.

(iii) The equations (3.1) and (3.4) have to any ϱ for sufficiently large c an $\omega(c)$-periodic solution such that $\omega(c) \to \infty$ as $c \to \infty$.

From Corollary 3.4 we obtain immediately

THEOREM 3.5. Suppose the hypotheses (D_2), (F_1), (F_2) are valid. Let be

$\gamma := \|D\|$, where $\|.\|$ is any matrix norm,

$\delta := \max_{u \in G} |f(u)|$, where $|.|$ is any vector norm.

Then we have

(i) The equation (2.1) has to any velocity c a wave train if γ to given f or δ to given D is sufficiently small. In case $\gamma \ll 1$ the period $\omega(\gamma)$ of the wave train satisfies $\omega(\gamma) \to c\,\tau_0$ as $\gamma \to 0$, in case $\delta \ll 1$ we have $\omega(\delta) \to \infty$ as $\delta \to 0$.

(ii) The equation (2.1) has to any γ, δ a wave train if c is sufficiently large. The corresponding period tends to infinity as $c \to \infty$.

REMARK 3.5.1. In case $\gamma \to 0$, that is $\|D\| \to 0$, it follows from (3.1) that the wave train $\tilde{u}(x \cdot \nu - ct)$ of the equation (2.1) tends to a wave train with period $c\,\tau_0$ of the system

$$c\,\frac{d\tilde{u}}{dz} + f(\tilde{u}) = 0 .$$

Using $dz = -c dt$ we can say that the wave train $\tilde{u}(x \cdot \nu - ct)$ tends to a wave train with period τ_0 of the kinetic system (2.2). Does there exist such a structure? From assumption (F_2) it follows that $p_0(t)$ is a τ_0-periodic solution of (2.2). Since (2.2) is an autonomous system, $p_0(t+\alpha)$ with any real α is also a τ_0-periodic solution of (2.2). If we consider x as an parametervector in (2.2) then we may put $\alpha = \frac{x \cdot \nu}{-c}$. Therefore, $p_0(\frac{x \cdot \nu - ct}{-c})$ is a wave train solution of (2.2) with period τ_0, that means, all points of the hyperplane $x \cdot \nu$ = const oscillate with the

same phase.
In cases $\gamma \to \infty$ and $c \to \infty$ we can conclude that the wave train $\tilde{u}(x\cdot\nu - ct)$ tends to the spatially homogeneous solution $u(x,t) = p_0(t)$ of (2.1).

REMARK 3.5.2. The crucial assumption of Theorem 3.5 is the existence of a non-degenerated limit cycle of the kinetic system (2.2). To check such an assumption we can take into consideration that the function f generally depends on some parameters. Then the Hopf bifurcation theorem may be an useful tool to prove the existence of a non-degenerated limit cycle with small amplitude.

The basic tool to prove Theorem 3.5 was the method of continuation. Now we replace the hypothesis on the existence of a limit cycle of the kinetic system (2.2) by the assumption of the existence of an equilibrium point u_0 of (2.2) and use the method of bifurcation theory to establish a wave train of (2.1) with small amplitude.

In order to be able to apply Theorem 5.6 we need the assumption

(F_3). $f(u_0) = 0$ (Without loss of generality we may assume $f(0) = 0$).

(F_4). $f \in C^2(G,R^n)$ where G is a neighbourhood of the origin.

The matrices L_0 and L_1 defined in (5.6), (5.7) of the appendix reads in case of the system (3.5)

$$L_0 = - D^{-1}f_u(0),\; L_1 := - Df_u(0)D^{-1}f_u(0).$$

Thus, the hypothesis (H_7) in the Theorem 5.6 can be formulated as follows (F_5). The matrix $- D^{-1}f_u(0) + cDf_u(0)D^{-1}f_u(0)$ has for sufficiently small ε a pair of simple complex conjugate eigenvalues $\alpha(\varepsilon) \pm i\beta(\varepsilon)$ satisfying $\alpha(0) = 0$, $\alpha'(0) \neq 0$, $\beta(0) \neq 0$.

PROPOSITION 3.6. Assume the hypotheses (F_3)-(F_5), (H_6), (H_8) hold. Then there is a positive number ε_3 such that for each ε with $0 < \varepsilon < \varepsilon_3$ the system (3.5) has a $\tau(\varepsilon)$-periodic solution with $\tau(\varepsilon) \to \frac{2\pi}{\beta}$ as $\varepsilon \to 0$.

From Proposition 3.6 it follows immediately.

THEOREM 3.7. Suppose the hypotheses of Proposition 3.6 hold. Then we have

(i). The equation (2.1) has to any velocity c a wave train with small amplitude if $\|D\|$ is sufficiently small where the period $\omega(\|D\|)$ of the wave train tends to $c\frac{2\pi}{\beta}$ as $\|D\| \to 0$.

(ii) The equation (2.1) has to any D a wave train with small amplitude if c is sufficiently large. The corresponding period tends to infinity as $c \to \infty$.

REMARK 3.8. The obtained results about the existence of wave trains of the system (2.1) can be extended to equations of the form

$$\frac{\partial u}{\partial t} = \sum_{i=1}^{m} \frac{\partial}{\partial x_i}\left(k(u)\frac{\partial u}{\partial x_i}\right) + f(u),$$

where $k : R^n \to R$ is positive and differentiable.

REMARK 3.9. A theorem analogous to Theorem 3.7 can be proved for equations of the type

$$\frac{\partial u}{\partial t} = D \Delta u + f(u(t-\sigma))$$

using results of R. Purfürst [12] and V.V. Strygin [19].

5. APPENDIX

Consider the autonomous differential system

$$\frac{dz}{dt} = q(z) \tag{5.1}$$

where $q : G \subset R^n \to R^n$ is assumed to be continuously differentiable on G.

DEFINITION 5.1. The equilibrium point $z_o \in G$ of (5.1) is said to be structural stable if $q_z(z_o)$ has no eigenvalue on the imaginary axis.

Let $p : R \to G$ be an τ-periodic solution of (5.1). The system

$$\frac{dy}{dt} = q_z(p(t))y \tag{5.2}$$

is called the linear variational system of (5.1) to the periodic solution p. Let Y(t) be the fundamental matrix of (5.2) satisfying Y(0)=E.

DEFINITION 5.2. The eigenvalues of the matrix $Y(\tau)$ are called the Floquet-multipliers of the system (5.2).

Any periodic solution p of (5.1) has 1 as a Floquet multiplier.

DEFINITION 5.3. The τ-periodic solution p of (5.1) is said to be nondegenerated if the linear variational system (5.2) has 1 as a simple Floquet-multiplier.

We now consider the singularly perturbed differential system

$$\frac{d\bar{u}}{dt} = g(\bar{u}, v, \varepsilon), \qquad \varepsilon \frac{dv}{dt} = h(\bar{u}, v, \varepsilon) \tag{5.3}$$

where $\bar{u} \in R^{n1}$, $\bar{v} \in R^{n2}$, $\varepsilon \in (-\varepsilon_o, \varepsilon_o)$, ε_o small. We assume

(H_1). There are regions $G_1 \subset R^{n1}$, $G_2 \subset R^{n2}$ and a positive number ε_1 such that g and h are continuous and differentiable with respect to $(\bar{u}, \bar{v})$ on $G := G_1 \times G_2 \times (-\varepsilon_1, \varepsilon_1)$ where the derivatives continuously depend on $(\bar{u}, \bar{v}, \varepsilon)$.

(H_2). The equation $h(\bar{u}, \bar{v}, 0) = 0$ has a unique isolated solution $\bar{v} = \varphi(\bar{u})$ on a region $G_1^{(1)} \subset G_1$.

Setting $\varepsilon = 0$ in (5.3) we obtain under the hypothesis (H_2) the system

$$\frac{d\bar{u}}{dt} = g(\bar{u}, \varphi(\bar{u}), 0) \tag{5.4}$$

which is called the degenerated system of (5.3). To guarantee the existence of a periodic solution of the singularly perturbed system (5.3) for $\varepsilon \neq 0$ we have essentially two different methods: the method of continuation and the method of bifurcation. Concerning the method of continuation we still need the following hypotheses.

(H_3). The equilibrium point $\bar{v} = \varphi(\bar{u})$ of the system

$$\frac{d\bar{v}}{dt} = h(\bar{u},\bar{v},0)$$

is structural stable for all $\bar{u} \in G_1^{(1)}$.

(H_4). The degenerated system (5.4) has a non-degenerated τ_o-periodic solution (limit cycle) $p_o(t)$, $\Gamma_o := \{ \bar{u} \in R^{n_1} : \bar{u} = p_o(t), 0 \le t \le \tau_o \}$.

THEOREM 5.4. (D.V. Anosov [1]). Assume the hypotheses (H_1)-(H_4) hold. Then there are positive numbers ε_2 and $\delta(\varepsilon_2)$ such that for each $\varepsilon \in (-\varepsilon_2, \varepsilon_2)$ the system (5.3) has a $\tau(\varepsilon)$-periodic solution which entirely lies in a δ-neighbourhood of $\Gamma := \{(\bar{u},\bar{v}) : \bar{u} = p_o(t), \bar{v} = \varphi(p_o(t)), 0 \le t \le \tau_o\}$ and satisfies $|\tau(\varepsilon) - \tau_o| < \rho$ where ρ tends to zero as $\varepsilon_2 \to 0$.

Concerning the bifurcation method we consider the phenomenon of singular Hopf bifurcation [13,18]. Let us assume

(H_5). $g(0,0,\varepsilon) \equiv 0, \quad h(0,0,\varepsilon) \equiv 0 \qquad \forall\, \varepsilon \in (-\varepsilon_2, \varepsilon_2)$.

(H_6). $g_{\bar{v}}(0,0,0)$ has no eigenvalue on the imaginary axis.

Applying the theory of center manifolds [14] we may prove [15]:

THEOREM 5.5. Assume the hypotheses (H_1), (H_5), (H_6) are valid. Then there exists a local integral manifold $\mathcal{M}_\varepsilon$ of the singularly perturbed differential system (5.3) for sufficiently small ε with the following properties.

(i) $\mathcal{M}_\varepsilon$ has the representation $\bar{v} = \psi(\bar{u}, \varepsilon)$, $|\bar{u}| + |\varepsilon|$ sufficiently small, where ψ has the same regularity properties as g and h.

(ii) $\psi(0, \varepsilon) \equiv 0$, $\psi(\bar{u},0) \equiv \varphi(\bar{u})$.

(iii) All periodic solutions of (5.3) near the origin are located on $\mathcal{M}_\varepsilon$.

Using the integral manifold $\mathcal{M}_\varepsilon$ we can reduce the singular Hopf bifurcation problem for (5.3) to the Hopf bifurcation problem of the regular system

$$\frac{d\bar{u}}{dt} = g(\bar{u}, \psi(\bar{u},\varepsilon), \varepsilon) =: \tilde{g}(\bar{u}, \varepsilon). \tag{5.5}$$

Assuming

$(\tilde{H}_1)$. $g \in C^2(G,R^{n_1}), h \in C^2(G,R^{n_2})$

there exist the matrices

$$L_o := \frac{\partial \tilde{g}}{\partial \bar{u}}(0,0), \quad L_1 := \frac{\partial^2 \tilde{g}}{\partial \bar{u}\partial \varepsilon}(0,0).$$

It is possible to represent L_o and L_1 as functions of g and h [11]. To this end we introduce the abbreviations

$$A_1 := g_{\bar{u}}(0,0,0),\ A_2 := g_{\bar{v}}(0,0,0),\ B_1 := h_{\bar{u}}(0,0,0),\ B_2 := h_{\bar{v}}(0,0,0). \tag{5.6}$$

We have

$$L_o = A_1 - A_2B_2^{-1}B_1, \quad L_1 = \frac{\partial^2 g}{\partial u \partial \varepsilon}(0,0,0) - \frac{\partial^2 g}{\partial v \partial \varepsilon}(0,0,0)B_2^{-1}B_1$$
$$+ A_2B_2^{-1}\left[\frac{\partial^2 h}{\partial v \partial \varepsilon}(0,0,0)B_2^{-1}B_1 - \frac{\partial^2 h}{\partial u \partial \varepsilon}(0,0,0) - B_2^{-1}B_1L_o\right]. \tag{5.7}$$

Applying the Ljapunov-Schmidt method we obtain the following result about singular Hopf bifurcation [16]

THEOREM 5.6. Assume the hypotheses (H_1) and (H_5) to be valid. Furthermore, suppose

$(\tilde{H}_6)$. det $B_2 \neq 0$

(H_7). $L_o + \varepsilon L_1$ has a pair of simple complex conjugate eigenvalues $\alpha(\varepsilon) \pm i\beta(\varepsilon)$ for sufficiently small ε satisfying

$$\alpha(0) = 0, \quad \alpha'(0) \neq 0, \quad \beta(0) \neq 0.$$

$(\tilde{H}_8)$. L_o has no eigenvalue λ obeying $\lambda = ni\beta$, $n = 0, \pm 2, \pm 3, \ldots$.

Then there exists a family of non-trivial $\tau(\mu)$-periodic solutions $\hat{u}(\mu)(t)$, $\hat{v}(\mu)(t)$ of (5.3) with $\varepsilon = \varepsilon(\mu)$ for sufficiently small $|\mu|$ such that

$$\tau(0) = \frac{2\pi}{\beta}, \quad \varepsilon(0) = 0, \quad \hat{u}(0)(t) \equiv 0, \quad \hat{v}(0)(t) \equiv 0;$$

moreover $\tau, \varepsilon, \hat{u}, \hat{v}$ are continuously differentiable with respect to μ.

If we replace $(\tilde{H}_8)$ by the stronger hypothesis

(H_8). L_o has no eigenvalue on the imaginary axis except $\pm i\beta$

then we may reduce the system (5.5) by means of the center manifold [14] to a two-dimensional system

$$\frac{dw}{dt} = \tilde{L}(\varepsilon)w + \tilde{h}(w,\varepsilon), \quad \tilde{h}(w,\varepsilon) = o(|w|) \tag{5.8}$$

where $\tilde{L}(\varepsilon)$ has the eigenvalues $\alpha(\varepsilon) \pm i\beta(\varepsilon)$.

Using the hypothesis (H_8) we have the following result regarding the direction of the singular Hopf bifurcation

THEOREM 5.7. Suppose the hypotheses of Theorem 5.6 hold. If we replace the assumptions $(\tilde{H}_6)$ and $(\tilde{H}_8)$ by the stronger hypotheses (H_6) and (H_8) and if we assume the equilibrium point $w = 0$ of (5.8) to be globally asympt. stable as $t \to +\infty$ or $t \to -\infty$ for $\varepsilon \leq 0$, then we have $\varepsilon(\mu) > 0$, that is, the family of $\tau(\mu)$-periodic solutions $\hat{u}(\mu)(t), \hat{v}(\mu)(t)$ exists only for $\varepsilon > 0$.

[1] Anosov, D.V.: On limit cycles of differential systems with a small parameter at the highest derivatives. Mat. Sbornik 5o, 299-334 (196o) AMS Translations, Ser. 2, 233-275 (1963).

[2] Ebeling, W. und R. Feistel: Physik der Selbstorganisation und Evolution. Akademie-Verlag, Berlin 1982.

[3] Fife, P.C.: Mathematical Aspects of Reacting and Diffusing Systems. Lect. Notes in Biomathematics, Vol. 28, Springer-Verlag, Berlin 1979.

[4] Haken, H. (Ed.): Synergetics. A Workshop. Springer Series in Synergetics, Vol. 2, Springer-Verlag, Berlin 1977.

[5] Haken, H. (Ed.): Pattern Formation by Dynamic Systems and Pattern Recognition. Springer Series in Synergetics, Vol. 5, Springer-Verlag, Berlin 1979.

[6] Haken, H. (Ed.): Dynamics of Synergetic Systems. Springer Series in Synergetics, Vol. 6, Berlin 198o.

[7] Koppel, N. and L. Howard: Plane wave solutions to reaction diffusion equations. Stud. in Appl. Math. 52, 291-328 (1973).

[8] Nicolis, G. and I. Prigogine: Self-Organisation in Nonequilibrium Systems. From Dissipative Structures to Order through Fluctuations. Wiley-Interscience 1977.

[9] Othmer, H.-G.: Current problems in pattern formation. In Lect. on Math. in the Life Sciences, Vol. 9, Providence 1977.

[1o] Pacault, A. and C. Vidal (Eds.): Synergetics. Far from Equilibrium. Springer Series in Synergetics, Vol. 3, Springer-Verlag, Berlin 1979.

[11] Purfürst, R.: Hopf-Bifurkation bei singulär gestörten Systemen. Preprint P-29/8o des Zentralinstitutes für Mathematik und Mechanik der AdW der DDR, Berlin 198o.

[12] Purfürst, R.: Hopf-Bifurkation bei singulär gestörten Funktional-Differentialgleichungen. Preprint des Institutes für Mathematik der AdW. Berlin (to appear).

[13] Schneider, K.R.: Bifurcation of periodic solutions from critical points of dynamical systems. Proc. Int. Conf. Nonlin. Oscill. Prague 1978, Vol. II, 645-652.

[14] Schneider K.R.: On the application of integral manifolds to Hopf bifurcation. Math. Nachr. 97, 313-323 (198o).

[15] Schneider K.R.: Qualitative Untersuchung von Systemen singulär gestörter autonomer Differentialgleichungen. Wiss. Beiträge der Ingenieurhochschule Zwickau, 7, 62-67 (1981).

[16] Schneider, K.R. and B. Wegner: Hopf bifurcation of singularly perturbed autonomous differential systems. Math. Nachr. 112, 5-18 (1983).

[17] Smoes, M.L.: Chemical waves in the oscillatory Zhabotinskii system. A transition from temporal to spatio-temporal organization. In "Dynamics of Synergetic Systems" ed. H. Haken, Springer Series in Synergetics, Vol. 6, Springer-Verlag, Berlin 198o.

[18] Strygin, V.V.: Exchange of stability and bifurcation of small periodic solutions of systems of differential equations with a small parameter at the highest derivative. DAN SSSR 199, 33-35 (1971).

[19] Strygin, V.V. and E.M. Friedman: Asymptotic integral manifolds of singularly perturbed systems of differential equations with retarded argument. Math. Nachr. (to appear).

[2o] Tyson, J.J.: The Belousov-Zhabotinskii-Reaction. Lect. Not. in Biomathematics, Vol. 1o, Springer-Verlag, Berlin 1976.

[21] Vidal, C. and Pacault, A. (Eds.): Nonlinear Phenomena in Chemical Dynamics. Springer Ser. Synergetics, Vol. 12, Springer-Verlag, Berlin 1981.

MODELLING THE SPATIAL EVOLUTION OF POPULATION AND EMPLOYMENT - THE CASE OF THE UNITED STATES

by P.M.Allen and G.Engelen
C.P.231 Université Libre de Bruxelles
Brussels 1050
BELGIUM

Introduction

Over recent years ideas and concepts emerging from the discovery and study of 'dissipative structures' have been used in the modelling of human systems (1),(2). In particular, view of the 'self-organizing' behaviour of complex systems was developed, with practical applications to problems of regional and particularly urban evolution. In this way a new type of dynamic spatial model was proposed with the aim of elaborating a comprehensive tool for decisional aid in planning and policy making.Such a tool could hopefully take into account the complex responses and interplay of the multiple actors of the system which so often either counters or sidesteps the hoped for effects of plans.

In this paper we describe how these ideas have been extended to the modelling of very large systems such as the continental United States, and also how the application clarified important issues in such modelling, and led to several improvements.(3)

Our method consists in writing dynamical equations which are based on the idea that in the long term the population of a State must adjust itself to the employment potential there, which is related to its resources, its comparative advantages for industry, for distribution and for manufacture, and the possibilities for tertiary employment there, within the national hierarchy of cities. It is an ecological model in some sense, and the migratory flows are generated endogenously in the model by the adjustment process of population to the potential for employment. In many models, the variable 'employment' for a State would be given simply by the number of actives there, but in our model we are going to distinguish between the active population at i, x_i, the jobs available at i, J_i, (both filled and as yet vacancies on offer), and the 'potential employment' that could exist at i. It is,for us, precisely the differences between these that drives the system, and only at equilibrium with full employment will they be equal.

The space in which our study has been made is disaggregated into 42 points corresponding roughly to the continental United States, but with some small States such as New Hampshire, Vermont, Rhode Island, Delaware and Connecticut aggregated with more populous neighbours like Massachusetts, New York and Washington D.C, Virginia and Maryland.

We have used as the variables which characterize each State i, the active, employed population x_i, the non-active and unemployed population n_i, the available jobs both filled and vacant, J_i, and the potential jobs M_i. The total population of a State is $P_i = x_i + n_i$.

The Mathematical Model

The first equation of our model is that which governs the evolution of the number of actives at each point i,

$$\frac{dx_i}{dt} = \epsilon\, n_i(J_i - x_i) \qquad (1)$$

which says simply that the active population x_i changes at a rate ϵ times the number of non-actives multiplied by the number of jobs vacant, or for decline, the number of jobs that have become redundant. The equation describing the change in non-actives at i is,

$$\frac{dn_i}{dt} = \lambda_i P_i - \epsilon\, n_i(J_i - x_i) + MG_i^{in} - MG_i^{out} \qquad (2)$$

The non-actives grow because of the rate of natural increase λ_i, which each year adds school leavers to the local labour market, and also because of the in-migration of non-actives from elsewhere. The number of non-actives is diminished by the out migration of people in search of jobs elsewhere, and also by the number of non-actives at i that succeed in finding a job, and who therefore become actives at i.

This particular choice of equations is based on the idea that non-actives migrate for a variety of reasons, most of which give rise to a quite random dispersion. However, it is supposed here that there is a systematic tendancy for migration to be greater towards those States with job vacancies than towards those with high unemployment, and hence we suppose that non-actives migrate in response partly to the need for employment. Either a 'new job' or a'job transfer' requires a 'vacancy' to be created in another State, but may or may not leave a vacancy in turn in the State from which the immigrant comes. So migration is generated by the relative push/pulls of the actives/non-actives ratio in different States, and generates flows of non-actives. These are converted into actives in a State providing that there exist sufficient vacancies. Our model generates a dynamic evolution of actives and non-actives resulting from the disequilibria between demography and economic development, and from this between vacancies, actives and non-actives.

The calculation of natural increase is done using an average rate only, but clearly, it would be simple and desirable at a later stage to take into account the evolution of the age pyramids of each

State, and their resulting greater or lesser propensity to generate migrants, who in turn will modify the age pyramids.

The parameter ϵ reflects the efficiency of the flow of information within the State i concerning job vacancies available to the non-actives n_i, and also the degree of matching of these vacancies with the qualifications of the job-hunters.

The migration flows MG^{in} and MG^{out} are calculated according to the number of non-actives at each point n_i, who are susceptible to migrate. This depends on the population increase and the rate of local economic development, the rate at which vacancies can absorb these individuals, as well as on the perceived attractivity of possible destinations (itself reflecting the number of vacancies which are available there). This is expressed mathematically as,

$$MG_i^{in} = \sum_{j \neq i} \xi \, n_j \frac{V_{ij}}{\sum_{i'} V_{i'j}} \tag{3}$$

and

$$MG_i^{out} = \xi \, n_i \frac{\sum_{j \neq i} V_{ji}}{\sum_{l} V_{jl}}$$

where ξ is a parameter reflecting the 'ease' of migration flow per non-active for a given stimulus - that is the natural mobility or propensity to move of the population, the relative costs of buying and selling houses and other inhibiting (or favourable) factors.

The attractivity V_{ij} is supposed to be simply a function of the advantages and disadvantages of the State i as viewed by the non-actives (or potential migrants) at j. It is supposed here that there is only a dependence on the job 'availability' at i and perhaps the distance between j and i, as follows,

$$V_{ij} = \left(\frac{J_i}{x_i}\right)^{\alpha} P_i \, e^{-\phi d_{ij}} \tag{4}$$

In this equation, if J_i is greater than x_i (i.e. there are more jobs available than people employed at i), then i will be viewed by the non-actives n_j at j as being attractive proportionally to its size P_i, and depending on the clarity with which it is perceived. The latter aspect is reflected by the parameter α. If α is zero, then there is no information about the different job opportunities; if it is infinite then there is 'total' information and unanimous behaviour in that all the migrants from j go to the same state, that which is temporarily the most attractive. Also, the attractivity of State i to individuals in State j may be specified as decreasing with increasing distance d_{ij} as in equation(4).

In addition to the above equations describing the evolution of active and non-active populations respectively, we must specify the process by which the available jobs J_i and the potential jobs M_i are determined at each point i. Once again let us stress that the actives x_i, the available jobs J_i, and the potential M_i will only be equal at a full employment equilibrium. Otherwise, x_i will be different from J_i, and J_i from M_i.

The rate of total change in job supply at i has been modelled by supposing that entrepreneurs tend to adjust their job offers or redundancies to match the opportunities, M_i :

$$\frac{dJ_i}{dt} = K J_i(1 - J_i/M_i) \qquad (5)$$

where K is a constant reflecting the dynamism of employers and the speed with which they react to opportunities.

The actual number of jobs available is divided into two parts:

$$J_i = J_i^{cc} + J_i^t$$

where J_i^{cc} denotes the 'carrying capacity'or basic sectors of agriculture, mining and manufactures, while J_i^t denotes the tertiary or service sector.

In reconstructing the past we need to know how M_i changed over time, and this can only be inferred from the way in which x_i changed. In other words, M_i, which is a potential level, is at no time directly measurable. We simply that for instance in the period 1950 to 1970 the number of people actually employed in the basic sector of State i changed from $x_i^{cc}(50)$ to $x_i^{cc}(70)$. The average rate of change during the 20 years was:

$$\frac{x_i^{cc}(70) - x_i^{cc}(50)}{20} \qquad (6)$$

We assume that the 'potential' jobs for these goods changed proportionally to this, giving:

$$\frac{dM_i^{cc}}{dt} = \eta \gamma_i \qquad \text{where } \gamma_i = \frac{x_i^{cc}(70) - x_i^{cc}(50)}{20}$$

and η is a constant. Then the values of η and K must be such that during a simulation experiment representing this period 1950 to 1970 the M_i^{cc} generated the actual job opportunities J_i^{cc} that led to the actual changes observed in basic employment $x_i^{cc}(70) - x_i^{cc}(50)$.

For the changes in tertiary activity, since a correct description of the spatial competition between tertiary centres of the urban hierarchy is beyond the scope of this initial phase of study, then we have adopted a simple hypothesis concerning the induction rate of tertiary activity per inhabitant. We have made the assumption that, starting from the initial tertiary structure, the induction rate per inhabitant of each state changed uniformly with the average rate for the U.S. as a whole. For example, from 1950 to 1970 tertiary employment per head rose by 20.4%, and so we have supposed that each State increases its rate of tertiary induction by roughly 1% a year. Clearly, this creates vacancies according to the population of each State at each moment, and this in turn depends on its demography and on the migration flows which in their turn depend on the job opportunities relative to other States.

The evolution of the induction rate β_i of State i is given by the average U.S. rate of increase acting on the initial value of β_i. The average rate of increase r is therefore:

$$r = \frac{\beta^{us}(t_{fin}) - \beta^{us}(t_{in})}{20}$$

where $\beta^{us}(t_{fin})$ and $\beta^{us}(t_{in})$ are the final and initial induction rates for the U.S. as a whole.

This leads to the following equation for each State i:

$$\frac{d}{dt} = r\,\beta_i(t_{in}) \qquad (8)$$

The equation governing the evolution of the total employment potential is therefore given by the sum of the basic and the tertiary parts:

$$\frac{dM_i}{dt} = \eta(\gamma_i + r\,\beta_i(t_{in})P_i) \qquad (9)$$

This closes the model and permits us to start simulating the spatial evolution of the U.S. and from this to find a set of values of that lead to good agreement with reality.

Model Calibration 1950-1980

The information input to the model is, for 1950: for each state, the population, the basic employment, the tertiary employment. For 1970: we require the basic employment of each State, the average growth of tertiary employment for the U.S., a demographic hypothesis for the period of study. Using these inputs, the model can then trace the evolution of population and employment in each State on the basis of just 5 parameters ϵ, ξ, α K and η, all of which are independent of any particular State i. It corresponds to an assumption of 'uniform human responses' (in aggregate though and not necessarily at the individual level)over the whole of the U.S. The first question is whether or not such a model can generatean important fraction of the real changes observed.

As a first step, we have run the model for the period 1950 to 1970 and in figure (1) we show the percentage discrepancies between the

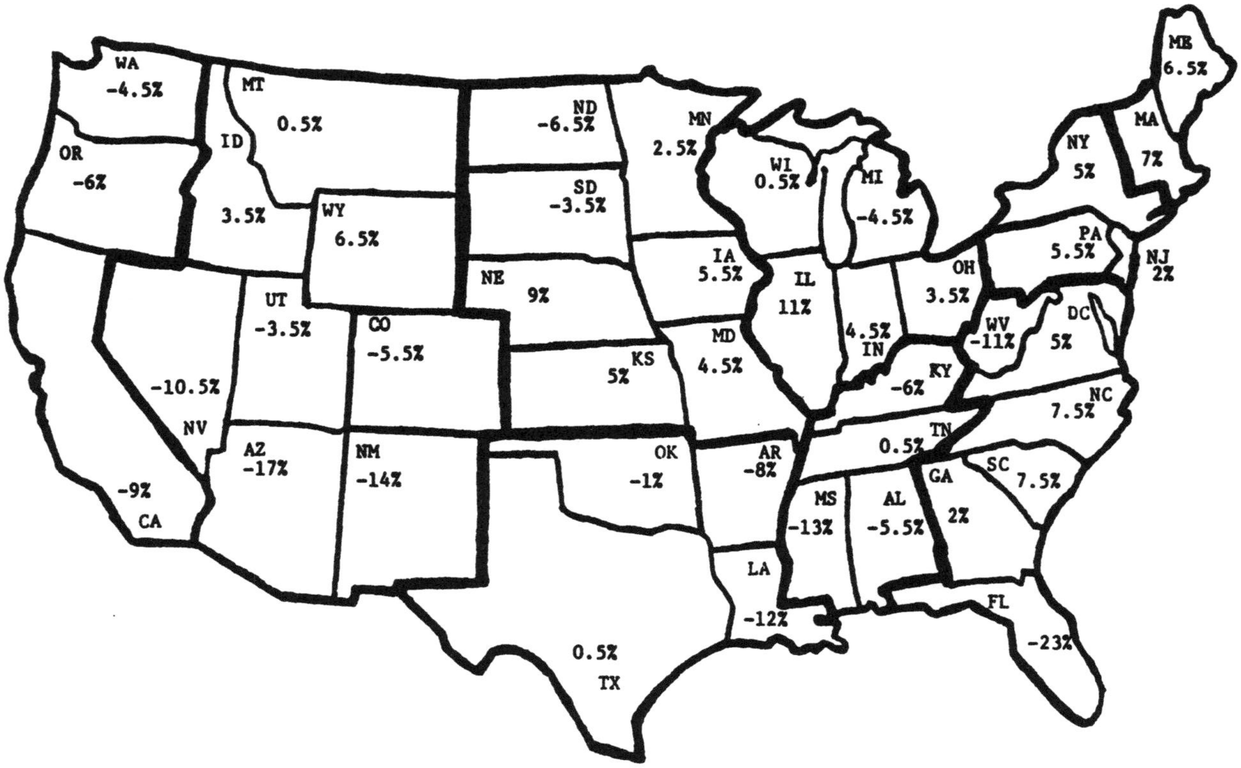

Figure (1): Difference in percentage between the populations observed in 1970 and those predicted by our model, running forward from 1950, using only global migration parameters.

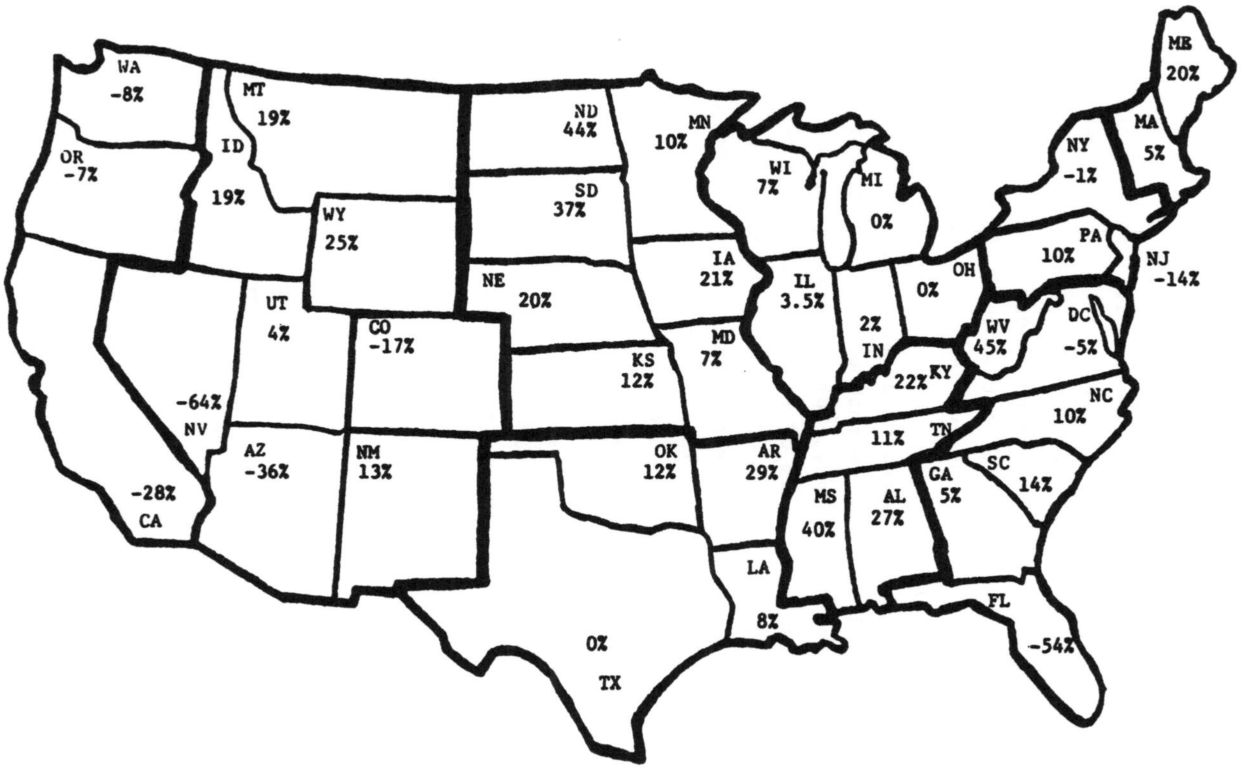

Figure (2) Difference in percentage of the observed populations per state in 1970 compared to those that would result from simply projecting the rates of natural increase.

population of each State generated by the simulation, and that really observed in 1970. In the map we see that the average error is roughly 5%. This seems a fairly satisfactory result considering that during the 20 years, the population of some States more than doubled, while that of others was halved, and our model generates some 95% of the structure in 1970. In order to judge the significance of this, let us compare it to the discrepancies that would have resulted from a model based only on the 'natural increase' observed in each State during the period. In the absense of the migration flows generated by our economic push/pull mechanism, the result after 20 years is that shown in figure (2). The average error is some 25%, showing that our 'economic redistribution' is acting very much in the right way.

However, if small average discrepancies are encouraging, it is also true that the residual differences themselves are interesting, since they are not distributed randomly, but are clearly spatially organized. This confirms the existence of spatial mechanisms not included in our simple model. A number of possibilities suggest themselves for the missing mechanisms: a) structural changes in the urban hierarchy b) evolution of the income distribution of the U.S.A. c) effects of changing age pyramids in different States d) anticipation on the part of actors e) stored growth in the system at the initial moment f) cultural differences between regions.

We have not examined all these possibilities yet, which will be the object of further studies, but two of these seem particularly important. The first is that concerning structural changes in the urban hierarchy which our presentsimulation has not modelled at all. Because we have imposed the national 'average' growth per head of tertiary activity, then clearly this corresponds to the assumption that the structure does not change its character. Our models are of course quite capable of treating such an evolution, as has been shown in our intra-urban models described elsewhere. (3). This would require more information than at present in order to describe the spatial competition between tertiary centres and this will be pursued further in future work.

A second important mechanism is that of 'stored growth'that may exist in the system, whereby delayed effects may still be occurring years after the events that triggered them. In order to examine this question we have examined the situation after 1970, at the end of a simulation during which the population grew from 150 to 202 millions, and tertiary activities grew by 60%. The amount of 'hidden change' stored in the system can be revealed by arresting population increase and the growth of tertiary activity per head, and allowing the simulation to run on. In this way we eliminate the changes caused by natural increase and by economic change, and we reveal those due solely to the differences between 'potential' and 'real' employment in the various

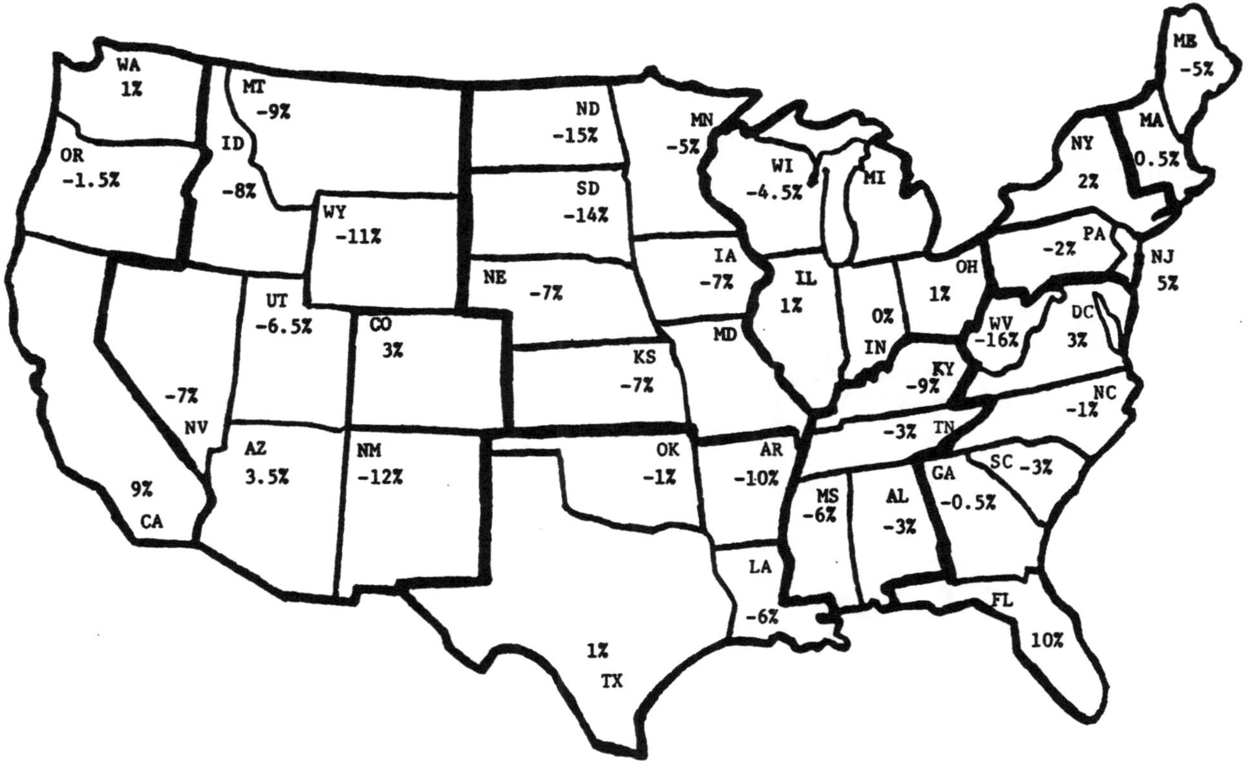

Figure (3): This shows the percentage changes of the population of each state stored purely in the disequilibria present at the end of our previous simulation 1950-1970. These are the "stored changes" 1970-1990, under constant total population, resulting from zero natural increase in each state, constant basic employment and no further increase in tertiary per head.

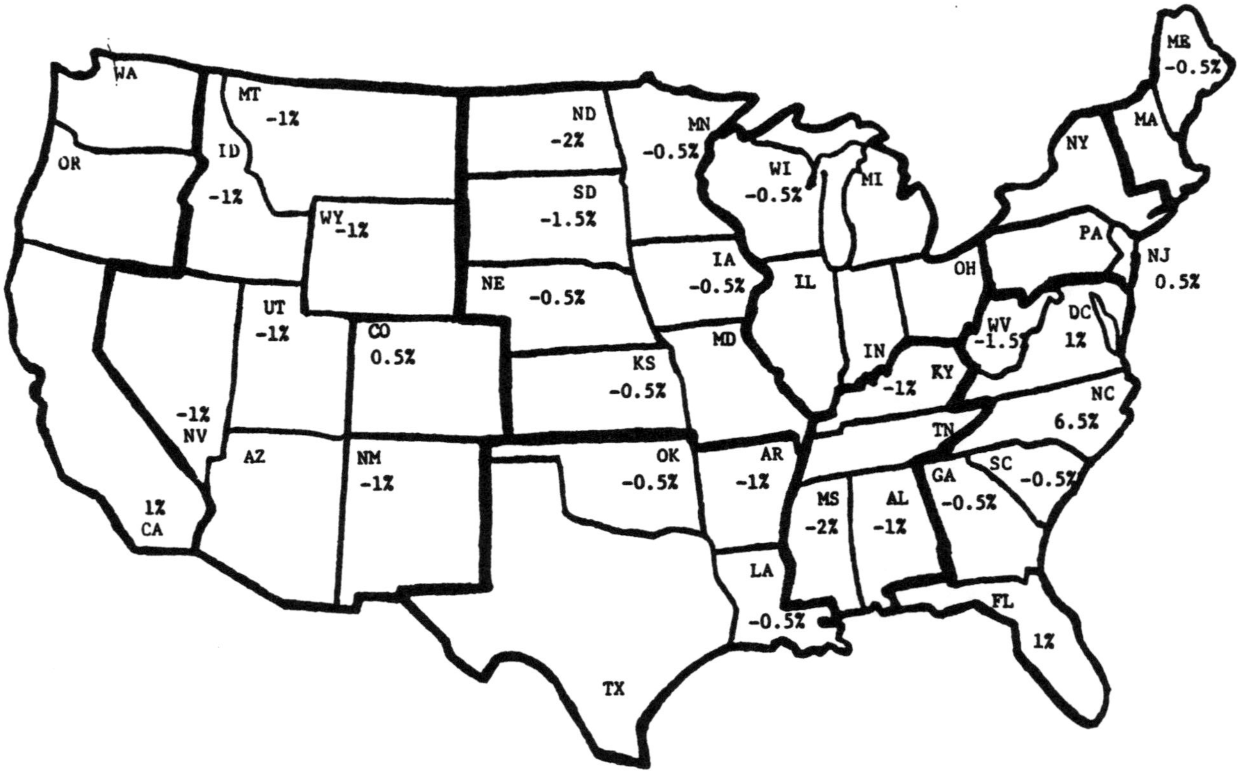

Figure (4): Remaining stored changes persisting in the system 1990-2010, after fixing zero natural increase, no change in basic employment or tertiary per head, per state, in 1970.

States in 1970. These differences can and do continue to produce migration flows, and these flows lead to the induction of new tertiary potentials, leading in turn to further migrations, as the 'non-equilibrium' of 1970 works its way out of the system. In figure (3) we see the population changes that occur during the first 20 years of this experiment, and in Figure (4), those occurring in the 20 years after that. We conclude that our equations express mechanisms which have a 'time-delay' of some 20 years, because most of this 'stored' evolution is dissipated by 1990. This seems to be in agreement with current ideas concerning the time taken for 'multipliers' to operate.

Of course, in reality, between 1970 and 1990 we shall not see the changes shown in Figure (3), but instead this evolution will be superimposed on that generated by economic and demographic changes taking place after 1970. In fact in 1990, there will again be another 20 years of change 'hidden 'in the potentials of the system. But of course, if that is true of the future, it was also true of the past and in particular of 1950, as a result of the phenomena which occurred prior to this date. We find that we have an 'initial conditiion' problem which is a profound one, suggesting that in order to 'fix' the 1950 values of J_i and M_i we must perform the simulation 1930-1950 and so on. This results from the fact that in a dynamic model we must always discuss some 'unobservable' variables, whose values are inferred from subsequent system behaviour. In order to have confidence in such a model it should best be calibrated over a period longer than that of the 'delay' which characterises its dynamics.

In order to actually use our model, however, there is still a further modification that must be carried out. Although a result with an average error of 5% is very satisfactory for such a model based on global mechanisms, it would still mean that we would begin any simulation of the future (i.e. after 1970) with initial conditions which were already seriously erroneous. However, to simply 'put in' the real values observed for 1970 would be false because the 'potentials' would still correspond to our final state, and for example Florida and Arizona would immediately start to decrease, and others overestimated to increase. In order to get around this problem, it is necessary to recognize first the extreme simplicity of our model, and the various mechanisms which we have listed above are missing. In that case, although 95% of the observed situation in 1970 is good, what we must do is to find what the attractivities of States really were in order to explain 100% of what occurred. Thus, we can multiply each attractivity by a factor f_i,

$$V_{ij} = \left(\frac{f_i J_i}{x_i}\right)^{\alpha} P_i e^{-\phi d_{ij}} \quad .$$

This factor f_i is centred on 1, and is adjusted per State until the observed flows are generated. This is a rather sensitive procedure

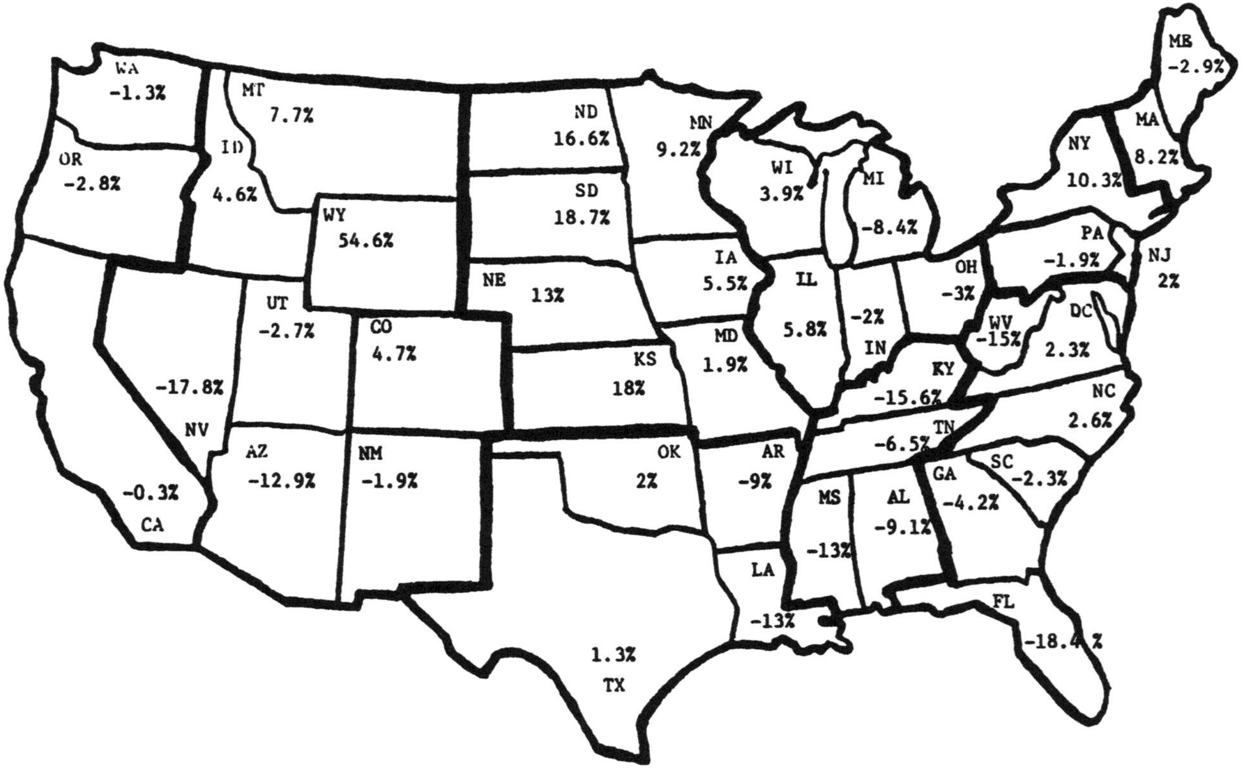

Figure (5) Differences in percentage of the observed populations per state in 1980 and those predicted by the model, using only global migration parameters.

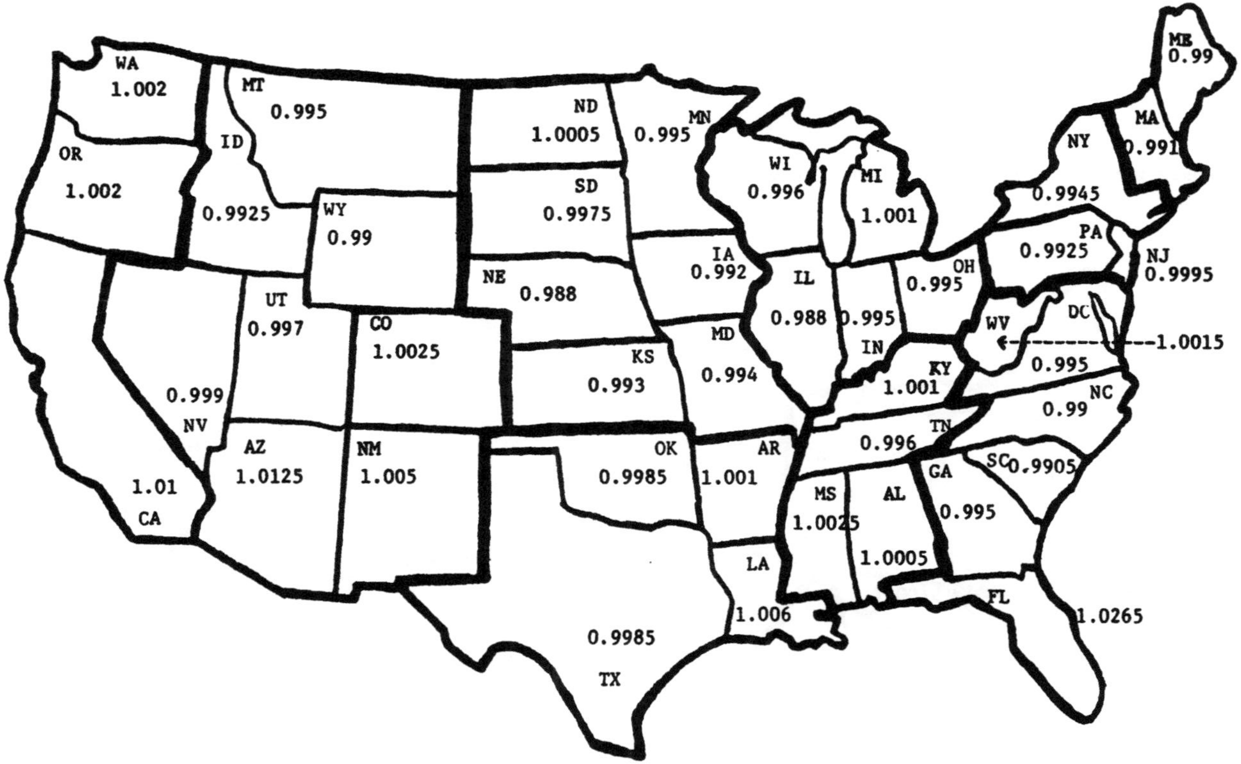

Figure (6): The adjustment factors f_i necessary for each state in order to reproduce exactly (0.5%) the changes observed in populations 1950-1970.

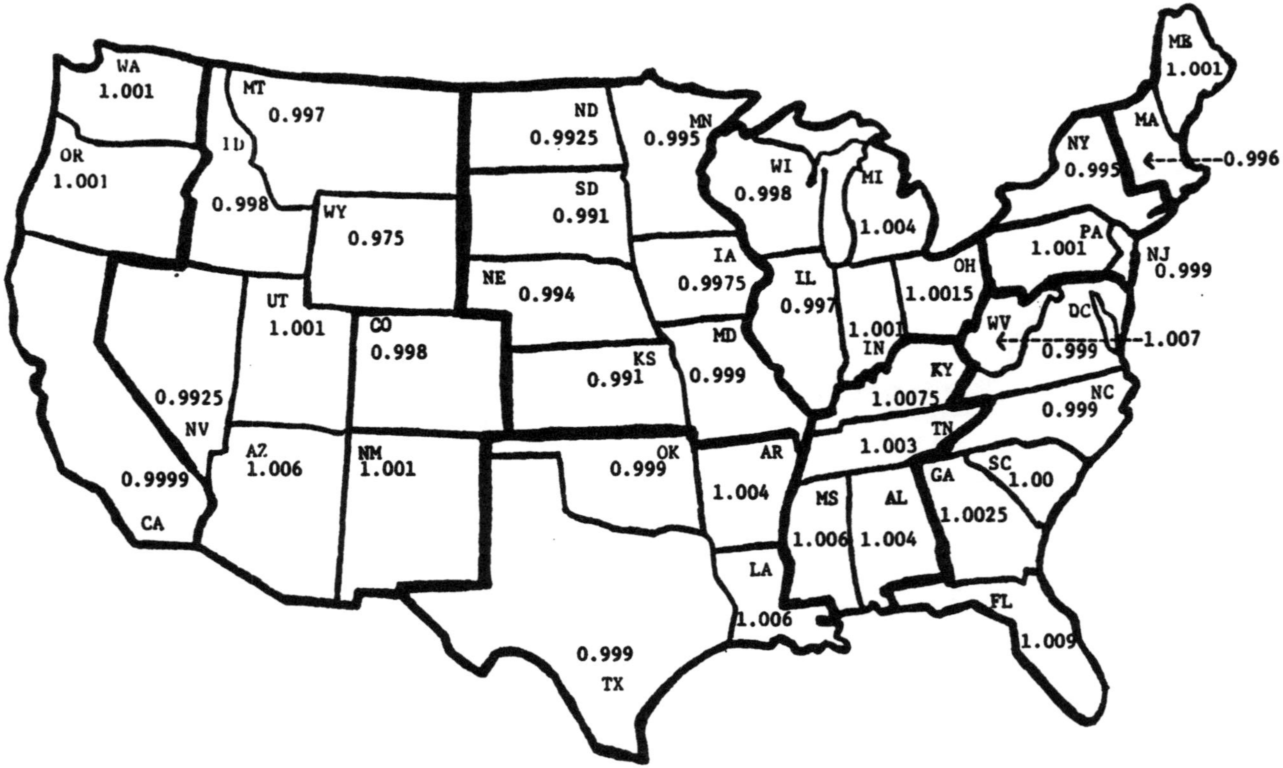

Figure (7): The adjustment factors f_i necessary per state in order to reproduce exactly (0.5%) the population changes 1970-1980.

because of the interconnected nature of all the variables . However, after a trial and error search it is possible to find a set of values, and to generate the situation in 1970 to within .5% permitting us to continue our simulation from 1970 to 1980.

Despite a prolonged search our global mechanismsstill lead to an average error of 5% for the 10 year period, which suggests that the basic 'rational' behaviour of 'economic push/pull' is less effective in describing migration patterns than it was over the earlier period. The map of the 'best result' is shown in figure (5) and it is interesting to note that the parameter values leading to this best fit are considerably different from those of the earlier period. They correspond to a higher mobility, and also much greater information concerning the relative pattern of job opportunities.

The two maps of the distortion factors f_i for the two periods are shown in figures (6) and (7). We notice a considerable stability in their spatial distribution, reflecting the persistent presence of factors outside those considered in our simple model. The interpretation of these values is again complex however, since there can be two possible reasons for either a 'high' or a 'low' score. For example, a higher attractivity than 1 can result either from growth driven by factors other than the strict economic rationality of our model (e.g. Florida) or from a rapid collapse of employment which has not yet led to out migration of the 'expected' magnitude (e.g. West Virginia). Again, a low score may result from employment growth'leading' the growth of population, (e.g. Texas1970-1980) or from people leaving a State faster than economics alone would dictate (e.g. New York 1970-1980). A further stage of this research would be to identify the underlying reasons for the pattern of factors f_i, and in this way include many other mechanisms in the model.

Exploring Possible Futures

Having calibrated the model for 1970-1980 we can now continue our simulations to explore possible future evolutions of the United States, under different assumptions. These assumptions concern the spatial distribution of investments in the basic sector, the growth of tertiary activity and demographic increase. In the first case we have assumed that the rate of increase or decrease of the basic sector of each State continues at the same rate as 1970-1980, and tertiary activity per head grows at the same rate. The results are shown in figure (8), and the growth league for 1980-2000 is: Texas 49.1%;Colorado 48.3%; Arizona 46.7%;Nevada 46.4%;Wyoming 44.3% and Florida 41.5%. The States exhibiting long term stagnation would be: Michigan 7.4% New Jersey 6.9% Ohio 6.6% and Pennsylvania 5.4%

Of course many questions arise when attempting to imagine the sequence of events which would occur in reality: could the water supply

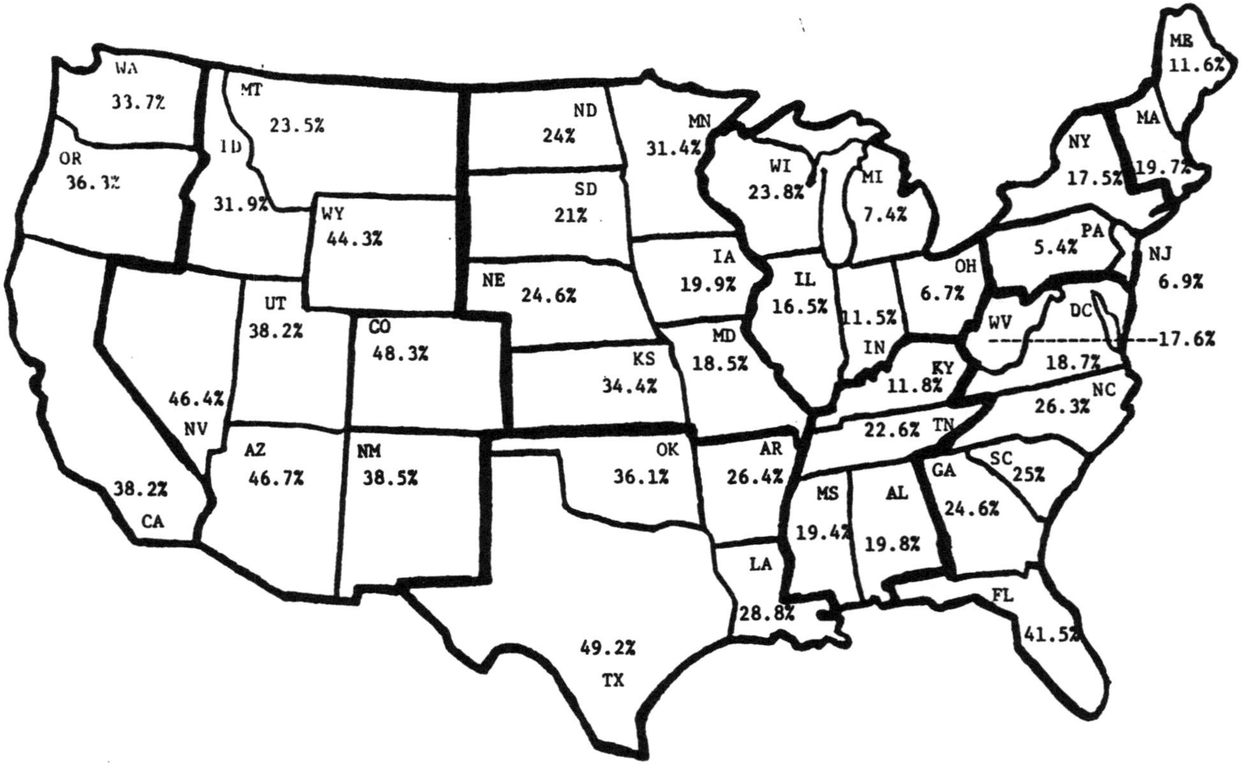

Figure (8) The changes generated in the populations of each state under the assumption that demography and the changes in basic employment follow the pattern of 1970-1980, and that tertiary per

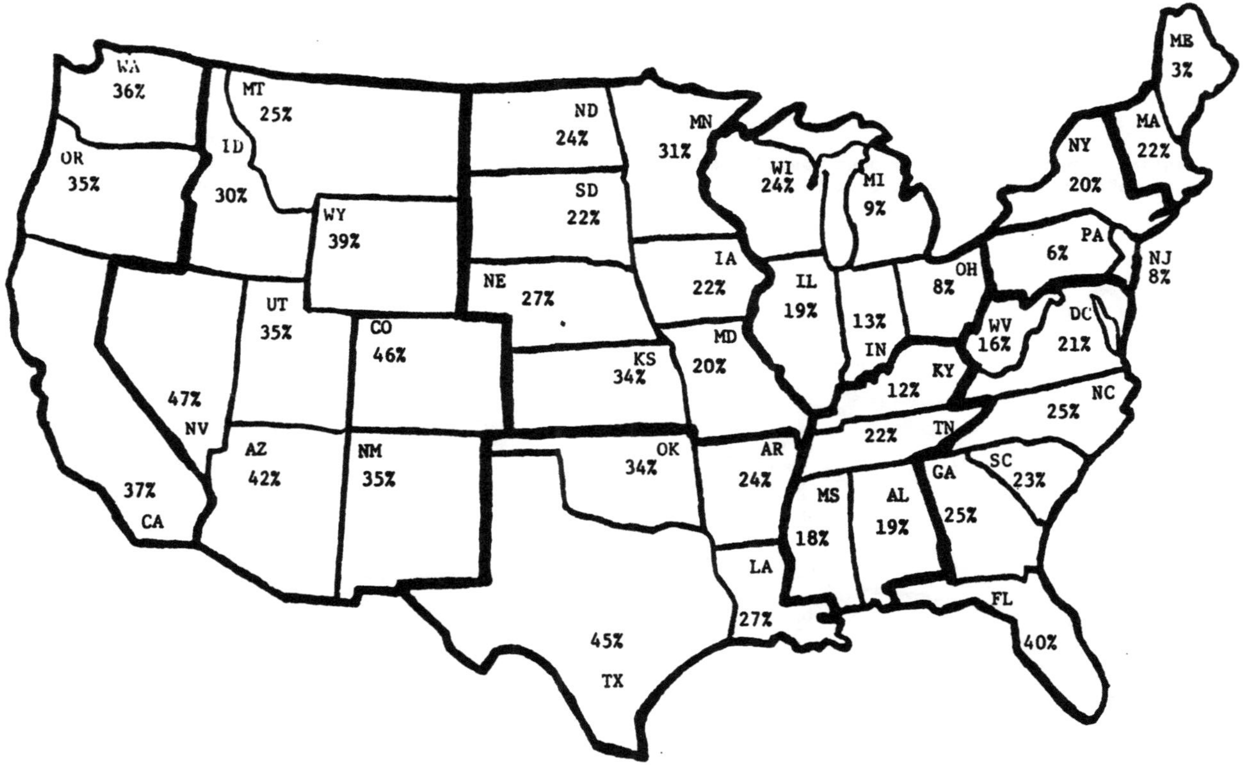

Figure (9): Changes predicted in the population, 1980-2000, supposing that demography follows the pattern of 1970-1980, that basic employment declines or increases at an annual rate which is 20% less

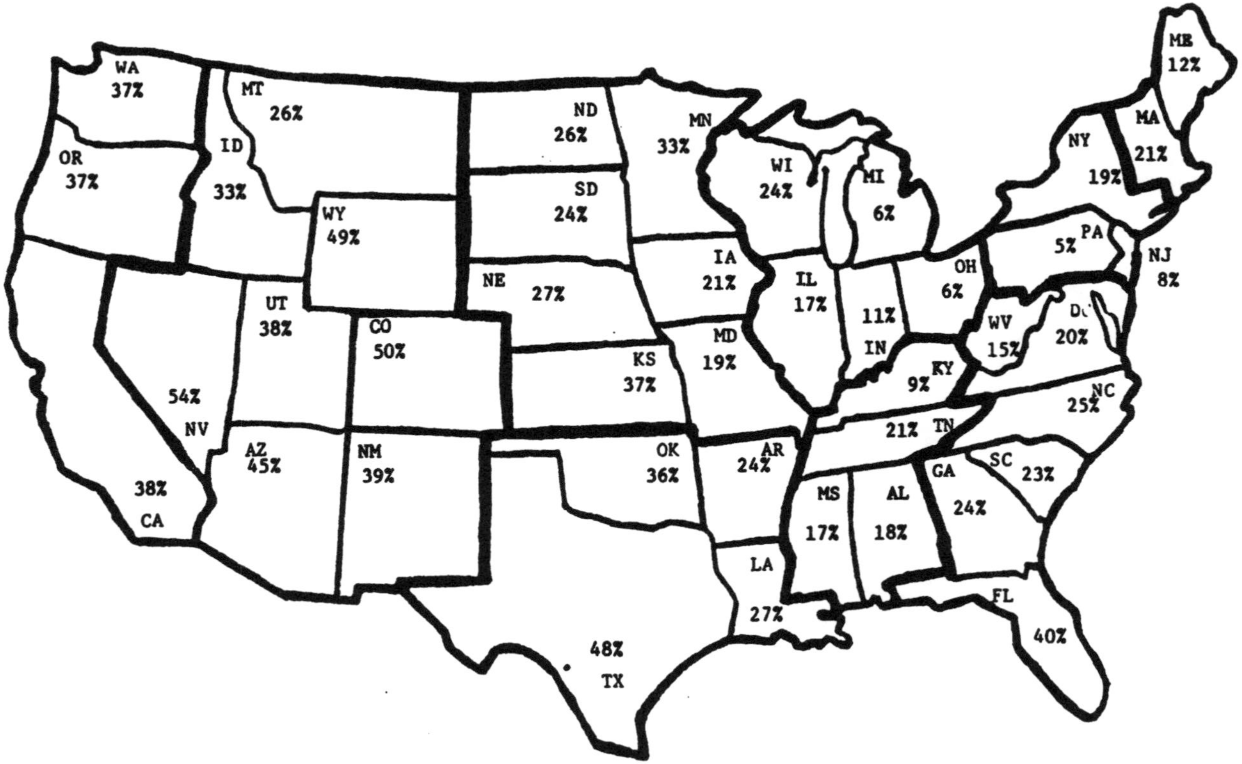

Figure (10): The population changes predicted for 1980-2000 supposing that demography and changes in basic employment follow the pattern 1970-1980. It is supposed that tertiary employment per head increases by 3% annually.

of Texas take another 7 million inhabitants ? Where could another 12 million Californians live ? These questions are very important because they may help us to decide how the 'adjustment factors' f_i may change in a new way reflecting the growing importance of some new factors which affects the attractivity of the State. Our model can then be used to explore how such a change in any particular State will modify the evolution of the other States. Our model furnishes us with a series of questions, however, concerning the water supply prospects for example, which when answered can then be fed back into the model to study the modified evolution. In this way the model is viewed as being useful in order to 'imagine' the conditions and pressures that may exist in given States at particular moments in the future, and the modifications which may occur as a result.

Several other possible scenarios have been examined using our model, but one of the most striking features is that there is a remarkable stability in the projections. They all show growth of over 40% for Texas, Colorado, Arizona, Nevada and Florida, and stagnation in States such as Michigan, Ohio and Pennsylvania. These results are presented in figures (8) to (10).

Our model thus has the potential of providing a useful tool for the exploration of future growth or decline of the various regions of the United States, under various scenarios of investment decisions, of tertiary growth and of demographic change. Clearly, such a model can be of great use in predicting housing, energy , transportation needs for example, and the spatial pattern of future areas of growth can be of assistance in investment plans, for both private and public investors, as well as in identifying the zones which will require improved transportation links on the national level.

Clearly, our model could now be developed further to take into account the various mechanisms which were shown earlier to be missing. This would then allow the support of policy evaluation for different regions using a dynamic framework within which the non-linear feedback interactions linking employment, transportation , migration flows and demographic parameters could be taken into account.

Hopefully then, the model presented here represents a first step towards the elaboration of dynamic spatial models which can offer a more solid base for policy and planning than present methodologies. Instead of being based on statistical extrapolation, or on essentially descriptive methods, our model attempts to treat causal relations, and to understand the future in terms of the interaction of the many actors of the system, each of whom is pursuing his own goals, but which are perceived, as is the system, with only a limited precision. In this way we move closer to a model which can anticipate structural changes in the urban hierarchy, changes in the age structure of each State, and the tertiary and basic employment that will be available.

References

(1) Allen P.M. and Sanglier M. (1981) 'Urban Evolution, Self-Organization and Decision Making', Environment and Planning A, Vol.13

(2) Allen P.M. (1982) 'Evolution and Design in Human Systems', Environment and Planning B.

(3) Allen P.M:, Chang G.L. Engelen G. Herman R and Mahmassani H. (1983) 'Modelling the Evolution of the U.S.Spatial Structure' Report by Purdue University, Indiana under contract DOT-CSSM-83-8 between the Depqrtment of Transportation of the United States and Purdue University, West Lafayette, Indiana 47909.

Acknowledgements.

The work presented here was performed at the University of Texas for a contract between Purdue University and The Department of Transport of the USA. The authors worked in close collaboration with Professors R.Herman and H.Mahmassani at Texas, whom they would like to thank for their many helpful, discussions, suggestions and efforts in the development of the ideas described here. Also, the authors wish to thank R.Crosby, of the Department of Transportation for his support for the project, and also Professor T.Sparrow of Purdue University for his help and support.

POSTER CONTRIBUTIONS

QUALITATIVE ANALYSIS OF VOLTERRA EQUATIONS AND ITS RICCATI TRANSFORMATION WITH SPECIAL REGARD TO BIFURCATION BEHAVIOUR

U. Grote [1)]

There are many problems in physics, biology and chemistry where the problem of bifurcation - e. g. basic structural changes of phase portraits - occurs. Here we consider the bifurcation of solutions of autonomeous differential equations

$$\dot{\underline{x}} = \underline{f}(\underline{x},\underline{\lambda}),\ x \in R^n,\ \lambda \in R^m \tag{1}$$

where the $f(\underline{x},\underline{\lambda})$ are analytical functions for all values of the variable $\underline{x} = (x_1,\dots,x_n)$ in a certain domain G which is independent of $\underline{\lambda}$. In the region $\lambda_1 < \lambda < \lambda_2$ with λ_1, λ_2 fixed $\underline{f}$ is assumed to be an analytical function of $\underline{\lambda}$ too. For simplicity we suppose here that $\underline{f}$ depends only on one parameter $\lambda \in R^1$.
Introducing a basis operator F = dln/dt it is possible to write equ. (1) in form of a Lotka-Volterra system [5]:

$$Fx_i = e_i - \sum_{j=1}^{n} G_{ij}\, x_j, \quad i = 1,\dots,n \tag{2}$$

with $\underline{x}(t) \geq 0$. The e_i and G_{ij} are the carrying capacity and interaction coefficients, respectively. Using the so-called Riccati-Transformation [5] eqn. (2) can be transformed as follows:

$$j \in A : \dot{u}_j = -G_{jj} \prod_{r \in A-j} u_r^{-G_{jr}/G_{rr}} \prod_{s \in \bar{A}} u_s^{G_{js}} \tag{3}$$

$$j \in \bar{A} : \dot{u}_j = u_j \prod_{r \in A} u_r^{-G_{jr}/G_{rr}} \prod_{s \in \bar{A}} u_s^{G_{js}}$$

with $j \in A$ if $G_{jj} \neq 0$ else $j \in \bar{A}$.
The aim of this paper is to discuss for the representation (2) and (3) of eqn. (1) the qualitative behaviour of trajectories in phase space in dependence on parameter λ. Possible bifurcation points are only those λ values for which singular points or trajectories are not crude [1].

1) Singular points
For Lotka-Volterra systems there exist crude equilibrium points, e. g. the characteristics of these equilibrium states cannot

1) Central Institute for Cybernetics and Information Processes of the Academy of Sciences of the GDR, Kurstraße 33, GDR 1086 Berlin

change at variations of the parameter $\lambda = f(G_{ij})$. Primary and secondary bifurcations of equilibrium points are not possible. The analysis of the Riccati-representation (3) shows that there are only singular points of higher order which cannot be investigated with methods of bifurcation theory.

2) Singular trajectories
For $n = 2$ there do not exist isolated periodic orbits in the case of Lotka-Volterra equations [2], however, for $n \geq 3$ the existence of Hopf bifurcations - e. g. branching of a periodic solution in a critical equilibrium point - has been proved theoretically [3].

According to the rules for the general Riccati transformation all u_j are greater than zero, e. g. only a monotonous behaviour is possible and no Hopf bifurcation may occur.
To allow for further discussions positive as well as negative values of u_j the equ. (3) has to be modified as follows:
- the exponents can take the integer values 0 or 1 only,
- in every equation only one u_j may have an exponent equal to one.

Then in the case $n = 3$ equ. (3) may be written in the following form:

$$\dot{u}_1 = a_1 \cdot u_2$$
$$\dot{u}_2 = a_2 \cdot u_3$$
$$\dot{u}_3 = a_3 \cdot u_1.$$

For the characteristic quantities follows

$$\sigma = -\left\{\sum \frac{\partial f_i}{\partial x_i}\right\} = 0$$

$$\Delta = \det \left\| \frac{\partial f_i}{\partial x_j} \right\| = a_1 \cdot a_2 \cdot a_3.$$

The characteristic equation $p^3 - a_1 \cdot a_2 \cdot a_3 = 0$ has the solution:

$$p = \sqrt[3]{a_1 \cdot a_2 \cdot a_3} = \pm \sqrt[3]{|a_1 \cdot a_2 \cdot a_3|} \cdot \begin{cases} 1 \\ \dfrac{-1+i\sqrt{3}}{2} \\ \dfrac{-1-i\sqrt{3}}{2} \end{cases}$$

The way of this solutions in the complex plane together with that way which would be necessary for the occurence of a Hopf bifurcation is shown in Fig. 1. Varying parameter λ - e. g. one of values a_1, a_2, a_3 - the imaginary part vanishes when crossing the imaginary axis and therefore no Hopf bifurcation is possible [4].
This result may be generalized for the case $n > 3$. The Jacobian then may be written as a product of a diagonal matrix with a permutation matrix. Therefore the solutions of the characteristic equation are

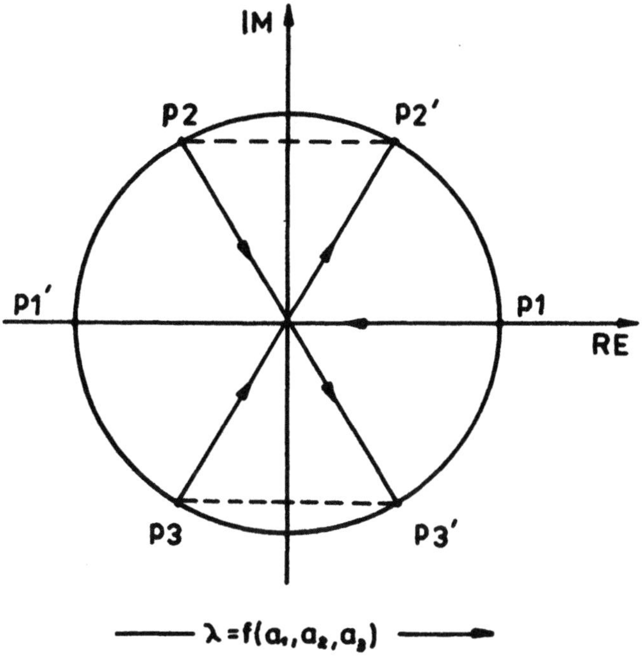

——— Way of the eigenvalues for the Riccati-representation

- - - - - Necessarily way of the eigenvalues for the Hopf bifurcation

Fig. 1

multiples of radicals of one and the same structure as given in Fig. 1. To summarize, it was shown

1) that branching of trivial solutions is not possible in either of the representations (2) and (3),

2) that Hopf bifurcation may occur only in the case of Lotka-Volterra systems with $n \geq 3$.

That means if one transforms equ. (1) in one of the representations (2) or (3) a lot of phenomena occuring in the original system (1) will vanish.

References

[1] Andronov, A. A. et al.: "Theorie der Schwingungen", Akademie-Verlag, Berlin, 1983.

[2] Andronov, A. A. et al.: "Qualitative Theory of Second Order Dynamic Systems", Wiley, New York, 1973.

[3] Hofbauer, J. in: Nonlinear Analysis, Theory, Methods & Appl., Vol. 5, No. 9, (1981), 1003 - 1007.

[4] Hopf, E. in: Berichte über Verhandlungen der Sächsischen Akademie der Wissenschaften - Math.-Phys. Klasse, Band 94 (1942), 3 - 22.

[5] Peschel, M.; Mende, W.: "Do we live in a Volterra World?", Akademie-Verlag, Berlin, 1984.

ROBUSTNESS OF THE SOLUTION OF STOCHASTIC VOLTERRA DIFFERENTIAL EQUATIONS

R. Funke [1)]

1. Introduction

It is known /3/ that the hypercycle solution of order r = 1 of an exponential chain is identical with the solution of an infinite long homogeneous exponential chain. Here we consider a stochastic variant of a generalized hyper-cycle of order r = 1, depending on three free parameters α, λ and μ.

Let $(\Omega, \mathcal{F}, P)$ be a complete probability space and $\{\mathcal{F}_t, t \geq 0\}$ an increasing family of σ-subalgebras of $\mathcal{F}$, containing all null subsets. Further let $\{w(t), t \geq 0\}$ be a standard Wiener process adapted to the family $\{\mathcal{F}_t, t \geq 0\}$. In order to study the behavior of a stochastic hypercycle solution of order r = 1 we consider the stochastic differential equation (with time homogeneous coefficient functions)

$$d\xi(t) = \lambda \xi^{1-\alpha}(t)\, dt + \mu\, \xi^{1-\frac{\alpha}{2}}(t)\, dw(t), \quad t \geq 0 \tag{1}$$

with the initial value $\xi(0) = x$, $x > 0$, where

$$I(t) = \int_0^t \varphi(s)\, dw(s)$$

is a stochastic Ito integral. For further considerations we put $\nu = \frac{2\lambda}{\mu^2}$ and assume $\mu \neq 0$.

The coefficient functions $a(x) = \lambda x^{1-\alpha}$ and $b(x) = \mu x^{1-\frac{\alpha}{2}}$ of equ. (1) don't satisfy the conditions of existence and uniqueness of the solution $\xi(t)$ of equ. (1) on an infinite interval (see /1/,/2/), because such processes $\xi(t)$ can explode in finite time. Nevertheless we can consider the process $\xi(t)$ as a stochastic process with absorption at the boundary /1/,/2/. We suppose $0 < \varepsilon \leq x \leq M < \infty$ and define $\xi(s) = \varepsilon$, $s \geq t$, if $\xi(t) = \varepsilon$ and $\xi(s) = M$, $s \geq t$, if $\xi(t) = M$ (Fig.1).

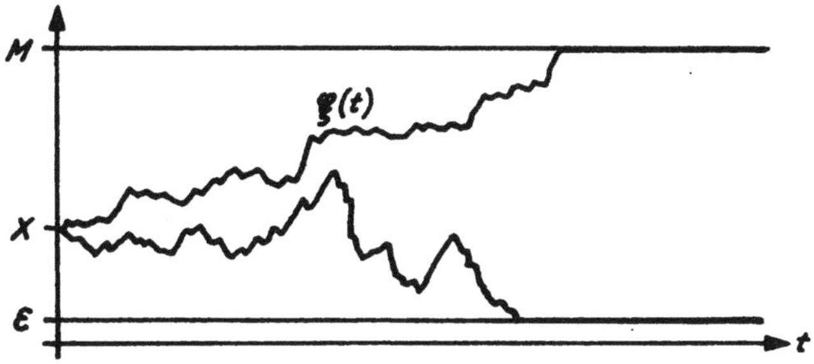

Fig.1. Sample paths of processes with absorption at the boundary

The functions a(x) and b(x) satisfy all conditions of existence and uniqueness of the solution $\xi(t)$ of equ. (1) on $\varepsilon \leq x \leq M$ (Lipschitz-continuity, boundness, differentiability) and $\xi(t)$ is a Markov process (see /1/,/2/). Further we define the probability that $\xi(t)$ has already been absorbed at the boundary ε and M, respectively, at the time t if the process $\xi(t)$ started at $\xi(0) = x$

[1)] Central Institute for Cybernetics and Information Processes of the Academy of Sciences of the G.D.R., Kurstraße 30-34, DDR-1086 Berlin

$$p_\varepsilon(t,x) = P\{\xi(t)=\varepsilon \,/\, \xi(0)=x\} \text{ and } p_M(t,x) = P\{\xi(t)=M \,/\, \xi(0)=x\},$$

the stopping time of the (first) reaching of one of both boundaries

$$\tau_x = \inf\{t:\ t \geq 0,\ \xi(t) \notin (\varepsilon, M)\}$$

and the probability that ε and M, respectively, is reached

$$p_\varepsilon(x;M) = P\{\xi(\tau_x)=\varepsilon \,/\, \xi(0)=x\} \text{ and } p_M(x;\varepsilon) = P\{\xi(\tau_x)=M \,/\, \xi(0)=x\}.$$

2. The probability distributions $p_\varepsilon(t,x)$ and $p_M(t,x)$

The probability distributions $u_1(t,x) = p_\varepsilon(t,x)$ and $u_1(t,x) = p_M(t,x)$ are solutions (in the sense of generalized functions) of the parabolic differential equation

$$\frac{\partial u_1(t,x)}{\partial t} = \lambda x^{1-\alpha}\frac{\partial u_1(t,x)}{\partial x} + \frac{1}{2}\mu^2 x^{2-\alpha}\frac{\partial^2 u_1(t,x)}{\partial x^2} \tag{2}$$

with the initial and boundary values

$$\lim_{t\searrow 0} p_\varepsilon(t,x) = 0,\ x>\varepsilon, \qquad \lim_{t\searrow 0} p_M(t,x) = 0,\ x<M,$$

$$\lim_{x\searrow\varepsilon} p_\varepsilon(t,x) = \lim_{x\nearrow M} p_M(t,x) = 1,\ t>0$$

$$\lim_{x\nearrow M} p_\varepsilon(t,x) = \lim_{x\searrow\varepsilon} p_M(t,x) = 0 \ /2/.$$

Unfortunately an analytically closed solution of equ. (2) can be obtained only in special cases. So for $\lambda = (1-\alpha)\mu^2/2$ and $\alpha \neq 0$ we obtain

$$\lim_{\substack{\varepsilon\to 0\\ M\to\infty}} p_M(t,x) = \exp\left\{\frac{x^\alpha}{\alpha^2\mu^2 t}\right\} \text{ if } \alpha<0 \text{ and}$$

$$\lim_{\substack{\varepsilon\to 0\\ M\to\infty}} p_\varepsilon(t,x) = \exp\left\{\frac{-x^\alpha}{\alpha^2\mu^2 t}\right\} \text{ if } \alpha>0.$$

3. The probability that either ε or M is reached

Formally we can take the limit $t\to\infty$ in equation (2). As proved in /1/ and /2/ $u_2(x) = p_\varepsilon(x;M)$ and $u_2(x) = p_M(x;\varepsilon)$ are nontrivial solutions of the equation (3)

$$\lambda x^{1-\alpha} u_2'(x) + \frac{1}{2}\mu^2 x^{2-\alpha} u_2''(x) = 0 \tag{3}$$

with the boundary values $p_\varepsilon(\varepsilon;M) = p_M(M;\varepsilon) = 1$ and $p_\varepsilon(M;M) = p_M(\varepsilon;\varepsilon) = 0$. The solutions of equ. (3) are

$$p_\varepsilon(x;M) = \frac{M^{1-\nu} - x^{1-\nu}}{M^{1-\nu} - \varepsilon^{1-\nu}} \text{ and } p_M(x;\varepsilon) = \frac{x^{1-\nu} - \varepsilon^{1-\nu}}{M^{1-\nu} - \varepsilon^{1-\nu}},\ \text{if } \nu \neq 1 \tag{4}$$

and $p_\varepsilon(x;M) = \dfrac{\ln M - \ln x}{\ln M - \ln\varepsilon}$ and $p_M(x;\varepsilon) = \dfrac{\ln x - \ln\varepsilon}{\ln M - \ln\varepsilon}$, if $\nu = 1$.

4. The first moment (expectation) of τ_x

The function $v_1(x) = E\,\tau_x$ satisfies the equation (5)

$$\lambda x^{1-\alpha} v_1'(x) + \frac{1}{2}\mu^2 x^{2-\alpha} v_1''(x) = -1 \tag{5}$$

with the boundary values $v_1(\varepsilon) = v_1(M) = 0$ (see /1/,/2/).

The solution of equ. (5) for $\alpha \neq 0$, $\nu \neq 1$ and $\nu \neq 1-\alpha$ is

$$v_1(x) = \frac{2}{\mu^2\alpha(1-\nu-\alpha)} x^\alpha + C_1\frac{x^{1-\nu}}{1-\nu} + C_2, \tag{6}$$

where $C_1 = C_1(\varepsilon, M)$ here and further are functions independent of x.

5. The second moment of τ_x

The function $v_2(x) = E\tau_x^2$ is the solution of the equation (7)

$$\lambda x^{1-\alpha} v_2'(x) + \frac{1}{2}\mu^2 x^{2-\alpha} v_2''(x) = -2 v_1(x) \qquad (7)$$

with the boundary values $v_2(\varepsilon) = v_2(M) = 0$ (see /1/,/2/).
The solution for $\alpha \neq 0$ and $\nu \neq 1-k\alpha$, $k = -1,0,1,2$ is

$$v_2(x) = \frac{4}{\mu^4\alpha^2(\alpha+\nu-1)(2\alpha+\nu-1)} x^{2\alpha} + C_1 \frac{4}{\mu^2\alpha(1-\nu)(\alpha+1-\nu)} x^{\alpha-\nu+1} + $$
$$+ C_2 \frac{4}{\mu^2\alpha(1-\nu-\alpha)} x^{\alpha} + C_3 \frac{x^{1-\nu}}{1-\nu} + C_4. \qquad (8)$$

6. Results

For $\nu<1$, $\nu = 1$ and $\nu>1$ different results are received. We obtain $\lim_{\substack{\varepsilon\to 0\\ M\to\infty}} p_\varepsilon(x;M) = 1$ and $\lim_{\substack{\varepsilon\to 0\\ M\to\infty}} p_M(x;\varepsilon) = 0$ from equations (4) if $\nu < 1$, i.e. an extinction of the process $\xi(t)$ occurs with probability 1. If $0<\alpha<1-\nu$ $\lim_{\varepsilon\to 0, M\to\infty} C_i = 0$, $i = 1,2$ in equation (6) and τ_x has a finite expectation

$$E\tau_x = \frac{2}{\mu^2\alpha(1-\nu-\alpha)} . \qquad (9)$$

If further $0<\alpha<\frac{1-\nu}{2}$ holds C_3 and C_4 disappear in equ. (8) too for $\varepsilon\to 0$ and $M\to\infty$ and the second moment and the dispersion of τ_x are finite:

$$E\tau_x^2 = \frac{4}{\mu^4\alpha^2(1-\nu-\alpha)(1-\nu-2\alpha)} x^{2\alpha} \qquad (10)$$

$$D\tau_x = \frac{4}{\mu^4\alpha(1-\nu-\alpha)^2(1-\nu-2\alpha)} x^{2\alpha}. \qquad (11)$$

The probability that extinction occurs and the probability that explosion occurs don' t exist as the limit $\varepsilon\to 0$ and $M\to\infty$ in the case $\nu = 1$, because it depends on how M depends on ε when ε converges to zero. All values from 0 to 1 can be obtained and all moments of τ_x are infinite. The case $\nu>1$ is similar to the first case, but an explosion occurs with probability 1. If $1-\nu < \alpha < 0$ holds than equ. (6) yields equ. (9) for $\varepsilon\to 0$ and $M\to\infty$. If further $(1-\nu)/2 < \alpha < 0$ holds than equ. (8) yields equ. (10) and (11) for $\varepsilon\to 0$ and $M\to\infty$, i.e. the process $\xi(t)$ explodes almost surely and the time of the explosion has finite expectation and dispersion. Both moments are infinite in all other cases not mentioned previously. Higher moments of τ_x can be calculated with the same method but they are unimportant for our considerations.

7. References

/1/ Gikhman,I.I. and A.V.Skorokhod: Introduction to the Theory of Random Processes. W.B.Saunders Co., Philadelphia 1969.

/2/ Gikhman,I.I. and A.V.Skorokhod: Stochastic Differential Equations. Springer-Verlag, Berlin,1972.

/3/ Peschel,M. and W.Mende: Do We Live in a Volterra World. Akademie-Verlag, Berlin 1984 and Springer-Verlag, Wien 1984.

AN ATTEMPT OF A FUZZY APPROACH TO VOLTERRA SYSTEMS

Bernd Straube [1])

Investigations of rough behaviour and robustness of Volterra systems are of special interest. For this purpose an attempt is made to describe Volterra systems like

$$\frac{\dot{x}_i}{x_i} = \sum_j G_{ij} x_j \quad , \qquad i=1(1)n, \quad j=1\,(1)n+1 \qquad (1)$$

$$x_{n+1} = \text{const.}$$

in a fuzzy manner.

First of all, both time and ranges of the x_i and the parameters G_{ij} are discretized.
Discretization of time is obtained by applying Euler foreward differences

$$x_i(k+1) = x_i(k) + \Delta x_i(k)$$
$$x_i(k) + h\cdot\dot{x}_i(k) = x_i(k)[1 + h\cdot\frac{\dot{x}_i(k)}{x_i(k)}] \qquad (2)$$

The ordered classes A,B,...,G,H are used for discretizing x_i whereas A means the fuzzy set 'very small' and H stands for 'very large'.

Fuzzy sets are also defined on the set of parameters G_{ij}. Here is a schematic representation of it:

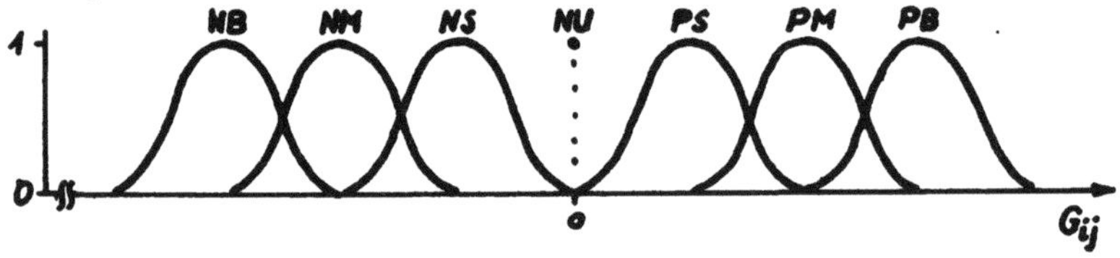

The fuzzy set NU is a (fuzzy) singleton with $NU(G_{ij})=1$ if $G_{ij}=0$ and $NU(G_{ij})=0$ if $G_{ij} \neq 0$.
Hence, a non-interaction between x_j and x_i (i.e., $G_{ij}=0$ in the original system) is preserved in the fuzzy model. All fuzzy sets introduced above have the following meaning NB (PB) 'negative (positive) big', NM (PM) 'negative (positive) median', NS (PS) 'negative (positive) small', and NU 'uncoupled'.

Considering the time k the known values of fuzzy classes of $x_i(k)$ have to be combined with the fuzzy values of G_{ij} for all j in a manner explained later. This leads to the relative growth rate $[\dot{x}_i/x_i](k)$. The partial relative growth rate caused by x_j via G_{ij} is denoted by

[1]) Akademie der Wissenschaften der DDR,
Zentralinstitut für Kybernetik und Informationsprozesse

$$\frac{\dot{x}_{ij}}{x_i} = G_{ij}x_j$$

Then, (1) is read

$$\frac{\dot{x}_i}{x_i} = \sum_j \frac{\dot{x}_{ij}}{x_i}$$

Now, fuzzy sets for the relative growth rates are defined on the set of $\dot{x}_{ij}/x_i$ (see also the fuzzy parameters for G_{ij} given above):

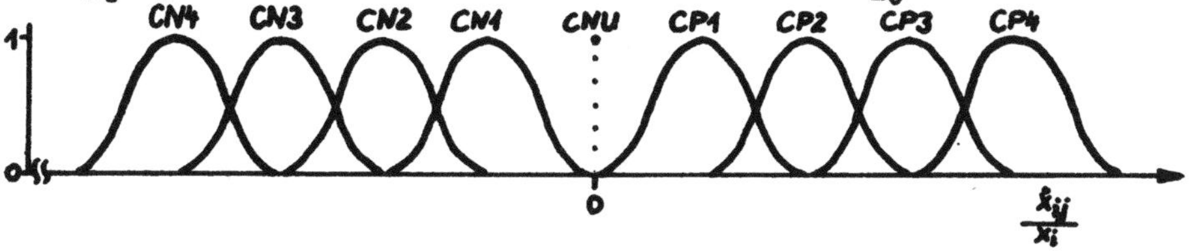

The interrelationship between the fuzzy classes of x_i, the fuzzy values of G_{ij}, and the fuzzy partial relative growth rates $\dot{x}_{ij}/x_i$ is seized by the fuzzy algorithm GX whose underlying operational table is shown in Table 1.

Next, the partial relative growth rates $\dot{x}_{ij}/x_i$ have to be combined to the relative growth rate $\dot{x}_i/x_i$ for all j. In order to prevent the addition of fuzzy sets this combination shall be carried out in another way.

The partial relative growth rate $[\dot{x}_{ij}/x_i](k)$ is assumed to be known. Using the Euler approximation (2) the single effect $x_{ij}(k+1)$ is caused by $x_j(k)$.

The pertinent fuzzy sets of $x_i(k)$, $[\dot{x}_{ij}/x_i](k)$, and $x_{ij}(k+1)$ are composed by the fuzzy algorithm RATE (see Table 2). (Note that the classes A and H are extented to A0,A1,...,A4 and H0,H1,...,H4, respectively, to prevent saturation effects).

Instead of an addition, $x_i(k+1)$ is calculated by a sequence of single effects $x_{ij}(k+1)$. This procedure is carried out according to the following calculation scheme:

step 0: choose row i; put j=0

step 1: put $x_{io}(k+1)=x_i(k)$

step 2: j=j+1

step 3: $(x_j(k), G_{ij}) \xrightarrow{GX} [\dot{x}_{ij}/x_i](k)$

step 4: $([\dot{x}_{ij}/x_i](k), x_{ij-1}(k+1)) \xrightarrow{RATE} x_{ij}(k+1)$

step 5: if j<n+1 go to step 2 else

step 6: put $x_i(k+1)=x_{ij}(k+1)$

step 7: unless all rows have been chosen go to step 0.

It should be noted that the used fuzzy algorithms GX and RATE always lead to fuzzy values like A,B,..,H of the variables x_i.

In the following five examples of Volterra systems treated in the described manner are presented.

The fuzzy simulations are carried out by means of our fuzzy simulation system FUZSIM [1].

In all examples x_5 has a constant value. Furthermore, only the class having the greatest membership value is depicted for a variable.

The treated original Volterra systems are characterized as follows.

Example 1 (Fig. 1): The system has a fixpoint at (1,1,1,1,1) and a stable cycle around it.

Example 2 (Fig. 2): $x_1, x_2, x_3 \longrightarrow \infty$, x_4 goes to a large constant.

Example 3 (Fig. 3): $x_1, x_2, x_3 \longrightarrow 0$, x_4 goes to a small constant.

Example 4 (Fig. 4): The system has fixpoints on a straight line which goes through (1,1,1,1,1). The trajectories go to one of these fixpoints.

Example 5 (Fig. 5): The system is a hypercycle and has a stable cycle around (1,1,1,1,1).

The fuzzy simulation results show these pertinent qualitative behaviours. Hence, the presented approach seems to be a promising means for getting answers for qualitative behaviours and robustness with respect to changes of the parameters of the system considered.

Reference

[1] Straube, B.: FUZSIM - A simulation program package for fuzzy systems ZKI-Informationen 2/1983 (Techn. report of the Zentralinstitut für Kybernetik und Informationsprozesse).

Table 1: Operational table for GX

GX	A	B	C	D	E	F	G	H
PB	CP1	CP1	CP2	CP2	CP3	CP3	CP4	CP4
PM	CP1	CP1	CP1	CP1	CP2	CP2	CP3	CP3
PS	CP1	CP1	CP1	CP1	CP1	CP1	CP2	CP2
NU	CNU	CNU	CNU	CNU	CNU	CNU	CNU	CNU
NS	CN1	CN1	CN1	CN1	CN1	CN1	CN2	CN2
NM	CN1	CN1	CN1	CN1	CN2	CN2	CN3	CN3
NB	CN1	CN1	CN2	CN2	CN3	CN3	CN4	CN4

Table 2: Operational table for fuzzy algorithm RATE

RATE	A4	A3	A2	A1	A0	B	C	D	E	F	G	H0	H1	H2	H3	H4
CP4	A0	B	C	D	E	F	G	H0	H1	H2	H3	H4	H4	H4	H4	H4
CP3	A1	A0	B	C	D	E	F	G	H0	H1	H2	H3	H4	H4	H4	H4
CP2	A2	A1	A0	B	C	D	E	F	G	H0	H1	H2	H3	H4	H4	H4
CP1	A3	A2	A1	A0	B	C	D	E	F	G	H0	H1	H2	H3	H4	H4
CNU	A4	A3	A2	A1	A0	B	C	D	E	F	G	H0	H1	H2	H3	H4
CN1	A4	A4	A3	A2	A1	A0	B	C	D	E	F	G	H0	H1	H2	H3
CN2	A4	A4	A4	A3	A2	A1	A0	B	C	D	E	F	G	H0	H1	H2
CN3	A4	A4	A4	A4	A3	A2	A1	A0	B	C	D	E	F	G	H0	H1
CN4	A4	A4	A4	A4	A4	A3	A2	A1	A0	B	C	D	E	F	G	H0

Example 1: $\frac{\dot{x}_i}{x_i} = \sum_j^n G_{ij} x_j$, $i = 1(1)n$, $j = 1(1)n+1$, $G = \begin{pmatrix} -0.4 & 1.0 & -0.6 & -1.0 & 1.0 \\ -1.0 & -0.4 & 1.0 & -0.6 & 1.0 \\ -0.6 & -1.0 & -0.4 & 1.0 & 1.0 \\ 1.0 & -0.6 & -1.0 & -0.4 & 1.0 \end{pmatrix}$

$x_{n+1} = \text{const.} = 1$

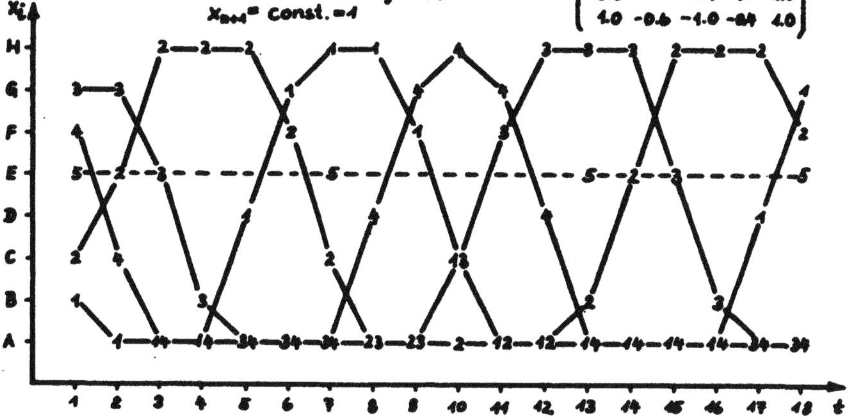

Ex.2: $G = \begin{pmatrix} -1 & -1 & 2 & 1 & 0 \\ 2 & 1 & 1 & 0 & 1 \\ 1 & 0 & 2 & 1 & 1 \\ 3 & 2 & 0 & -1 & 1 \end{pmatrix}$

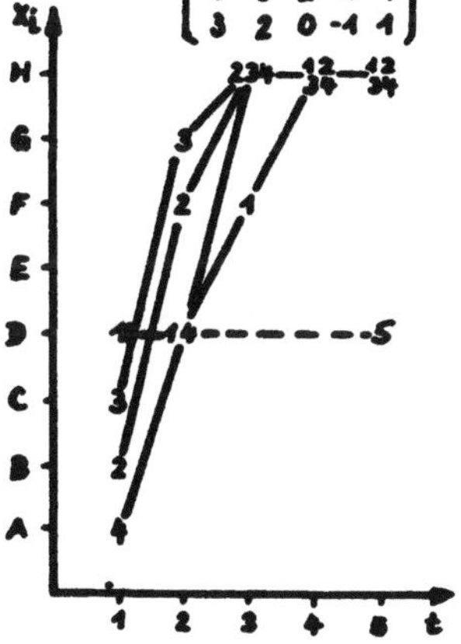

Ex.3: $G = \begin{pmatrix} 1 & 1 & -2 & -1 & 0 \\ -2 & -1 & -1 & 0 & -1 \\ -1 & 0 & -2 & -1 & -1 \\ -3 & -2 & 0 & 1 & -1 \end{pmatrix}$

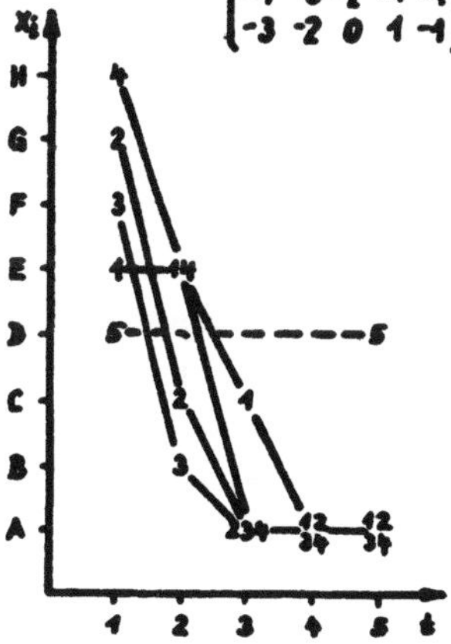

Ex.4: $G = \begin{pmatrix} -3 & 1 & 1 & 0 & 1 \\ 1 & 0 & -3 & 1 & 1 \\ 4 & -1 & -4 & 1 & 0 \\ -5 & 2 & 5 & -1 & -1 \end{pmatrix}$

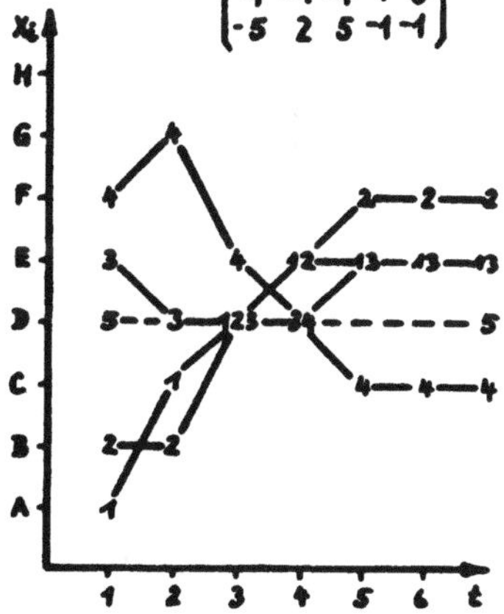

Example 5: $G = \begin{pmatrix} -K & K & 0 & 0 & 0 \\ -K & 0 & K & 0 & 0 \\ -K & 0 & 0 & K & 0 \\ -K & 0 & 0 & 0 & K \end{pmatrix}$

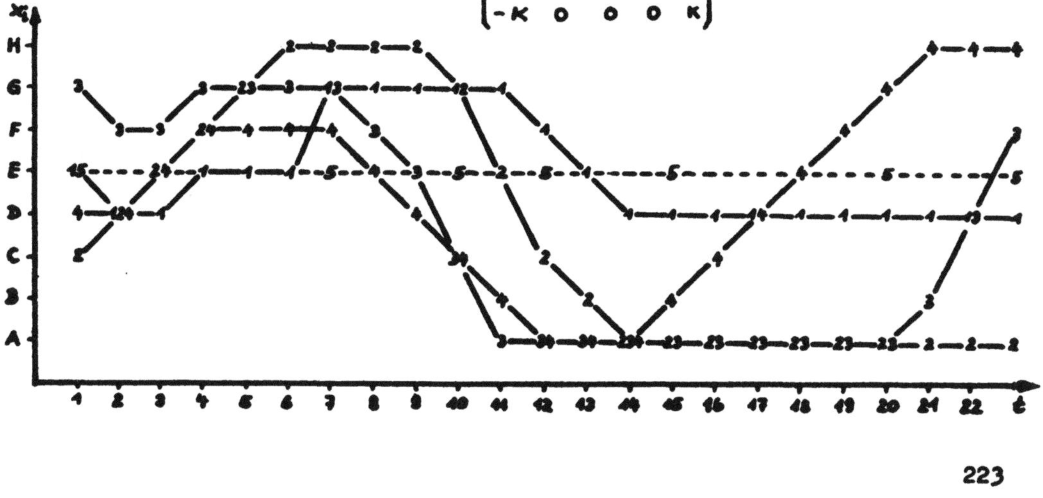

CHAOS IN A LOTKA-VOLTERRA-SYSTEM - A NUMERICAL STUDY

Thomas Schulmeister and Wolfgang Meiske[1)]

Consider the following normalized system:

$$\begin{aligned} \dot{x} &= 1 - bx - xy^2 - exy + z \\ \dot{y} &= a(xy^2 + d - y) \\ \dot{z} &= f(exy - z) \end{aligned} \qquad (1)$$

with x, y, z > 0 and parameters a, b, d, e, f > 0,
It represents a mathematical model of an autocatalytic, second-order in product reaction with reversible deposition of the substrate. The existence of non-periodic motion has been suggested in [2] . An extensive numerical simulation has been performed to analyze the onset of chaos in the parameter space. A topological explanation of the chaos-generating effects on the basis of simulations alone is, of course, impossible but detailed numerical investigations may improve the knowledge of these bifurcation phenomena and help to find chaos in other models. For the considered particular set of parameters (a=5, d=0.1, f=0.1) in system (1) up to three stationary states exist. Stable and unstable periodic solutions branch by Hopf bifurcation [2] . For values of a certain parameter region within the rectangle $0.34 < b < 0.37$ and $0 < e < 2.0$ motions appear which do not show any periodicity. This region is interspered with stripes generating only stable limit cycles. The existence of chaotic attractors is confirmed by positive values of the corresponding maximal Liapunov exponents [3]. Varying the parameters we observe different ways into chaos:

1. The Feigenbaum sequence

This route starts with a stable limit cycle. At the end of an infinite cascade of period doubling bifurcations attractive chaos arises. This chain of supercritical bifurcations is well-known from other chaos-generating systems [1,4,5]. Besides the pure Feigenbaum cascade in system (1) two special types of this route were observed. In the first case a second stable limit cycle emerges from a multiple limit cycle (tangential bifurcation). The old limit cycle disappears and the cascade acts on the new orbit. In the other case a period-3-orbit arises. Then, on the basis of a mixture of further tangent-type bifurcations and Feigenbaum sequences parameter regions, generating stable limit cycles, prechaos or attractive chaos, appear.

Academy of Sciences of the GDR, Central Institute of Molecular Biology
1115 Berlin-Buch Robert-Rössle-Straße 10, GDR

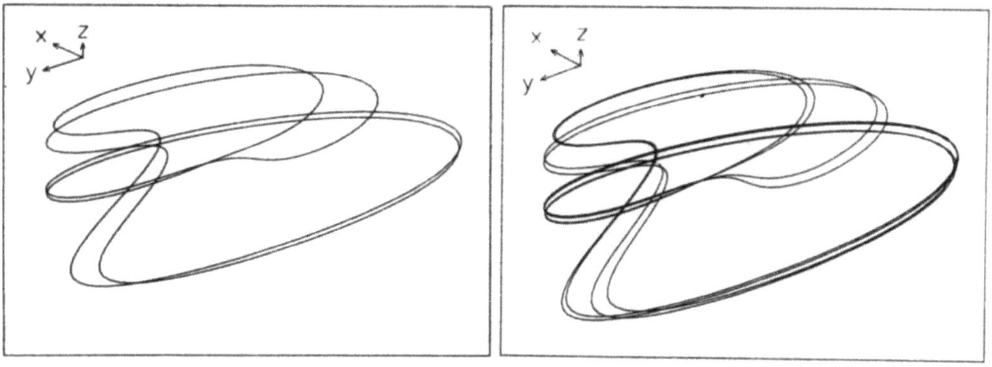

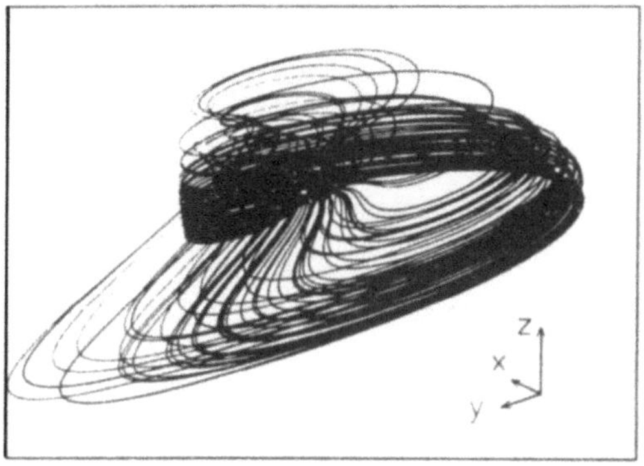

Fig.1 Feigenbaum sequence: Phase portraits for the exceptional case a=4, b=0.35, d=0.1, f=0.2 generating very intelligible graphs. Above: folded stable limit cycles after two (left-hand part, e=1.54) and three (e=1.53) period doubling bifurcations. Below: attractive chaos (e=1.4). Liapunov exponent= 0.051 . Initial value: 0.3445; 1.8085;1.0652 .
singly periodic orbits: e > 1.7
double periodic orbits: 1.57<e<1.69
fourfold period. orbits: 1.54<e<1.56
eightfold periodic orbits: e ~ 1.53
attractive chaos: e < 1.52

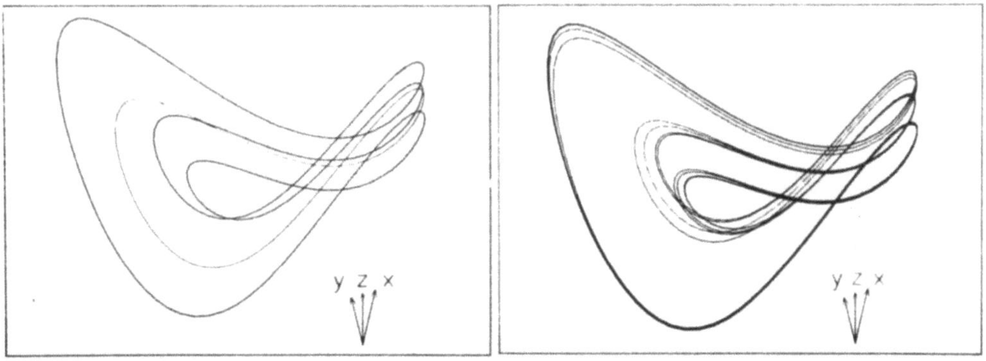

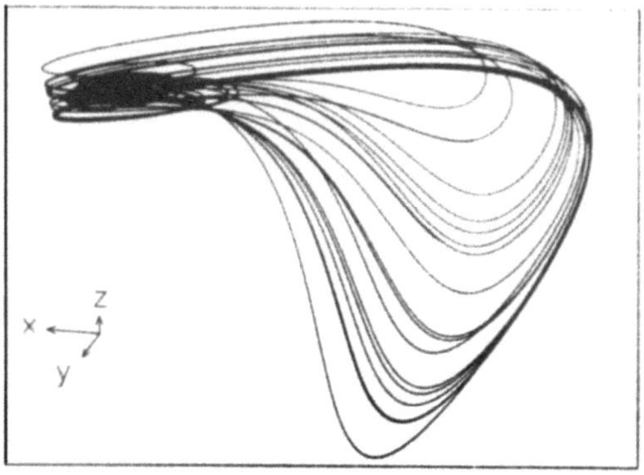

Fig.2 Above: Phase portraits of stable limit cycles in the period-3-region. Simultaneous existence of a singly periodic oscillation and a period-3-orbit (left-hand part, b=0.365, e=1.76) and a period-3-orbit after two period doubling bifurcations (b=0.365, e= 1.756).
Below: Attractive chaos near the line of coalescing foci (b=0.353, e=1.3). Liapunov exponent=0.044 . Initial value: 2.5077;0.2225;0.7515.

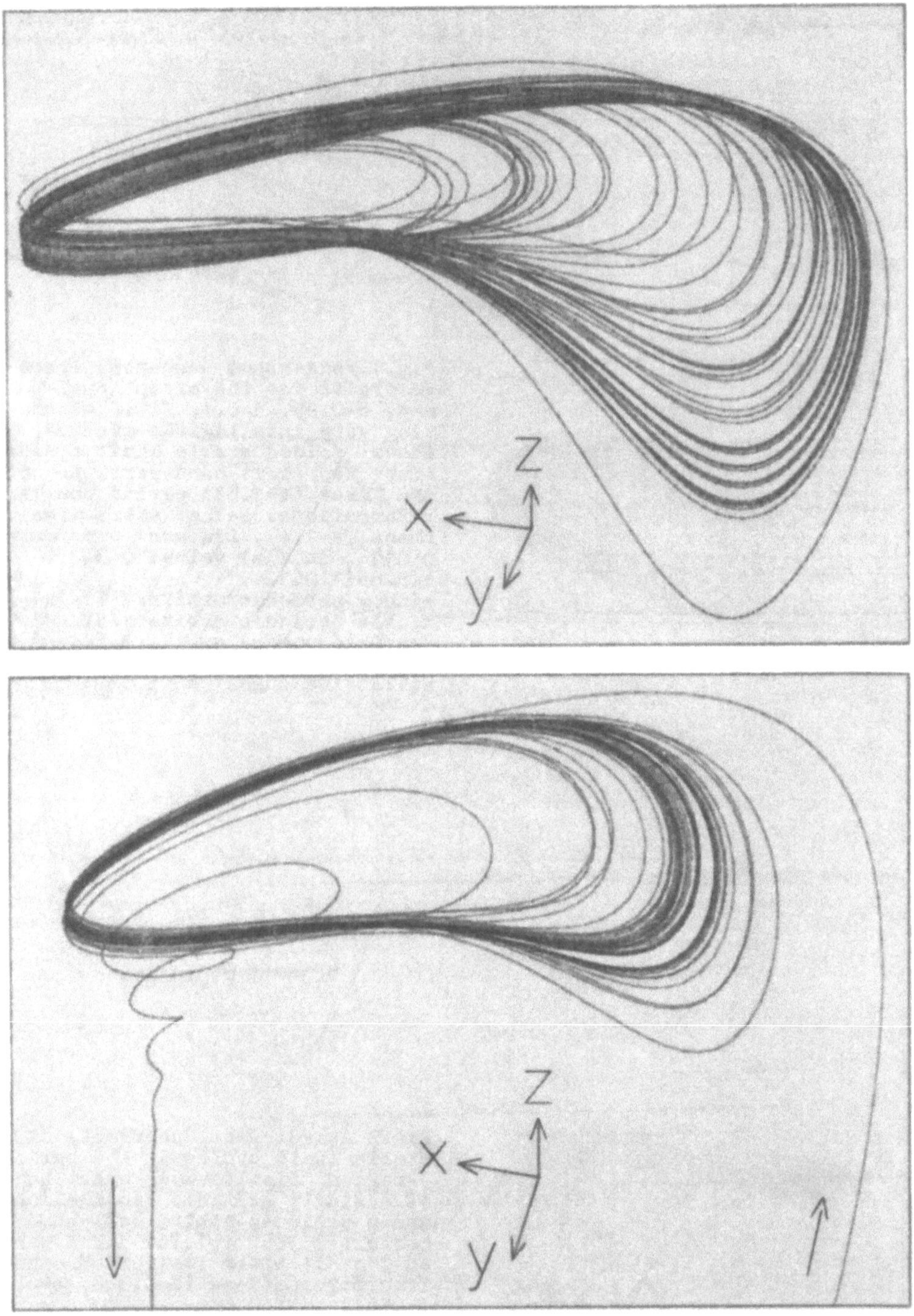

Fig.3 The prechaos route. Above: Phase portrait of a typical attractive chaos (b=0.36, e=1.647). Initial value: 1.6516;0.1564;1.0337 . Liapunov exponent=0.072 . Below: Prechaotic behaviour (b=0.367, e=1.6905). Initial value: 0.224;2.030;1.168 . The solution tends to the stable focus at 2.4878;0.1870;0.7863 .

2. The route with bifurcations of stationary states

We observed two ways into chaos starting with a bifurcation of stationary states. In the first case a stable focus loses stability by Hopf bifurcation and in the second case a pair of focus-type stationary states coalesces and disappears. In both cases the global behaviour of the stable and unstable manifolds of the stationary states leads to homo-clinicity, that is, into chaos.

3. The prechaos route (explosion)

At certain parameter values the existing chaotic attractor disappears suddenly leaving "metastable chaos", that is, chaotic behaviour persists for a long but finite time [1]. Later on all solutions tend to the other (now only) existing stable mode, a stationary state in a distant region of the state space.

We emphasize that the numerical methods used in this work cannot prove the appearance of chaos in system (1) in a unique manner. The same is true for the presented onset mechanisms. Special theoretical investigations are necessary to complete the understanding of this interesting attractor. Note that system (1) may serve as a rough model of the glucose metabolism of E. coli [2].

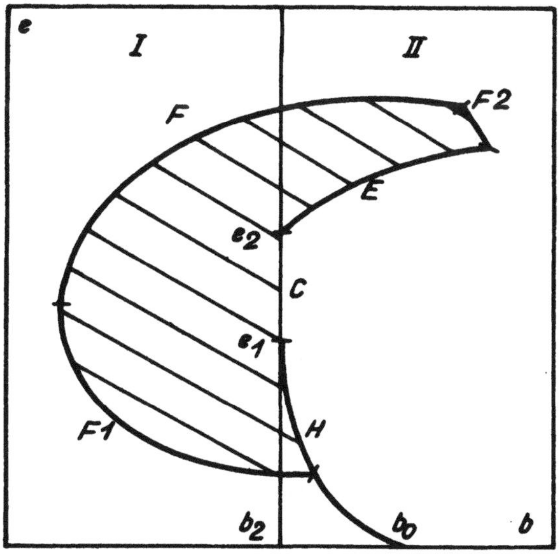

Fig.4 Schematic diagram of the ways into chaos in the b-e-plane (a=5, d=0.1, f=0.1). Within the shaded region system (1) generates attractive chaos or stable folded limit cycles. F, H, C and E indicate the start of the Feigenbaum sequence, Hopf bifurcation curve (b_o=0.3553, e_1=1.223), curve of coalescing foci (b_2=0.35382) and the change of prechaos into attractive chaos (e_2=1.35), respectively. F1 and F2 denote the two special types of the Feigenbaum route. In region I exists only one (unstable) stationary state, in region II there are three stationary states.

[1] Kaplan,J.L. and J.A.Yorke: The Onset of Chaos in a Fluid Flow Model of Lorenz. Ann.N.Y.Acad.Sci. 316 (1979) 400-417.
[2] Schulmeister,Th.: Chaos in a Lotka-Scheme with Depot. studia biophysica 72 (1978)3, 205-206.
[3] Shimada,I. and T.Nagashima: A Numerical Approach to Ergodic Problems of Dissipative Dynamical Systems. Prog.Theor.Phys. 61 (1979) 1605-1616.
[4] Sparrow,C.T.: The Lorenz Equations: Bifurcations, Chaos, and Strange Attractors. Springer, New York 1982.
[5] Tomita,K.: Chaotic Response of Nonlinear Oscillators. Physical Reports 86 (1982)3, 113-167.

STRUCTURING AND EVOLUTION PROCESSES

Hans-Michael Voigt [1]

Introduction

Evolution processes are closely related to structuring. We are postulating that structuring is very closely connected to the features of evolving system.
In the following we use the so called replicator equation, a special type of a Volterra-Equation, to verify this postulate. Since the replicator equation is autonomous in the state variables it represents by the stationary features only a spotlight of the richness of the underlying structure. Therefore we introduce naturally well-defined control variables, which are possible to interprete in the language of evolving systems.

The replicator equation

The replicator equation plays a central role in population genetics, but obviously this model is more far reaching. So it is used e.g. in analyzing the selfregulation in animal societies or in general in ecological systems.
Introducing the relative frequency x_i of species within a population yields the replicator equation

$$x_i = x_i \left(\sum_j G_{ij} x_j - \sum_k \sum_j x_k G_{kj} x_j \right) \quad i = 1,2,\dots,n.$$

The phase space is the simplex $S_n = \left\{ (x_1,\dots,x_n)/x_i \geqslant 0, \sum_i x_i = 1 \right\}$.

The above notion leads to an interesting game-theoretic interpretation of the replicator equation as a non-zero sum game where

$$\phi = \sum_k \sum_j x_k G_{kj} x_j$$

represents the value of the game.
At this point we sketch some remarkable features of the replicator equation, which will be useful for the further discussions [4] :

i) Adding any constant to a column of G does not change the replicator equation at all, i.e. it can be always transformed to a standard form such that all diagonal elements of the pay-off matrix G are

1) GDR Academy of Sciences, Mathematics and Cybernetics Division, 1199 Berlin, Rudower Chaussee 5

equal to zero.

ii) For each n there are only a finite number of stable classes of fixed points on the simplex S_n.

The controlled replicator equation

The replicator equation may be generalised in such a way that environmental influences are taken into account contrary to the uncontrolled equation, where the only connection to the environment consists in the value ϕ of the differential game.

We introduce control variables u_i, $i = 1,2,\dots,n$, acting on the reproduction rates of the species. Than the pay-off matrix of the standardised controlled replicator is given by

$$\bar{G} = G + \mathrm{diag}\,\{u_i\}_1^n$$

and the replicator equation may be written in the form

$$\dot{x}_i = x_i\,(\sum_j G_{ij}\,x_j - \phi) + x_i\,(u_i\,x_i - \sum_k u_k\,x_k^2)$$

The stationary solutions of this equation have the property that for $u_i \to -\infty$ the evolutionary stable strategy $e_i = 1/n$ will be reached, i.e. the barycentre of the simplex S_n is the only stable attractor. For $u_i \to +\infty$ we observe the reverse flow on the simplex S_n, i.e. all vertices of the simplex are stable attractors. This is shown in fig. 1 for n = 3, where 0 represents a repellor and ● represents an attractor. For $u_i \to -\infty$ the resulting global state will be defined as the state of uniform cooperation and for $u_i \to +\infty$ as the state of uniform competition. This implies the scaling of the global state of the controlled replicator equation with respect to the control variables u_i in the following way

$$u_i \to -\infty \quad \xrightarrow[\textstyle\xleftarrow[\text{between species}]{\text{increasing cooperation}}]{\text{increasing competition between species}} \quad u_i \to +\infty$$

From the above it is seen that the underlying structure of the system given by the pay-off matrix G does not play any role for the two extremal values of the environmental controls $u_i \to -\infty$ and $u_i \to +\infty$, respectively. The interdependence of the species in these cases is completely determined by the external control variables u_i. For $u_i \to -\infty$ the value of the game ist given by $\phi \to -\infty$, i.e. the environment represents a total looser. For $u_i \to +\infty$ the opposite holds.

In the following we will discuss the simplified case with $u_i = u_j$ for all i and j. Than the pay-off matrix is given by

$$\bar{G} = G + u\,I$$

with I the identity matrix.
A fixed point in the interior of the simplex S_n is defined by the equation $x(u) = \operatorname{adj} \bar{G}(u) / (e, \operatorname{adj} \bar{G}(u)\, e)$. Fixed points on the boundary of the simplex S_n may be determined by discarding the corresponding columns and rows from the controlled pay-off matrix $\bar{G}(u)$.
At this point we formulate the thesis:

> Stable structures are represented by the set of stable attractors of the controlled replicator equation on the simplex S_n. Each stable attractor represents a cluster of the graph given by the matrix $\bar{G}(u)$.

Example: We consider an undirected graph with three objects as nodes having the structure

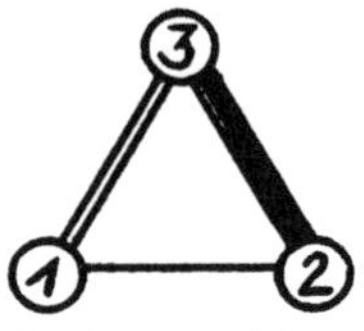

$$G = \begin{bmatrix} 0 & 1 & 2 \\ 1 & 0 & 4 \\ 2 & 4 & 0 \end{bmatrix}$$

where the interaction between the objects is given by the pay-off matrix G as values of the edges of the graph. The variable u controls the forced degree of cooperation and competition, respectively, between the objects of the given structure. Depending on the degree of cooperation we observe the structures as shown in fig. 1.

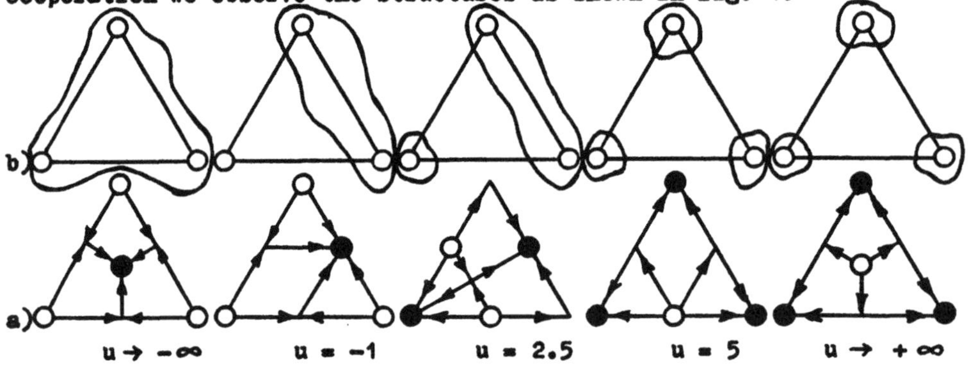

c)

Fig. 1 a) Set of fixed points of the controlled replicator equation on the simplex S_3, b) Structures of the graph corresponding to the pay-off matrix G, c) Values of the control variable u

References

/1/ Crow, J.F. and M. Kimura: An Introduction to Population Genetics Theory. New York: Harper and Row 1970

/2/ Eigen, M. and P. Schuster: The Hypercycle - A Principle of Natural Selforganization. Berlin, Heidelberg, New York: Springer 1979

/3/ Maynard-Smith; J.: The Theory of Games and the Evolution of Animal Conflicts. J. Theor. Biol. 47, 209-221 (1974)

/4/ Zeeman, E.C.: Population Dynamics from Game Theory. Proc. Int. Conf. on Global Theory of Dynamical Sysstems. Berlin, Heidelberg, New York: Springer 1980

/5/ Hofbauer, J., P. Schuster and K. Sigmund: Competition and Cooperation in Catalytic Selfreplication. J. Math. Biol. 11, 155-168 (1981)

/6/ Peschel, M. and W. Mende: Leben wir in einer Volterra-Welt. Berlin: Akademie-Verlag (1983)

/7/ Voigt, H.-M.: Evolution Processes and Structuring. in: Report R-MATH-01/83, ISSN 0232-5136, Institut für Mathematik der AdW der DDR

INVESTIGATION OF WORLD ENERGY CONSUMPTION AND WORLD POPULATION IN TERMS OF A HYPERLOGISTIC EVOLUTION MODEL

U.Kriegel[1], W.Mende[2] and M.Grauer[3]

Under the assumption that evolution is a smooth deterministic process - e.g. describable by differential equations - and that large scale systems are hierarchically organized, a hyperlogistic evolution model (HLEM)

$$\frac{dx(t)}{dt} = A\cdot\left(\frac{x}{\bar{x}}\right)^k\cdot\left(\frac{B-x}{\bar{x}}\right)^l \qquad (1)$$

has been formulated [1], which describes the transition from one stationary level ($x=0$) of the system variable to the next one ($x=B$) as a process with a very smooth dynamics - there are no sudden changes in growth rates, rates of rates etc. In eqn.(1) k and l describe the strength of co-operativity of driving and saturation processes, respectively and A is the scale factor. The renormalization constant $\bar{x}$ has been introduced to get reasonable units of the system variable x(t). In order to get a reasonable description of historical data of world primary energy consumption and world population eqn.(1) was modified in the following manner:

- Since there is no information on the saturation process in either of these data sets the saturation level B was expressed as a function of the co-operativities and the value of the state variable at the time t_M when the maximum relative growth rate occurs.
- There is an irregular pattern in historical data between 1914 and 1949 which is probably due to the effects of World War I, the Depression, World War II and post-war depressions.It is assumed that these events only slow down the evolutionary process without changing its structural parameters, e.g. the exponents. Therefore the

[1]Central Institute for Cybernetics and Information Processes of the Academy of Sciences of the GDR,Kurstraße 30-34,1080 Berlin

[2]Institute for Geography and Geoecology of the Academy of Sciences of the GDR, Rudower Chaussee 5, 1199 Berlin

[3]International Institute for Applied Systems Analysis, A2361 Laxenburg, Austria

scale factor A was reduced by a factor $0 < \varepsilon \leq 1$ between 1914 and 1949.

- The co-operativity of driving and saturation process was assumed to be equal, e.g. k=1.

A close agreement between HLEM-curves and historical data was achieved. From the value of the slow-down factor ε it was estimated that the evolution of both population and energy systems has been delayed by about 20 years. World primary energy consumption and population were then projected up to the year 2100, assuming that the maxima of the corresponding relative growth rates lie between 1970 and 1975. The HLEM curves given in Figs.1 and 2 show a smooth transition to saturation levels between 27 TWyr/yr and 35 TWyr/yr (for energy consumption) and of 10^{10} people (for population). For comparison the corresponding predictions from IIASA studies [2] are also given, showing a much steeper rise for the time after the year 2000 (annual growth rates between 1.68% and 2.53% compared with 0.63% to 0.84% for the period 2000-2030).

Fig.3 shows the projection of per capita world primary energy consumption. After a steep rise to a maximum of between 3.3 kW per capita and 4.2 kW per capita around the year 2010 the curves decrease smoothly to their plateau values between 2.7 kW per capita and 3.5 kW per capita. In contrast to the smooth behaviour the IIASA study predicts an increase of per capita energy consumption with annual growth rates of 0.8% to 1.6% for the period 2000-2030. The corresponding HLEM rates lie between -0.1% and 0.1%.

References

1 Kriegel,U.,Mende,W. and Grauer,M.: Syst.Anal.Model.Simul.1(1984), 3,223.

2 Häfele,W.:"Energy in a Finite World",Ballinger Publ.Comp.,Cambridge/Mass.,1981.

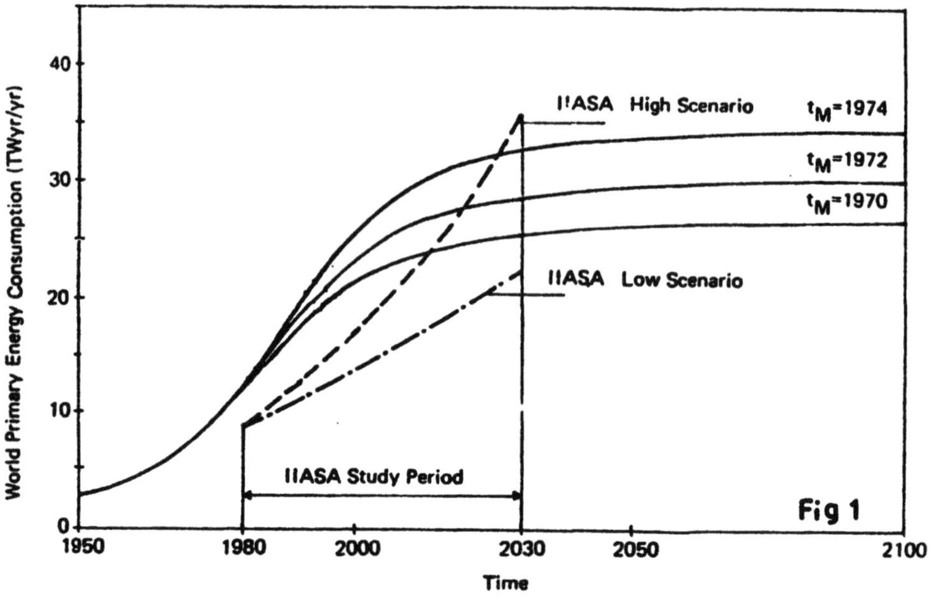
World Primary Energy Consumption (TWyr/yr)
40
30
20
10
0
IIASA High Scenario
t_M=1974
t_M=1972
t_M=1970
IIASA Low Scenario
IIASA Study Period
1950
1980
2000
2030
2050
2100
Time

Fig 1

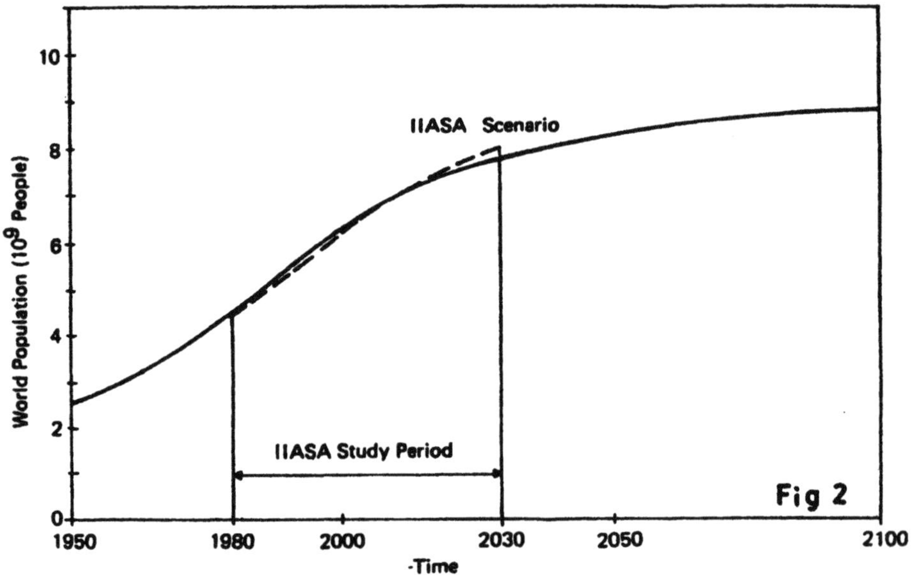
World Population (10^9 People)
10
8
6
4
2
0
IIASA Scenario
IIASA Study Period
1950
1980
2000
2030
2050
2100
Time

Fig 2

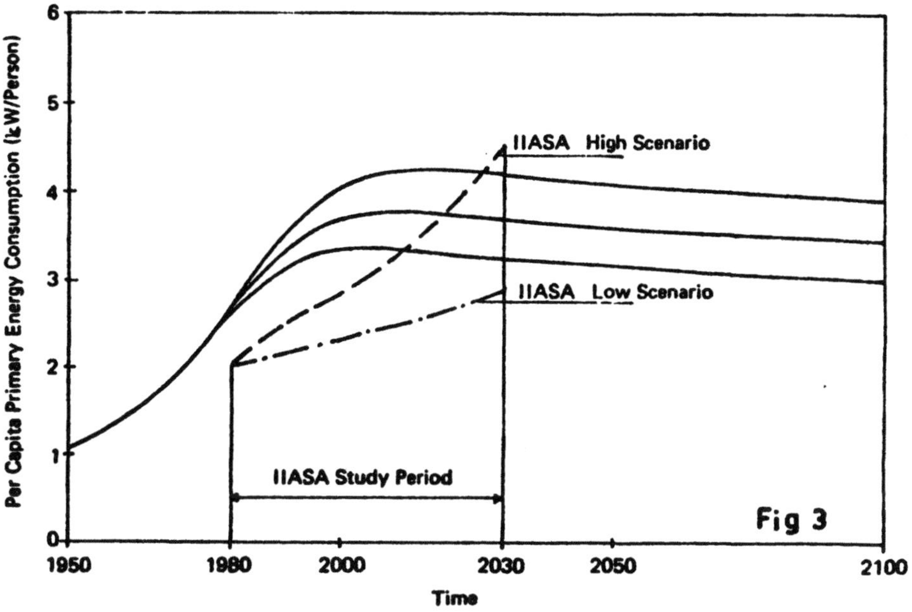
6
5
4
3
2
1
0
Per Capita Primary Energy Consumption (kW/Person)
IIASA High Scenario
IIASA Low Scenario
IIASA Study Period
1950
1980
2000
2030
2050
2100
Time

Fig 3

CHAIN STRUCTURES IN 1-ST ORDER DIFFERENTIAL EQUATIONS

K.-F. Albrecht[1], W. Mende[2], M. Peschel[1]

In a chain structure every element x_i is influenced by its predecessor x_{i+1} and influences its successor. Such chain structures often appear in reality for instance in energy transmission processes, food chains, chemical reaction chains, radioactive decay, diffusion processes, scattering processes and individual developement of organisms because the growth of the predecessor is as a rule the supposition for the developement of the successor. The most impressive example of a chain is the sequence of generations in the biological evolution.

We investigated chain structures, which can be **described** by a special kind of a first order differential equation system. The general form of the considered chain structures is:

$$\frac{dx_i}{dt} = f_i(k_i, x_i, x_{i+1}) \text{ with } i=0,\ 1,\ 2 \ldots N \text{ and } x_{N+1} = x_s$$

where x_{N+1} is the input- and $x_o(t)$ is the output signal of the chain. The chain structure is a homogenous one if the f_i differ only in the parameters k_i. In this case $f=f_i$ is called basic module. A homogenous chain structure is called a tower if all initial values $x_i(0)$ are equal. We consider here the case $x_i(0)=x_{N+1}=\text{const}$.

For the considered special chain structures a pointwise convergence of the output signal (element x_{oN}) to a function $x_{o\infty}(t)$ has been found. On the other hand it is possible for a given smooth signal $x(t)$ to contruct a special chain structure, which produces this signal: $\lim_{N\to\infty} x_{oN} = x(t)$ (Signal expansion principle arround $x(0)$). In practice a relative small number N gives an already good approximation $x_{oN} \to x(t)$.

A tower with a linear basic module (linear tower)

$$\dot{x}_i = K_i + x_{i+1}$$

generates a signal $x_o(t)$. On the other hand this linear tower may serve as a signal generator for a given sufficient frequently differentiable signal $x(t)$. The coefficients are then uniquely determined by $K_i = x_o^{(i+1)}\big|_{t=0}$. Stepwise integration yields nothing else than the well known Taylor expansion. But it puts a new light on it as the natural additive decomposition of a given function (superposition principle).

Another basic module namely

$$\dot{x}_i = K_i x_i x_{i+1}$$

leads to a new signal generation by rate coupling. The corresponding basic operator is the logarithmic derivative which determines the K_i recursively: $K_i = F^{(i+1)} x_o\big|_{t=0}$ with $F = \frac{d\ln}{dt}$. This chain of simple

[1] Central Institute for Cybernetics and Information Processes, Academy of Sciences, Berlin (G.D.R.)

[2] Institut for Geography and Geocology, Academy of Sciences, Berlin (G.D.R.)

bilinear basic modules we call exponential tower. It is a decomposition of a given function into its relative growth rates, growth rates of growth rates and so on.

The convergency conditions of the exponential tower are discussed in /1/ (e.g. $K_i \to K$ or the K_i are finally periodic is sufficient). If the exponential tower has periodic coefficients $(K_0 \ldots K_{r-1})$ one gets for $N \to \infty$ the solution for the r-order hypercycle $x_s = x_{or}$ /1/.

In the case of convergence for a given smooth function coefficients K_i can be determined, so that $\lim_{N\to\infty} x_{oN} = x$ yielding the Taylor expansion for the linear tower and for the exponential tower a new principle of function expansion /1/. For that tower the influence of fluctuations of the coefficients K_i on the mean relative quadratic deviations s has been invetigated. s depends on t in a nonlinear way (fig. 2) /2/.

Many other modules are possible (e.g. logistic ones). The corresponding signal expansion principle allows nonlinear models of a given system in the following sense: We choose for the module on the lowest level an already good nonlinear approximation (local model) for the real behaviour of the system. The higher levels will be used for corrections of this local model.

A large number of real transition processes can be described by the so called Evolon transition model x(t) given by:

$$\dot{x} = K(x-A)^k((B-A)^w - (x-a)^w)^l$$

Therefore it is interesting to investigate chains of such transitions, for which the transition process in the i-th level is controlled by the reached state in the i+1-st level.

Table 1 contains some characteristics of 3 different towers. For these special towers we obtain (apart possible singularities) already a good approximation $x_{oN} \to x$ for a small number N. This is illustrated in fig 1.

The behaviour of a 'logistic tower' $k_i = k = l_i = l = 1$ for 3 periodic coefficients ($K_0 = 0.5$, $K_1 = 1$, $K_2 = 2$) is illustrated in fig 1c for $x_1(0) = x_{N+1} = 0.5$. Fig. 1d illustrates how the relative deviation of x_{oN} in respect to $x_{o\infty}$ decreases with $N \to N+1$.

The exponential and logistic towers are only special cases of a signal expansion which abandons the linear superposition principle. But it gives the possibility to decompose a function in most suitable nonlinear components. The convergence of the tower is reflected in the more and more weakened influence of the higher levels of the tower (approximations property).

By the so called structural design principle /1/, which generates additional feedbacks in the tower, the approximation character vanishes but the towers become finite.

Literature

/1/ Peschel, M., Mende, W.: Do We Life in a Volterra World, Berlin 1984
/2/ Albrecht, K.-F., Chetverikov, V., Ebeling, W., Funke, R., Mende, W., Peschel, M.: 'Random Phenomena in Nonlinear Systems in Connection with the Volterra approach', Proceedings of the 4-th Meeting of the UNESCO's working group on Cybernetics, Plattnitz (Austria) 1983

	kind of tower		
	linear	exponential	hyperlogistic
basic module	$\dot{x}_i = x_{i+1} + K_i$	$\dot{x}_i = x_{i+1} \cdot K_i x_i$	$\dot{x}_i = (x_{i+1} - A_{i+1}) K_i (x_i - A_i)^{k_i} \cdot$ $\cdot ((B_i - A_i)^{w_i} - (x_i - A_i)^{w_i})^{l_i}$
coupling	additive	multiplicative (coupling of rates)	multiplicative
coefficients for the generation of a given signal $x = \lim_{N \to \infty} x_{ON}$	$K_i = F^{(i+1)} x(0)$ = Taylorcoef. with $F = \frac{d}{dt}$	$K_i = F^{(i+1)} x(0)$ with $F = \frac{d \ln}{dt}$	for $\dot{x} = K(x-A)^k (B-x)^l$ and $\dot{x}_i = (x_{i+1} - A) K_i (x_i - A)(B-x)$ see footnote x)
convergence of x_{ON} with rising N	identical with convergence of the Taylor series	uniform convergent in every closed interval out of $-\infty < t < 1/k$, with $K = \text{supr.} K_i$	uniform convergent in every closed interval for $K_i, k_i, l_i > 0$ and $A < x_i(0) < B$
standart special case	$x_i(0) = x_s = 1, K_i = K = 1$	$x_i(0) = x_s = 1, K_i = K = 1$	$x_i(0) = x_s, K_i = K = 1, w_i = w = 1$ $k_i = k,\ l_i = l,\ A_i = A,\ B_i = B$
$x = \lim_{N \to \infty} x_{ON}$ for the simplified special case	$x = Ke^{-t}$	$x = \frac{1}{(1-Kt)}$	$\dot{x} = K(x-A)^{k+1} (B-x)^l$

x) A hyperlogistic function $\dot{x}_0 = K(x-A)^k (B-x)^l$ can be generated for instance by logistic basic modules $\dot{x}_i = (x_{i+1} - A) K_i (x_i - A)(B - x_i)$ with $K_i = \frac{\dot{x}_1}{(x(0)-A)^2 (B - x(0))}$, x_i recursive from $x_i = \frac{\dot{x}_{i-1}}{K_{i-1}(x_{i-1}-A)(B-x_{i-1})}$

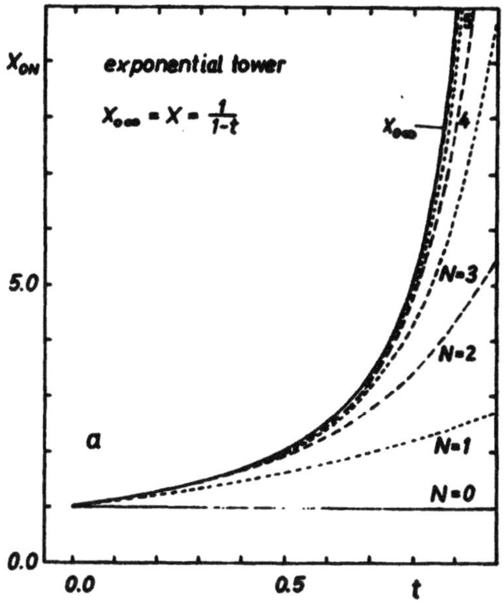

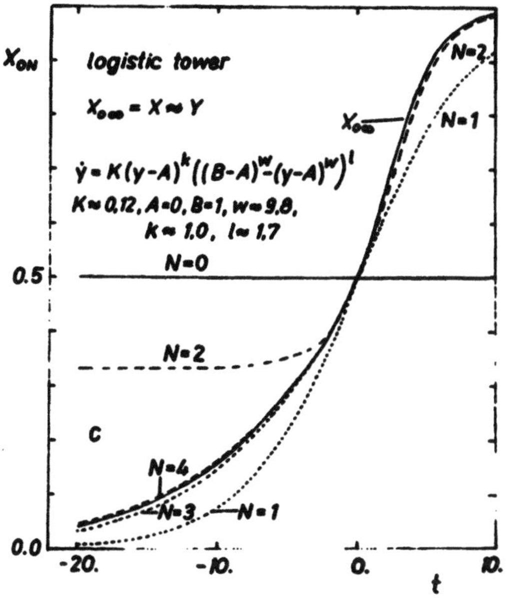

a) Exponential tower with $x_i(0) = x_{N+1} = 1$, $K_i = K = 1$

c) Logistic tower with 3 periodic coefficients $K_0=0.5$, $K_1=1$, $K_2=2$ and $x_i(0) = x_{N+1} = 0.5$

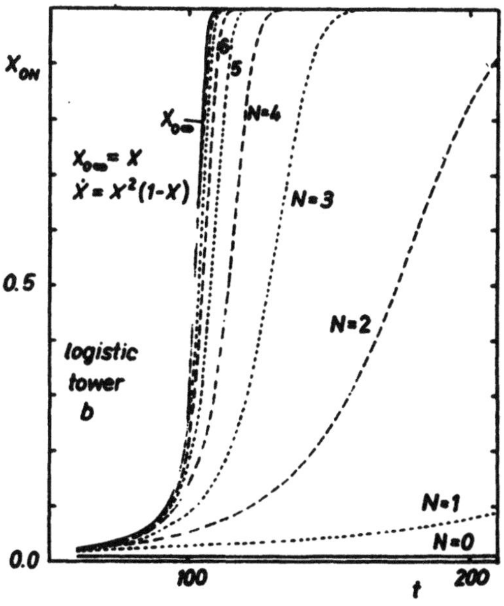

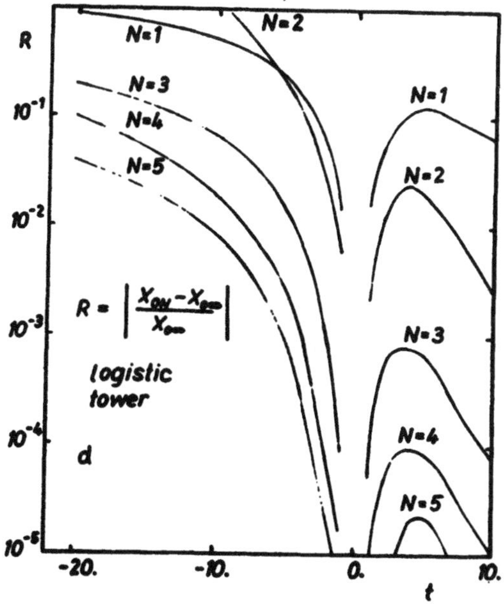

b) Logistic tower with $x_i(0) = x_{N+1} = .01$, $K_i = K = 1$

d) Relative deviation of x_{oN} from the limit function x_{Ooo} for the logistic tower of fig. 1c)

Fig. 1 Behavior of the output signal x_{oN} of towers in dependence of time t and the number of tower levels N

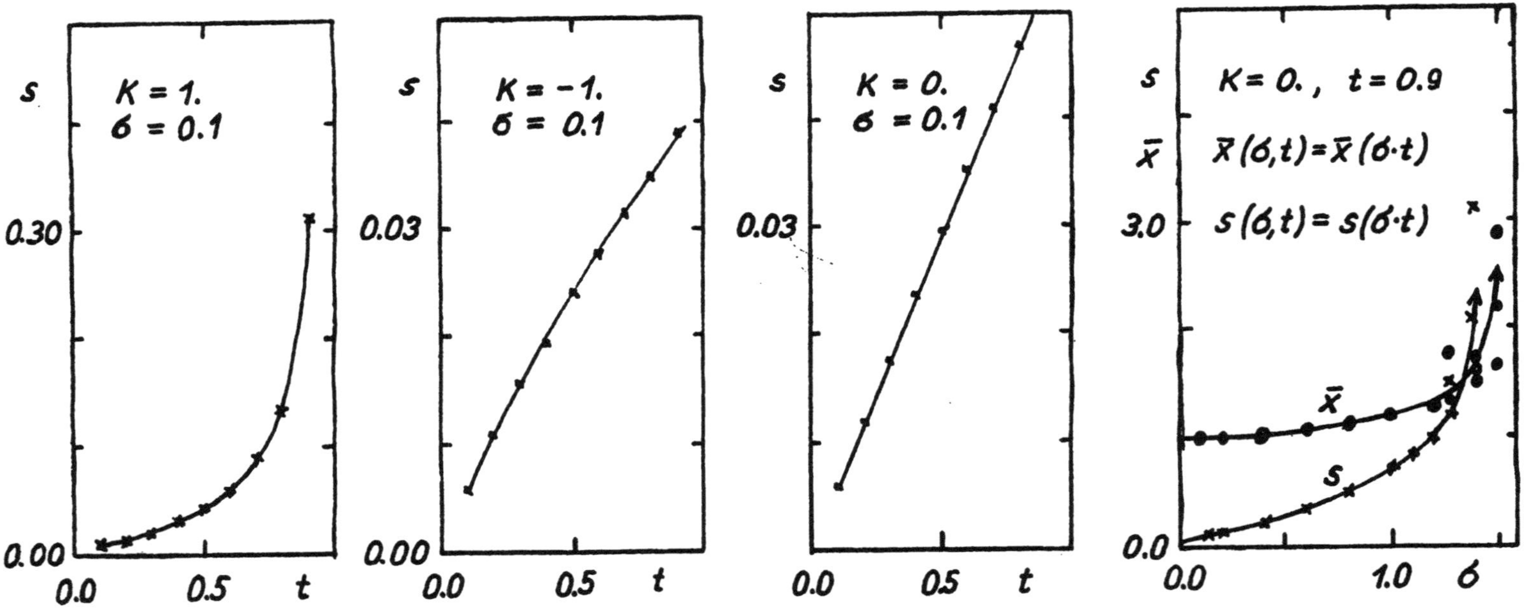

Fig. 2 The influence of noisy coefficients on the relative mean quadratic deviation s between a tower with diturbed coefficients K_i and a tower with undisturbed coefficients K in the case of the standard exponential tower (see table). $K_i = K + 2\sigma\xi_i - \sigma$ (ξ_i = random numbers, $0 < \xi_i < 1$), N = 10 number of tower levels, n=1000 number of integrations for different samples K_i

STATE MODELS - AN APPROPRIATE TOOL FOR IMPROVING THE EFFICIENCY OF BIOTECHNICAL PROCESSES

Brigitte Heinritz and Thomas Bley [1]

Microbial growth in technical systems gets more and more interest for modelling and investigations in systems analysis. The reasons for this may be these two :

- Microbial growth is an essential for most processes of biotechnology.
- Microbial growth in a controllable technical system allows us to study characteristic properties of the dynamics of ecological systems.

Generally the microbial population is described from the point of view of chemists or biochemists : The variables of the models correspond to an unstructured biomass or to some biochemical components of the cells (e.g. protein, nucleic acid, ATP ..). The results of such a way of modelling are not always satisfactory. The reason for this is that the physiological state of microbial cells during their course of growth and proliferation is not considered. Therefore we use a model structure in which the variables correspond to the states of the cells deduced from the growth cycle. This approach of modelling is supported by the results of biological experiments /4/ as well as by theoretical considerations /2/.

In the simplest case investigated a cell passes two states during its cycle : one with and another without biomass growth. In both cases substrate is consumed, but with different kinetics. These states are connected with different morphological features.

The type of differential equations for the model of a continuous fermentation process is the following (cf./1/):

$$\dot{x}_1 = (\mu(s) - k_1 - D)x_1 + k_2(s)x_2$$
$$\dot{x}_2 = k_1 x_1 - (k_2(s) + D)x_2$$
$$\dot{s} = (s_0 - s)D - \alpha_1(s)x_1 - \alpha_2(s)x_2$$

1) Academy of Sciences of the GDR, Institute of Technical Chemistry DDR 7050 Leipzig, Permoserstr. 15

x_i : concentration of cells of different states
s, s_o : concentration of (carbon)substrate
$\mu(s)$: specific biomass growth rate
D : dilution rate
k_i : transition rates
α_i : stoichiometric coefficients, i=1,2

Using a proper choice of $\mu(s)$, $k_2(s)$, $\alpha_1(s)$, and $\alpha_2(s)$ it is evident that this type of model is connected closely with a VOLTERRA model and it can be transformed easily by using the suitable formalism (cf./6/).

With models of this type a broad spectrum of behaviour of fermentation processes, for instance the synchronization by substrate pulses or oscillations by low dilution rates, can be described and investigated theoretically.

With the consideration of further cell states and the introduction of minimum retardation times of staying in some states which is biologically founded by SMITH and MARTIN /7/ we get a model which describes the behaviour of microbial growth in a technical system in a very realistic way /3/.

Simulations with the model described showed the possibility of increasing the efficiency of microbial carbon-compound conversion or production of special metabolites by a periodic control conception using synchronized populations /4/. The heading "Dynamic Process Control" is used for this type of fermentation processes /5/.

For illustration the result of a model simulation and a technological example are demonstrated in the two figures.

<u>References</u>

/1/ BLEY,Th., SCHMIDT,A., stud.biophys. 78 (1980) 11-12
/2/ BLEY,Th., BOGNER,H., stud.biophys. 83 (1981) 217-224
/3/ BLEY,Th., HEINRITZ,B., SCHMIDT,A., stud.biophys. 98 (1984)119-124
/4/ HEINRITZ,B. et al., Z. Allg. Mikrobiol. 21 (1981) 583-588
/5/ HEINRITZ,B. et al., Acta Biotech. 3 (1983) 125-131
/6/ PESCHEL,M., MENDE,W., Syst. Anal. Model. Simul. 1 (1984) 17-26
/7/ SMITH,J.A., MARTIN,L., Proc. Natl. Acad. Sci. USA 70 (1973) 1263-1267

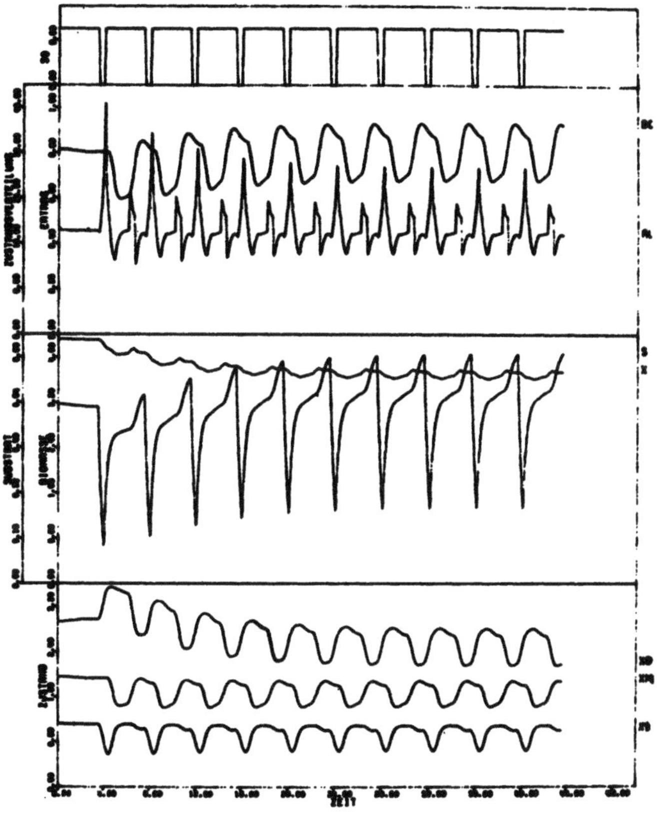

Fig.1 Simulation of periodically varying carbon substrate feeding with a 3-state model

Symbols: SO : substrate concentration in the incoming medium
BC : percentage of budding cells (XM)
AL : specific biomass yield of the whole population
S : substrate concentration in the fermentor
X : biomass concentration
XO, X1, XM : concentration of cells of the different states

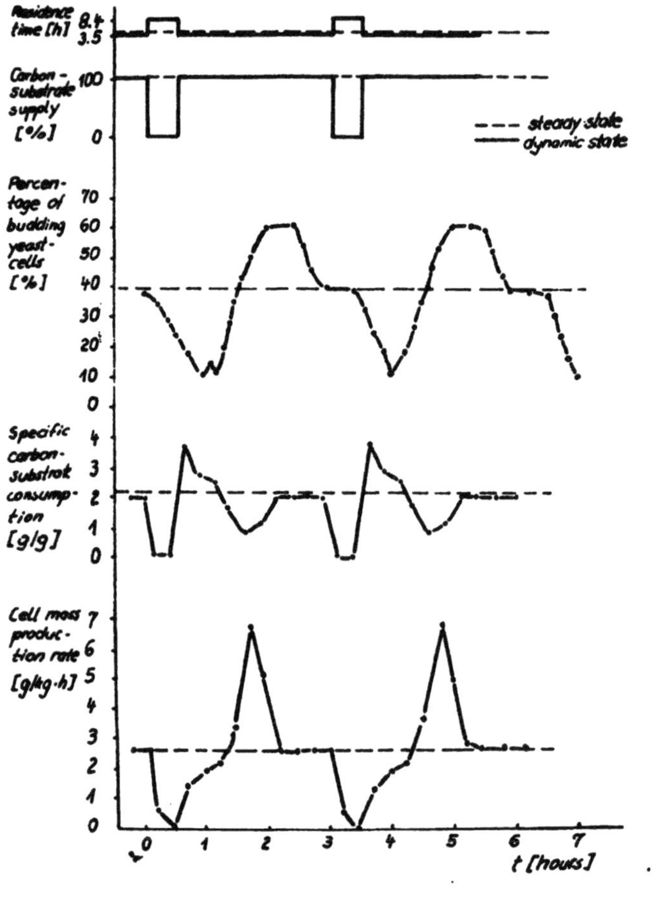

Fig.2 Experimental result of a dynamic process control
The efficiency of carbon compound conversion was increased by about 15% compared to a stationary process.

Immigrating and Emigrating Organisms in a Finite Volterra World

Heinz Schilling[1)]

The investigations concern with two kinds of diffusing organisms, predator X and prey Y. Their concentrations are assumed to be a function of position coordinate r and of time t :

$$X = X(r,t), \quad Y = Y(r,t) \quad . \tag{1}$$

The densities of their currents may be written as

$$J_X = -a_X \nabla X + b_Y \nabla Y , \quad J_Y = -b_X \nabla X - a_Y \nabla Y \tag{2}$$

/1/, where ∇X and ∇Y define gradients of concentration, a_X and a_Y coefficients of self diffusion, b_X and b_Y such of mutual diffusion respectively. Positive values of coefficients characterise natural behaviour, i.e. $b_X > 0$ escaping prey, $b_Y > 0$ persecuting predator, $a_X > 0$, $a_Y > 0$ competition within the same species. The resource C_A for the prey reproduction is assumed to be finite and may therefore be written as

$$C_A(r,t) = \frac{C_{A0}}{1 + \gamma Y(r,t)} , \tag{3}$$

where C_{A0} denotes the maximum concentration of the resource and the parameter γ characterizes their decrease due to the existing prey.

The reproduction velocity of prey is supposed to be proportional to the concentrations of resource and of prey /2/ :

$$\left(\frac{\partial Y}{\partial t}\right)_{\text{reproduction}} = k_1 C_A Y \quad . \tag{4}$$

k_1 is the rate constant of the reproduction.
Using normalized coordinates the equations describing the source processes of birth and death have the following form /3/ :

$$\left(\frac{\partial X}{\partial t}\right)_s = \frac{1}{t_2}(-X + XY) , \quad \left(\frac{\partial Y}{\partial t}\right)_s = k_1 C_A Y - k_1 C_{A0} XY , \tag{5}$$

where t_2 is the mean lifetime of the predator.

---

1) Central Institute for Cybernetics and Information Processes (ZKI/AdW), 1080 Berlin, Kurstr. 33

According to the equations of local balance /4/

$$\frac{\partial X}{\partial t} = - J_X + \left(\frac{\partial X}{\partial t}\right)_s \quad , \quad \frac{\partial Y}{\partial t} = - J_Y + \left(\frac{\partial Y}{\partial t}\right)_s \quad , \tag{6}$$

(2) and (5) give the following foundamental equations for the co-operation of current effects and source processes in a finite Volterra world

$$\frac{\partial X}{\partial t} = a_X \Delta X - b_Y \Delta Y + \frac{1}{t_2} (-X + XY) \ , \tag{7}$$

$$\frac{\partial Y}{\partial t} = b_X \Delta X + a_Y \Delta Y + \frac{k_1 C_{A0} Y}{1 + \gamma Y} - k_1 C_{A0} XY \ . \tag{8}$$

On the boundary of the flow field the concentrations may change abruptly to X_U of predator and to Y_U of prey. The density of their currents may be there /5/

$$J_X(L,t) = \alpha_X [X_U - X(L,t)] \quad , \quad J_Y(L,t) = \alpha_Y [Y_U - Y(L,t)] \ , \tag{9}$$

where α_X and α_Y are velocities of transition, controlling emigration and immigration.

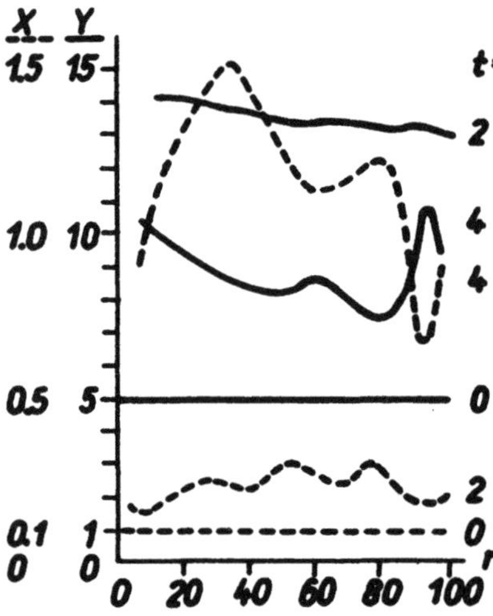

Fig. 1 . $b_Y = 100.0$, $b_X = a_X = a_Y = 0.1$

For the numerical solution the differential equations (7), (8) and (9) are approximated by difference equations.
In the following figures the settlement for various properties of the organisms is illustrated. The length of the area to be settled is $L = 100$. Organisms immigrate from border at $r = 100$. Constant concentrations of the environment are $X_U = 1.25$, $Y_U = 1.0$. The coefficient of transition is $\alpha_X = 0.1$. Prey may not pass the border.
The initial state is $X(r,0) = 0.1$, $Y(r,0) = 5.0$.

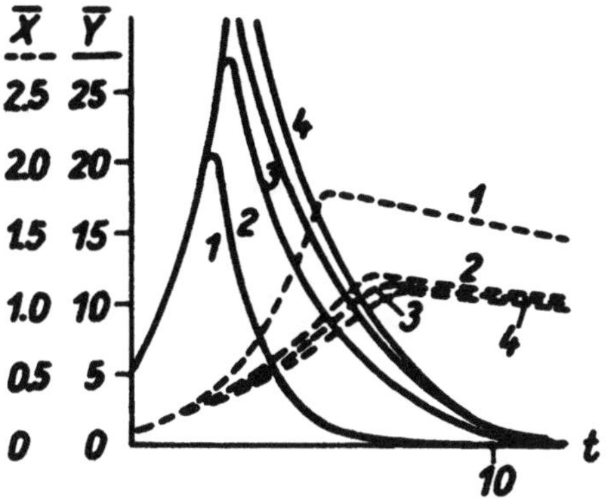

Fig. 2 . 1) b_Y = 100. , 2) a_X = 100. , 3) b_X = 100. , 4) a_Y = 100. ; other coefficients 0.1

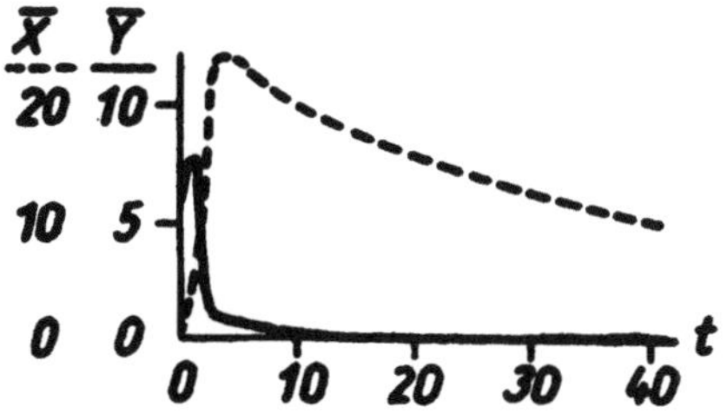

Fig. 3 . a_Y = −100. , $a_X = b_Y = b_X$ = 0.1

Parameter of resources are C_{A0} = 200, γ = 1.6 , prey's rate k_1 = 0.01 , predator's mean lifetime t_2 = 50 . Fig. 1 shows the distributions of predator and prey for the initial state, after two and after four time units in the case of a high agility of predator in persecuting prey. Fig. 2 represents the mean concentrations for high agility of persecuting predator (curves 1), high mobility of predator in competition (curves 2), high agility of escaping prey (curves 3) and high mobility of prey in competition (curves 4). The result for unnatural prey crowding in presence of predator is shown in fig. 3 .

Literature

/1/ Fife, P. C.: Comparision principles for reaction-diffusion-systems, J. Diff. Equ. 40 (1981), 2, 168

/2/ Peschel, M., and W. Mende: Do We Live in a Volterra World . Springer New York, Akademie-Verlag Berlin 1984

/3/ Ebeling, W., and R. Feistel: Physik der Selbstorganisation und Evolution . Akademie-Verlag Berlin 1981

/4/ Kluge,G., and G. Neugebauer : Grundlagen der Thermodynamik. Deutscher Verlag der Wissenschaften Berlin 1976

/5/ Pao, C. V.:On nonlinear reaction-diffusion systems . J. M. A. A. 87 (1982), 165

DISCRETNESS IN A CONTINUOUS MODEL OF BIOLOGICAL EVOLUTION

By A. Engel

1. Evolution as hill-climbing in the phenotypspace

One of the simplest but most succesful models of pre-biological evolution is the model due to Eigen [3]:

$$\partial_t n_i = \left(E_i - \langle E \rangle\right) n_i + \sum_{j \neq i} \left(a_{ij} n_j - a_{ji} n_i\right) \qquad ; \langle E \rangle = \frac{\sum_i E_i n_i}{\sum_i n_i} \tag{1}$$

n_i and E_i denote the concentration and fitness of the species i respectively and the a_{ij} describe mutation between the species i and j. Feistel and Ebeling [4] have proposed a continuous generalization of this model:

$$\partial_t n(q,t) = \left(E(q) - \langle E \rangle\right) n(q,t) + D \nabla^2 n(q,t) \; ; \tag{2}$$

$$\langle E \rangle = \frac{\int E(q) n(q,t)\, dq}{\int n(q,t)\, dq} \; , \; D = \frac{1}{2} \int dq' (q-q') a(q,q') (q-q')$$

Now a species is represented by a point $q \in R^d$ of the highdimensional phenotypspace. The coordinates of q fix all phenotypic properties of the species q. The value function E(q) giving the fitness as function of this phenotypic properties may be expected to be very complicated and is therefore modelled by a random field [2]. The evolution process can now be pictured as hill-climbing in this random landscape, i.e. regions of lower values of E(q) will be left in favour of those with high-valued E(q). Our aim here is to investigate some details of this hill-climbing process.

2. Formal analogy to the localization problem

We may simplify eq. (2) with the help of the transformation

$$y(q,t) = n(q,t) \exp\left\{ \int_0^t \langle E \rangle (t')\, dt' \right\} \tag{3}$$

to get

$$\partial_t y(q,t) = E(q)\, y(q,t) + D \nabla^2 y(q,t) \tag{4}$$

1) A. Engel, Sektion Physik, Bereich 04 der Humboldt-Universität zu Berlin, 1040 Berlin, Invalidenstr. 42

Eq. (4) is formally equivalent to a Schrödinger equation for a quantum particle in a random potential and a lot of the results obtained for this important and extensively studied problem can be used for our purposes [2] . A formal solution of (4) can be written as

$$y(q,t) = \sum_n e^{E_n \cdot t} c_n \psi_n(q) \tag{5}$$

where the ψ_n and E_n are determined by the eigenvalue problem

$$D\nabla^2 \psi_n + E(q)\psi_n = E_n \psi_n \tag{6}$$

The solutions ψ_n of (6) relevant to our problem are localized [1], i.e. concentrated around some point q_n and exhibiting for large $|q-q_n|$ the asymptotic form

$$\psi_n(q) \sim \exp\left\{-\frac{|q-q_n|}{l(E_n)}\right\}$$

where $l(E_n)$ is the localization length. The $\psi_n(q)$ are nothing but the continuous generalizations of the Eigen-Schuster quasi-species.

3. Mean transition time and mean transition distance

We consider now the case in which at t = 0 the sum in (5) is dominated by a master (-quasi-) species ψ_o at q_o with fitness eigenvalue E_o and ask for the mean transition time $\bar{t}(E_n)$ to the next quasispecies ψ_n with $E_n > E_o$ and the mean distance $\bar{r}(E_n) = |q_o - q_n|$ as a function of E_n (see Fig. 1). The bar denotes configurational averaging over the realizations of the random field E(q).

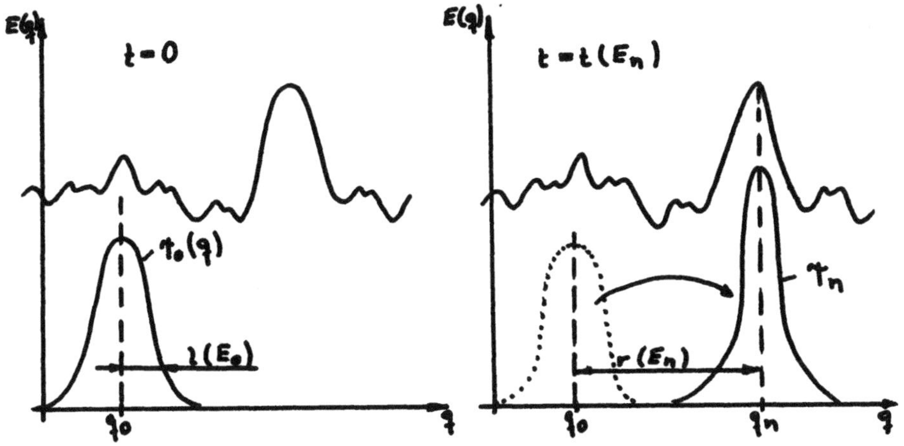

Fig.1 Definition of $\bar{t}(E_n)$ and $\bar{r}(E_n)$ for a d = 1-dimensional phenotyp-space.

Proceeding along the lines of ref. (3) but in addition taking into account correlations between the different localized states [5] we obtain:

$$\bar{r}(E) = \left[\frac{d}{\bar{\varrho}(E)\, S_d}\right]^{1/d} \exp\left\{\frac{S_d}{d}\,\bar{\varrho}(E)\, R^d(E)\right\} \cdot \int_{\frac{S_d}{d}\bar{\varrho} R^d}^{\infty} u^{1/d}\, e^{-u}\, du \tag{7}$$

$$\bar{t}(E) = \frac{\bar{r}(E)}{l(E)(E - E_0)} \tag{8}$$

$S_d = \frac{2 \cdot \pi^{d/2}}{\Gamma(d/2)}$ is the area of the d-dimensional unit surface, $\bar{\varrho}(E)$ is the density of states giving the number of states ψ_n with eigenvalue E per unit volume and

$$R(E) = \frac{l(E)\, l(E_0)}{l(E) + l(E_0)}\, \ln \frac{(E \cdot E_0)^{1/2}}{|E - E_0|}$$

gives the minimal distance between states with eigenvalues E_0 and E [5]. From (7) we get

$$\text{for } E \to \infty \qquad \bar{r}(E) \sim [\bar{\varrho}(E)]^{-1/d} \to \infty \tag{9}$$

$$\text{and for } E \to E_0 \qquad \bar{r}(E) \sim \left| \ln\left(\frac{E}{E_0} - 1\right)\right| \to \infty \tag{10}$$

Eq. (9) follows from the fact, that $\bar{\varrho}(E)$ goes exponentially to zero for $E \to \infty$ [3,5].
(9) and (10) ensure the existence of a r_{min} with $0 < r_{min} < \bar{r}(E)$ and using (8) we find a similar result for $\bar{t}(E)$ (see fig. 2). The quickest and therefore mostly realized transitions will thus occur to states with $E \approx E_{min}$ which have a mean distance $r(E_{min}) \approx r_{min}$ from q_0.

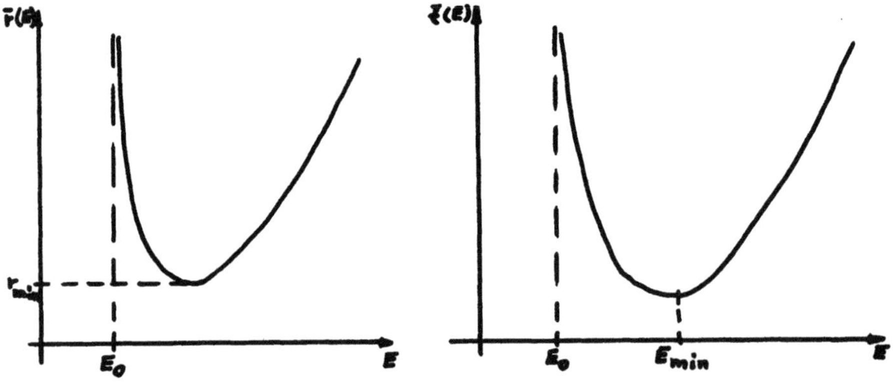

Fig. 2 Qualitative plot of the functions r(E) and t(E) as given by eqs. (7) and (8).

4. Conclusion

Despite of the continuous character of our model the evolutionary process has been shown to proceed via small but discrete steps both in the phenotypspace and in the fitness increase (see fig. 3). This may serve as indication that the stepwise character of evolution as observed in nature is not a direct consequence of the discrete action of mutations but a rather general feature of selection-diffusion processes

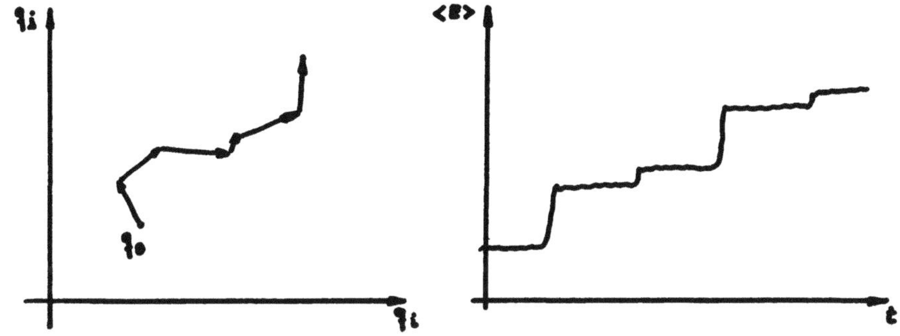

Fig. 3 Qualitative plot of the evolution pathway in a two-dimensional phenotypspace (left) and the increase of the ensemble fitness with time (right)

Acknolegdments: The author is indebted to Prof. W. Ebeling and M. Suhrke for stimulating discussions.

References

[1] Anderson, P.W.; Phys. Rev. 109 (1958) 1492
[2] Ebeling, W., A. Engel, B. Esser, R. Feistel; subm. to J. Stat. Phys.
[3] Eigen, M.; Naturwiss. 58 (1971) 465
[4] Feistel, R., W. Ebeling; Biosystems 15 (1982) 291
[5] Mott, N.F., E.A. Davis; Electronic Processes in Non-Crystallnie Materials 2 nd ed., Oxford, 1978

DISSIPATIVE STRUCTURES IN AN IONIC LOTKA-VOLTERRA TYPE CHEMICAL REACTION SYSTEM

By H. Malchow[1)]

Various chemical reaction models have been considered recently where the ionic character of the reactants has been neglected, see e.g. [1], [4], [6], [7]. Here we deal with a system of one neutral and two charged reactants using the concept of ambipolar diffusion by Nernst and Schottky as an approximation [3], [5].
We have

$$\dot{x}_i = f_i(x_1, x_2, x_3) - \nabla j_i \;; \quad i = 1,2,3; \tag{1}$$

with

$$j_i = -D_i \nabla x_i - z_i M_i x_i \nabla \varphi \;; \quad i = 1,2,3; \tag{2}$$

where $z_1 = 0$, $z_2 = -1$, $z_3 = +1$ are the charge numbers of the reactants i with concentration x_i, nonlinear reaction function f_i, diffusion coefficient D_i and mobility M_i; φ is the electric potential satisfying the Poisson equation. It must hold $f_2 = f_3$ because of charge conservation [2]. The system (1) can be reduced [5] to

$$\dot{x}_1 = f_1(x_1, x) + D_1 \Delta x_1 , \tag{3}$$

$$\dot{x} = f_2(x_1, x) + D_A \Delta x , \tag{4}$$

with the mean concentration $x = x_2 \approx x_3$ and the ambipolar diffusion coefficient $D_A = 2D_2D_3/(D_2 + D_3)$.
The potential gradient is

$$\nabla \varphi = \frac{D_3 - D_2}{D_3 + D_2} \cdot \frac{kT}{e} \cdot \nabla \ln x . \tag{5}$$

Now we consider the special reaction scheme

$$\begin{aligned} A + X_1 &\xrightarrow{k_1} 2X_1 , \\ B + 2X_1 &\xrightarrow{k_2} 3X_1 , \\ X_1 + X_2^- &\xrightarrow{k_3} 2X_2^- + X_3^+ , \\ X_2^- + X_3^+ &\underset{k_5}{\overset{k_4}{\rightleftharpoons}} C . \end{aligned} \tag{6}$$

[1)] Humboldt-Universität, Sektion Physik, Bereich 04, DDR-1040 Berlin, Invalidenstr. 42

with the kinetic equations

$$\dot{X}_1 = X_1 + \alpha X_1^2 - X_1 X_2 - \nabla j_1 , \tag{7}$$

$$\dot{X}_2 = X_1 X_2 - \beta X_2 X_3 + \gamma - \nabla j_2 , \tag{8}$$

$$\dot{X}_3 = X_1 X_2 - \beta X_2 X_3 + \gamma - \nabla j_3 , \tag{9}$$

where α , β and γ are functions of the reaction rates k_i.
The concentrations of the substances A, B and C are kept constant. The system (7) - (9) can be reduced to

$$\dot{X}_1 = X_1 + \alpha X_1^2 - X_1 X + D_1 \Delta X_1 , \tag{10}$$

$$\dot{X} = X_1 X - \beta X^2 + \gamma + D_A \Delta X . \tag{11}$$

Bistability is obtained for $\alpha\beta < 1 < 2\alpha\beta$ and $\beta < \gamma < [4\alpha(1-\alpha\beta)]^{-1}$.
The 3 homogeneous stationary solutions are

$$X_1^o = 0 \quad , \quad X^o = (\gamma/\beta)^{1/2} ; \tag{12}$$

$$X_1^{\pm} = [2\alpha(1-\alpha\beta)]^{-1} \cdot \left\{ 2\alpha\beta - 1 \pm \left[(2\alpha\beta-1)^2 - 4\alpha(\gamma-\beta)(1-\alpha\beta) \right]^{1/2} \right\} , \tag{13}$$

$$X^{\pm} = 1 + \alpha X_1^{\pm} ;$$

where (X_1^o, X^o) corresponds to a stable node and (X_1^+, X^+) to a stable node and a focus respectively. (X_1^-, X^-) is always a saddle point. In the focus regime a Hopf bifurcation is possible by varying γ as bifurcation parameter. A stable limit cycle bifurcates from (X_1^+, X^+) crossing a critical value of γ .
There is a variety of possible solutions in the distributed system. Some cases are listed below:

a) Inhomogeneous time-independent substance distributions as a result of a Turing instability of (X_1^+, X^+) in the two-node-case by crossing critical ratios of D_1 and D_A, bearing of direct-current voltages, compare Fig. 1;

b) moving or standing interfaces between the two "phases" (X_1^o, X^o) and (X_1^+, X^+) for suitable boundary conditions in the two-node-case, bearing of a direct-current voltage, compare Fig. 2;

c) homogeneous time-independent solutions (X_1^o, X) or (X_1^+, X^+) in the two-node-case and homogeneous time oscillations in the limit cycle region respectively, no bearing of voltages, compare Fig. 3;

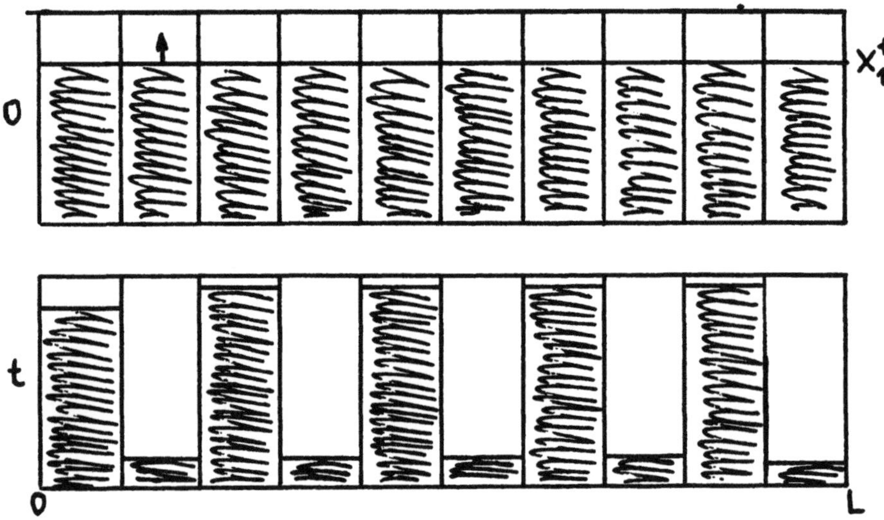

Fig. 1. Formation of a time-independent pattern as a result of a Turing instability of (X_1^+, X^+) for Neumann boundary conditions. Parameters: $\alpha = 0.4$, $\beta = 2.0$, $\gamma = 3.0$, $D_1 = 0.01$, $D_2 = 0.3$, $D_3 = 0.35$

d) one obtains limit cycle oscillations in the bistable region for

1.) $\alpha\beta < 1 < 2\alpha\beta < \alpha + 1$, (14)

2.) $\beta < 1$, (15)

3.) $\alpha < \alpha^-$; $\alpha < \alpha_2^+$, (16)

or 3!) $\alpha_1^+ < \alpha < \alpha_2^+$, (16')

4.) $X_1^+ = 2\beta / (1 + \alpha - 2\alpha\beta)$ (17)

5.) $\gamma < \gamma_c = \beta + X_1^+ [2\alpha\beta - 1 - \alpha X_1^+ (1 - \alpha\beta)]$, (18)

with $\alpha_1^\pm = [4\beta(1-\beta)]^{-1} \cdot \{1 - 2\beta \pm [1 + 4\beta(1-\beta)]^{1/2}\}$ (19)

and $\alpha_2^+ = (4\beta)^{-1} \cdot [1 + (1 + 8\beta)^{1/2}]$. (20)

Oscillations in time and space, i.e. traveling waves, are observed in the distributed system connected with the bearing of alternating-current voltages, compare Fig. 4.

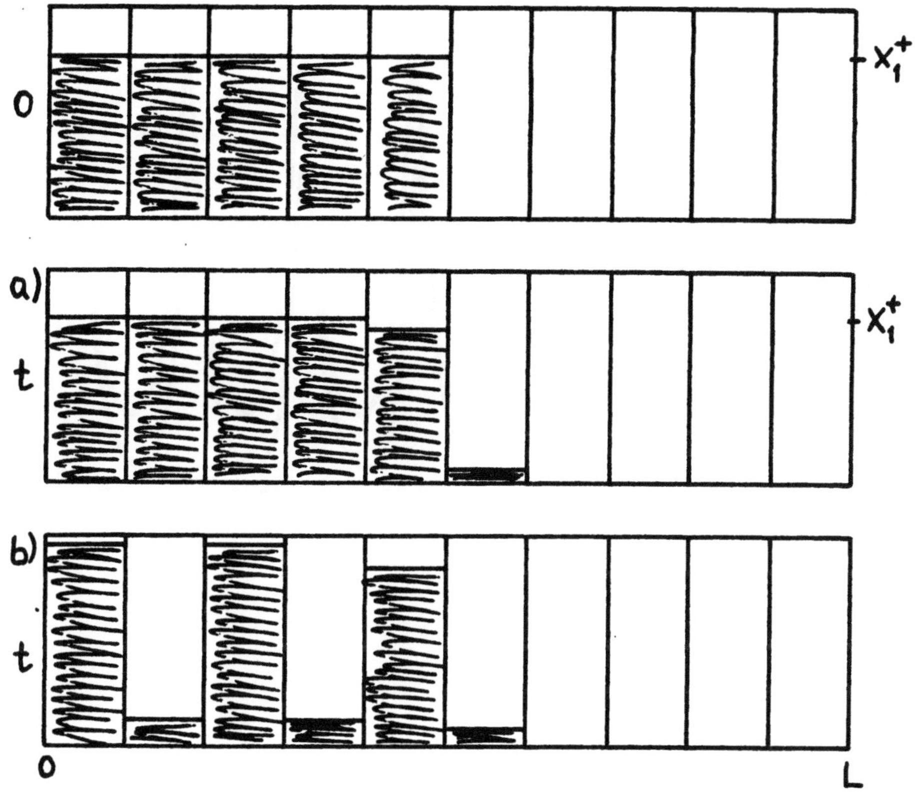

Fig. 2. Standing interface between (X_1^+, X^+) and (X_1^o, X^o) for Neumann boundary conditions in the stable two-node-case (a) and in the region of Turing instability of (X_1^+, X^+) (b).
Parameters: $\alpha = 0.4$, $\beta = 2.0$, $\gamma = 3.$, $D_1 = 0.01$, $D_2^a = 0.01$, $D_3^a = 0.02$, $D_2^b = 0.3$, $D_3^b = 0.35$

It is imaginable that at least in principle electrical dissipative structures serve as a new source of electrical energy. However electric potential differences are not only important in electrochemistry but for the regulation of processes in and on biological cells too.

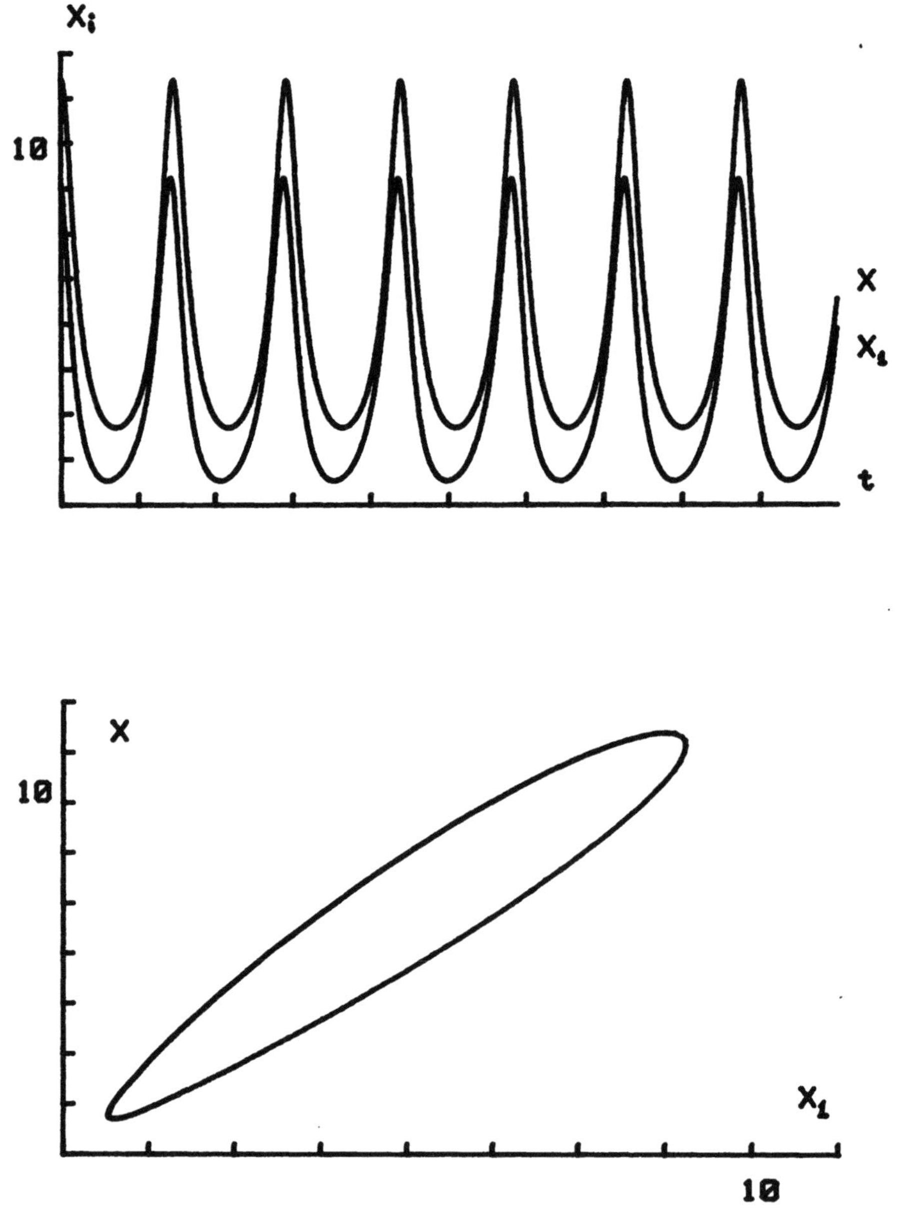

Fig. 3. Homogeneous time oscillations without bearing of voltages for homogeneous initial conditions.
Parameters: $\alpha = 1.1$, $\beta = 0.8$, $\gamma = 1.35$, $D_1 = 0.01$, $D_2 = 0.01$, $D_3 = 0.02$

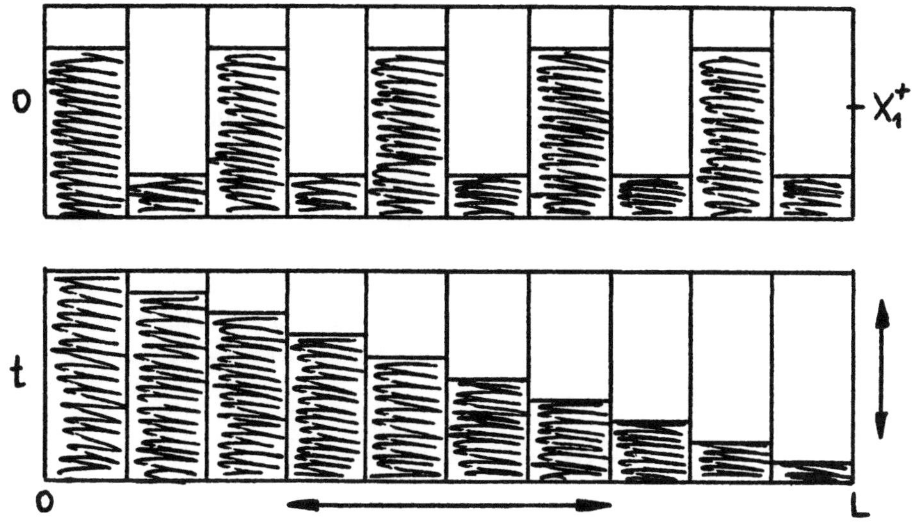

Fig. 4. Traveling waves in the limit cycle region for Neumann boundary conditions. Bearing of alternating-current voltages. Parameters: $\alpha = 1.1$, $\beta = 0.8$, $\gamma = 1.35$, $D_1 = 0.01$, $D_2 = 0.01$, $D_3 = 0.02$

References

[1] Ebeling, W. and Feistel, R.: Physik der Selbstorganisation und Evolution, Akademieverlag, Berlin 1982
[2] Ebeling, W. and U. Feudel: Ann. Physik (Leipzig) 40 (1983) 1, 68 - 78
[3] Falkenhagen, H.: Theorie der Elektrolyte, Hirzel Verlag, Leipzig 1971
[4] Haken, H.: Synergetics. An introduction, Springer-Verlag, Berlin (W.) etc. 1978
[5] Malchow, H., W. Ebeling and R. Feistel: studia biophysica 97 (1983) 3, 231 - 236
[6] Nicolis, G. and I. Prigogine: Selforganization in nonequilibrium systems, Wiley, New York etc. 1977
[7] Polak, L.S. and A.S. Mikhailov: Samoorganizatsiya v neravnovesnykh fiziko-khimicheskikh sistemakh, Nauka, Moskva 1983

EFFECT OF FLUCTUATIONS ON FRONT PROPAGATION IN NONEQUILIBRIUM SYSTEMS

by L. Schimansky-Geier[1] and A.S. Mikhailov[2]

A theory of plane front propagation in bistable stochastic media was developed in recent papers [1]. The effect of fluctuations was shown to be twofold. They lead (i) to random variations of the front shape and result (ii) in a random motion of the front position. The general expression of the dispersion of the position for arbitrary stochastic noises were given. This result is generalized here. Small deviations of the front in a second space direction will be taken into account.
We assume that the density of a nonequilibrium system follows the reaction diffusion equation [2]

$$\frac{\partial n}{\partial t} = W(n) + D\Delta n(x,y,t) \tag{1}$$

and that (1) possesses a solution $n^0(x - vt)$ with stationary profile which moves with constant velocity in the x-space-direction. If $\xi = x-vt$ then n^0 is given by boundary conditions and by

$$W(n^0) + v\frac{d}{d\xi}n^0 + D\frac{d^2}{d\xi^2}n^0 = 0. \tag{2}$$

Suppose further that the nonlinear reaction function $W(n, K_1 \ldots K_e)$ depends on fluctuating values

$$k_i^0 \longrightarrow k_i(x,y,t) = k_i^0 + \delta k_i(x,y,t) \tag{3}$$

where $|\delta k_i| \ll k_i^0$ and

$$\langle \delta k_i \rangle = 0 \; ; \; \langle \delta k_i(x',y',t')\, \delta k_j(x,y,t) \rangle = \delta_{ij} S_i(|x-x'|, |y-y'|, |t-t'|). \tag{4}$$

These variations will induce small fluctuations of the density profile near the steady $n^0(\xi)$

$$n(x,y,t) = n^0(\xi) + \delta n(x,y,t). \tag{5}$$

[1] Sektion Physik der Humboldt-Universität Berlin, 1086 Berlin, GDR

[2] Department of Physics, Moscow State University, 117234 Moscow, USSR

In linear approximation we get for the evolution of δn

$$\frac{\partial}{\partial t}\delta n = \hat{\tau}\delta n + \mathcal{F}(\xi, y, t; n^0) \tag{6}$$

with the random field

$$\mathcal{F} = \sum_{i=0}^{\ell} \left.\frac{\partial W}{\partial k_i}\right|_0 \delta k_i(\xi + vt, y, t) \tag{7}$$

and the linear operator

$$\hat{\tau} = \left.\frac{\partial W}{\partial n}\right|_{n^0(\xi)} + v\frac{\partial}{\partial \xi} + D\frac{\partial^2}{\partial \xi^2} + D\frac{\partial^2}{\partial y^2}\,. \tag{8}$$

$\hat{\tau}$ together with its hermitically adjointed operator possess two complete biorthogonal sets of eigenfunctions $\phi_{jq}(\xi, y)$ and $\phi^{jq}(\xi, y)$. If δn as well as $\mathcal{F}(\xi, y, t)$ are expanded in serieses of the eigenfunctions with coefficients $c_{jq}(t)$ and $\mathcal{F}_{jq}(t)$ we obtain

$$\dot{c}_{jq} = -Dq^2 c_{jk} + \mathcal{F}_{jq}(t). \tag{9}$$

This is a linear approximation and holds as long as curvature are smaller than the critical one [2]. Invariance under transformation of the front which does not modify its profile in x-direction is described by eigenfunctions

$$\phi_{0k} = \frac{dn^0}{d\xi}e^{iqy} \qquad \text{with the eigenvalue} \qquad \lambda_{jq} = -Dq^2. \tag{10}$$

Since

$$\begin{aligned} n(x,y,t) &= n^0(\xi) + \frac{dn^0}{d\xi}\sum_q c_{0k} e^{iqy} + \sum_{q,j\neq 0} \phi_{jq} c_{jq}(t) \\ &\approx n^0(\xi + c_0(y,t)) + \delta n_1(\xi, y, t)\,, \end{aligned} \tag{11}$$

with $c_0(y,t) = \sum_q c_{0q}(t)\exp(iqy)$ we get a random shift of the front position

$$\Delta x(y,t) = c_0(y,t). \tag{12}$$

δn_1 describes stochastic deviations of the shape of the profile which we suppose small as long as $n^o(\xi)$ is structural stable.
The statistic characteristics of Co(y, t) may be calculated from eq. (9) and

$$C_o(y,t) = \frac{1}{4\pi^2 Z}\int e^{\frac{v\xi}{D}} \frac{dn^o}{d\xi} \frac{\mathfrak{F}_{ok,\omega}}{i\omega + Dk^2} e^{iqy+i\omega t} d\xi\, dq\, d\omega \tag{13}$$

where

$$Z = \int e^{\frac{v\xi}{D}} \left(\frac{dn^o}{d\xi}\right)^2 d\xi. \tag{14}$$

$\mathfrak{F}_{ok,\omega}$ is the Fourier-component of $\mathfrak{F}_{ok}(t)$. Therefore we obtain

$$\langle C_o(y,t)\rangle = 0, \tag{15}$$

$$\langle C_o(y,t)\, C_o(y',t')\rangle = \frac{1}{Z^2}\int e^{\frac{v(\xi_1+\xi_2)}{D}} \frac{dn^o}{d\xi_1}\cdot\frac{dn^o}{d\xi_2} \times \tag{16}$$

$$\times \sum_{j=1}^{l} \frac{\partial W}{\partial k_i}\Big|_{n^o(\xi_1)} \frac{\partial W}{\partial k_i}\Big|_{n^o(\xi_2)} \frac{g(q,\omega,\xi_1-\xi_2)}{\omega^2+D^2q^4} e^{i\omega(t-t')+iq(y-y')} dq\, d\omega\, d\xi_1\, d\xi_2$$

where $g(q, w, \xi_1 - \xi_2)$ is the spectral-density of (4). Since no special assumption about Si was made the expression (16) is valid for wide classes of stochastic noises.

References

[1] Mikhailov, L. Schimansky-Geier, W. Ebeling, Phys. Lett. 96A (1983) 453; Ann. Phys. (Leipzig) in print

[2] Ebeling, W., R. Feistel: Physik der Selbstorganisation und Evolution, Akademieverlag, Berlin 1982

THE EVOLON GROWTH MODEL AND ITS APPLICATION TO NATURAL PROCESSES

W. Mende +

The face of our earth is the result of hudge accumulation of structure during several billion years of evolution, that is structure building. We ourselves are the result of an uninterrupted information accumulation process in the biosphere. In contrast to globally acting dissipation processes structure accumulation is a very local event but nevertheless also a very general feature of our world. Not only on earth but also in other cosmic regions structure building is at work on all hierarchical levels and on all scales of the universe. In figure 1 the known (or somewhat beyond) orders of magnitude of the mass and length scale are sketched. Two strong physical limits the Schwarzschild-radius (the limit of communication by light) and the Comptonwavelength (the limit of resolution) span an evolutionary cone (as one example of the cones of physically possible evolution processes from which us told Prof. Ebeling yesterday).

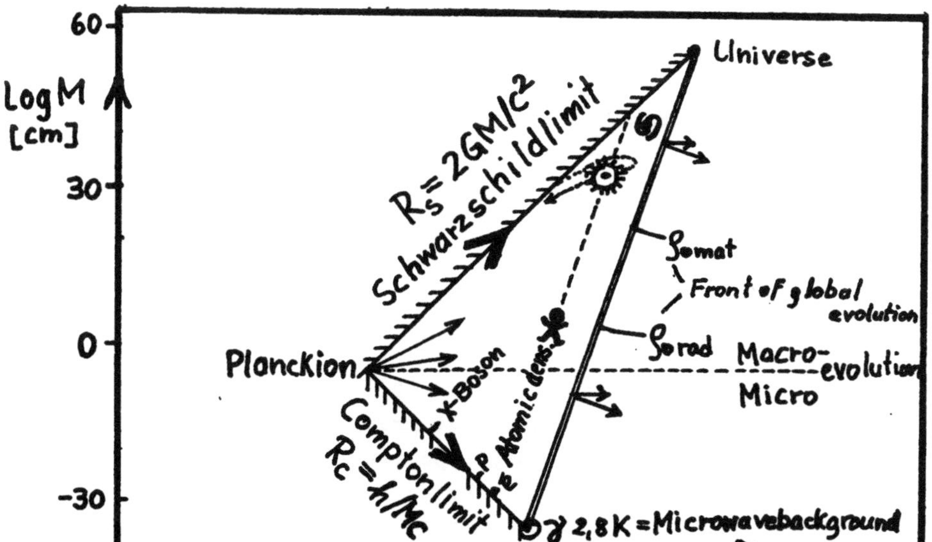

Fig.1: Evolutionary cone of R-M-scales: masses 10^{-37}- 10^{56}gr, lengthes 10^{-33}- 10^{28}cm, densities 10^{-34}- 10^{93}gr/cm^3.
A nonevolutionary M-R diagram is given in Carr and Rees [2].

+Institut for Geography and Geoecology of the Academy of Sciences, GDR

Reality explodes in this cone originating from a (still hypothetical) Planckion Antiplanckion Pair near the intersection of the two borderlines and spans the possible ranges of mass-length-scales. At very early times cosmic evolution branches in the two processes: the horizon-expansion along the Schwarzschild-limit and the particle decay along the Comptonline and the light connects the two processes. The recent front of global evolution process is the microwave background radiation with the faintest energy density $\varrho_{o\ rad}$ and the mean-density of matter $\varrho_{o\ mat}$.

Every concrete evolution process has his evolution path in the M-R evolution cone from the origin initially to lower densities conform to the general expansion process and eventually (if it is involved in local structure building) a reversion to higher densities according to a local eigenevolution. All interesting "objects" we observe in macro- and microcosmos are results of such a local structure building process ("space-cosmetics", [9]) which is strongly connected with growth processes of appropriate ordering parameters (i.g. masses, biomasses etc.). Many processes in the evolution cone, esp. the boundaries are coupled by simple power laws (straight lines in the diagram). If the one process grows according to a power law ($\dot{Y} = Y^m$) the other does also.

Hyperbolic growth of evolution processes

Not all growth processes lead to structure building, for example exponential growth has no potential for structure building. Also the simple introduction of a new mutant X_2 with an higher carrying capacity N_2 in a LV-system:

$$\dot{X}_1 = K_1 X_1 (N_1 - X_1 - X_2) - d_1 X_1$$

$$\dot{X}_2 = K_2 X_2 (N_1 - X_1 - X_2) - d_2 X_2$$

with the escalation condition $(N_2 - d_2 / K_2)\ N_1 - d_1 / K_1$

leads far from saturation to exponential growth and finally to exponential saturation for the successful species X_2, i.e. a logistic transition behaviour (see f.e.[11]). But already in the initial growth phase we observe in natural evolution processes considerable deviations from exponential growth. Exponential growth is the growth of uncoupled reproducing elements (with unlimited resources and complete waste disposal) and the deviation from exponential growth represents the inhibiting or stimulating interactions with each other or with their environment. The exponent m in the autonomical poweransatz $\dot{Y} = K\ Y^m$ is an highly compressed measure of these inhibiting ($m < 1$) or stimulating ($m > 1$) interactions. It is already known (see for ex. [13]) that evolving structure building systems typically show hyperbolic growth or hyperbolic transition behaviour as its macrodynamics. One of the most simple devices which generate a limiting cycle the Brüsselator for example

$$\dot{X} = 1 - X + X^m Y$$
$$\dot{Y} = + BX - X^m Y$$

has a hyperbolic growth term X^m, $m > 1$, which generates the oscillation behaviour after the Hopfbifurcation if the dissipation A - B is high enough. Essentially the same nonlinearity is responsible for the limiting cycle in the Sel'kov glycolysis model. The chaos generating systems have also hyperbolic terms or bilinear chains, which are sufficently long for generating a similar nonlinearity.
If our world is an evolving one and endowed with structure building processes we should look for hyperbolic growth processes in all levels of the cosmic hierarchy in all time scales as indicators for evolution processes with structural gain. For an example on cosmic scales we remember that the dynamics of the gravitational collaps and the condensation of a star by the Jeans instability is a hyperbolic one (measured by the mass which is accreted in a small volume around the gravitational centre ($dM \sim M^2$)).

Population growth as a hyperbolic process

Perhaps the most famous example for hyperbolic growth in intermediate dimensions (near the centre of the evolutionary cone surface)is the growth of the human population during the whole history of mankind. World population is a very complex system and its growth depends on a tremendous number of factors like food, housing, resources, fertility, climate and last not least on the level of production, technology, science and other socio-cultural determinants. Despite this complexity we observe in a long term perspective a stable macrodynamics with a relatively uniform increase of (the average of) growth rates with only relatively small deviations from the trend hyperbola (see Fig. 2)

$$P(t) = 1.1 \cdot 10^{12} / (2050 - t)^{1,3}$$

Considering the fact that the evolution of mankind suffered strong disturbances by ice ages, natural catastrophes, epidemics, wars and starvation the existance of such an uniform trend is the more remarkable. Mankind explodes already since its whole history of several million years not only in modern times. The increasing growth rates for such a long time are not accidentally, but a characteristic feature for selforganizing structure accumulating systems. The hyperbolic growth of mankind reflects the cooperative behaviour of the individuals within the society. In ecology these symbolic effects are known as Allees effect or simply social effects. A lot of other examples of hyperbolic growth (primate evolution, socialeconomic parameters, phylogenesis is discussed elsewhere [6], [7], [4], [1], [13]). All these examples corroborate an experience that hyperbolic (or more generally hyperexponential) growth and structure building in natural evolution processes are closely related.

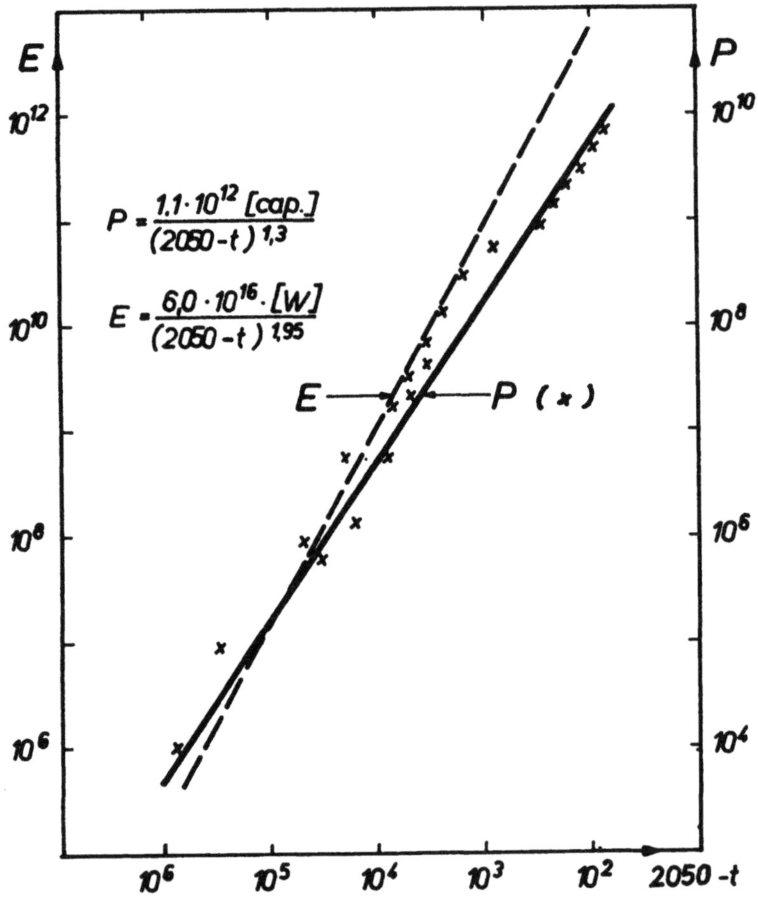

Fig. 2: Long term trend of earth population growth P and energy consumption E and hyperbolic trend lines [6], [7]

<u>The exponential tower as a tool for generating hyperbolic and parabolic growth</u>

The question arises how hyperbolic growth could be generated out of a world of locally linear or bilinear processes. In [6], [8] the exponential tower was introduced as a possibility to understand the nature of hyperbolic growth and its relation to linearity. An exponential chain is the most simple infinite Lotka-Volterra system consisting only of bilinear coupling terms:

$$\dot{X}_i = K_i X_i X_{i+1}, \qquad i = 0,1,2,\ldots$$

It becomes an exponential tower by the condition $X_i(0) = 1$ for all i. The latter condition generates a convergency property : The truncated tower T_n (1st index: level, 2nd: chain length)

$$T_n: \dot{X}_{in} = K_i X_{in} X_{(i+1)n} \qquad \text{for } i = 0,1,2,\ldots,n \text{ , where}$$

$X_{n+1,n}(t) \equiv 1$, the X_{on} converge under certain conditions, pointwise in an intervall to a certain continous function $X_o(t)$

A simple convergency condition is: If $K_i \rightarrow K$ then $X_{no} \rightarrow X_o$ in an intervall $[0,1/K)$ for $t > 0$. Many interesting properties of such a strong convergency structure are discussed elsewhere [7], [13] and ref. here. The basic operator F which leads from the i-th to the i+1-level is the logarithmic differential operator $df/f =: F$ only defined in $[0,1/K)$ on the set of all positive, differentiable functions with $f(0) = 1$. Appropriate K_i on each level garantee that the operator F could be iterated and inverted. F^{-1} is then the exponential integration. The hyperbola $X(t) = \frac{1}{1-t}$ is the unique positive solution of the operator-equation

$$F X = X = F^{-1}X \qquad (\text{and } F^2X = X = (F^{-1})^2 X)$$

This is completely analog to the exponential function as the fixed point for the common differential operator. The simplicity of the linear systems theory is essentially based on this fact, conversely: the hyperbola and the logarithmic differential operator are fundamental for the tower concept.

For example the tower yields analog to the Taylor expansion a new kind of expansion for a wide class of positive monoton C^∞-functions: "exponential analytic functions". A tool which could generate such a wide class of functions nevertheless shows especially good approximation properties for a very special class of functions, the parabolas and hyperbolas: a homogeneous tower $K_i = K$ with arbitrary K_o generates the solution of the equation $\dot{Y} = A Y^k$. Given any "exponential analytic" function so the first two steps of its tower expansion yield a local fitting hyperbola or parabola as a local approximation model. Because the higher levels act slower and slower and loose its influence on the output, a long chain decouples the output more and more from the processes at the input of the chain. This property we apostrophized as the "law of long chains".

Especially good approximations one gets if the levels have equal time constants K_i. Because the exponential integration F^{-1} is a continuous operator with respect to compact convergency the truncated homogeneous tower with constant input converges to a solution of $\dot{Y} = AY^m$. Thus the following expansion typs are of special interest, they reflect different organization typs of growth resp. declining processes and express the fact that all higher levels either could stimulate or inhibit the process on the basic level:

K_o, K, K, K,... growing hyperbolas

$-K_o$, -K, -K, -K,... declining hyperbolas

K_o, -K, -K, -K,... growing parabolas

$-K_o$, +K, +K, +K,... declining parabolas

We should expect to encounter these growth and declining types in a good approximation in nature, because rate-coupled chains of this kind are frequently met: f.ex. multiple layers in membranes, polymers, proteins, food chains in ecology, scattering processes and other kinds of homogeneous energy dissipations. The reason why rate-coupled structures with a prevailing chain and weak feedbacks are so commonly distributed may lie in the fact that energy transport is organized in chain structures and that mostly a main-chain-flow-path exists. Also in the very common case if we cannot identify the single levels, we could get qualitative inferences and an approximative macrodynamics: If the levels "eat" or dissipate or inhibit we should expect hyperbolic declines. If the basic system grows exponentially and the higher controlling levels have inhibiting or breaking influence parabolic growth should result. If all levels stimulate a basic autocatalytic process hyperbolic growth should be expected. In the following we will sketch some examples of these growth types from a phenomenological point of view without consideration of the very complex underlying mechanisms.

Hyperbolic decline

The case where a rate-coupled system has in every level an exponential decaying system generates hyperbolic decline of the tower output. Typically a sytem shows hyperbolic decline after it was highly excited and many degrees of freedom are rate-coupled ("food chains") and are dissipating simultaneously (provided a single degree of freedom declines exponentially which is very often the case). For example if a system is pushed to very high temperature and then cools down by many complex dissipation processes.

Consider for example the decay of radioactivity after a nuclear explosion: The initial explosion is very short. After a few milliseconds an inventory of several hundreds of radioactive isotops has been built up and the following decay process generates decay chains. Only the shortestliving isotopes contribute to the total radioactivity. They feed from the slower decaying isotopes (if there is a decay-path). Despite the fact that there are several chains and the total radioactivity is an additive superposition of all chains the overall macrodynamics of the complex feeding decaying process fits well to a hyperbolic decline. An identification of the very complex process with the levels of our structure seems to us difficult.

We found by a Čebyshev approximation to the bottom up computed curve of the total radioactivity R for a megaton nuclear explosion:

$$A = 1.8 \cdot 10^{16}/(t-t_0)^{1.3} \qquad \text{(t in s)}$$

with σ = 11 %, over 9 decimal orders of magnitude of the time.
A similar example but on a different scale is the evolution of luminosity of galaxies. Model calculations based on the local mass distribution of stars lead to an hyperbolic luminosity of the integral luminosity L

$$L(t) = 10/t^{0.898} \qquad \text{(t in } 10^9 \text{years)}$$

with an error of 1 % (Data from digitized model output curves from D.L. Meier [5]).
It is interesting to note that the evolution on the most global scale is also a hyperbolic declining process in the framework of the Friedman model: For example the sum of matter and energy density is given by the equation: $= -\dot{K}^{1.5}$

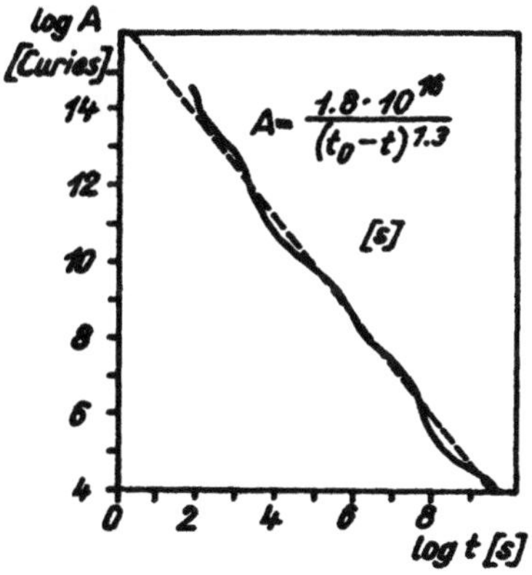

Fig.3: Hyperbolic decline of total radioactivity of a 1 megaton nuclear explosion

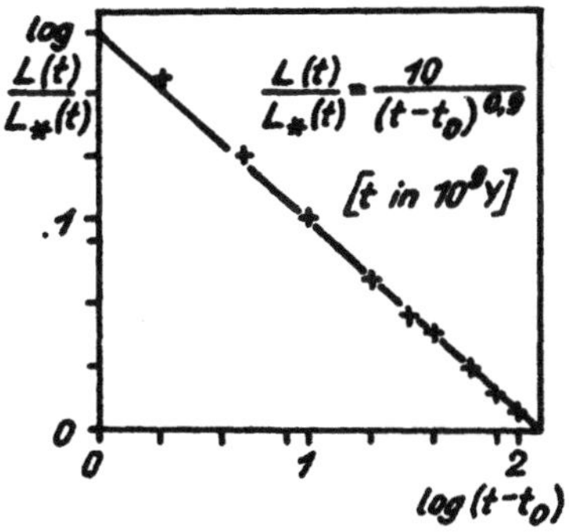

Fig.4: Hyperbolic decline of the luminosity of a model galaxy

Hyperbolic process and the dialectics local - global

Hyperbolic growth processes are the counterpart to hyperbolic decline processes. In the tower expansion-interpretation a basic growth process is stimulated in every level (i.e. in every higher growth rate). Hyperbolic growth is the dialectical counterpart to hyperbolic decline in the dialectics global - local. When global evolutions are hyperbolic declining certain local centres behave inversely and show hyperbolic growth (see [9]).
We cite here the case of our sun[7] . The photonoutput of the central-H-burning phase (the first 5 billion years) describes the hyperbola

$$L(t) = 10.48/(11.83 - t)^{1.006} \qquad \text{(t in } 10^9 \text{ years, L(t) relativ to now)}$$

with an error of the sun model relative to the hyperbolic model of

0.1 %. The sun model data are taken from I.Iben jr [3]. Without the central gravitational field of the sun the H-burning process would be a hyperbolic declining one because the velocity determing process is deuterium generation according $\dot{L} \sim -X^2$ (X concentration of hydrogen). The gravitation inverts the growth type because it establishes many stimulating couplings which finally lower the pressure, the rising fusion rate reestablishes the hydrostatic equilibrium.
Also such an exotic object as a black hole evaporates hyperbolically by the Hawking radiation (with an higher cooperativity exponent m = 25 as the sun):

$$L_{BH}(t) = 2.6 \quad 10^{23}/(t_o - t)^{2/3} \qquad (t \text{ in s, } L_{BH} \text{ in W}).$$

Parabolic growth

In parabolic growth processes an exponential process on the basic level is controlled by damping mechanisms in the higher levels. We meet this case if the growth process is (together with a certain environment) controlled by an inner program. In this case damping influences act on many levels of organization to reach the predetermined target. In ontogenesis such a situation is given. Here the target is the adult organism as the decoded message of the genetic program and this final attractor influences already very early stages of development. Indeed, only in the earliest stages of the embryonic development the cells divide independent of each other according to an exponential law. Already after the 4th division the so called micromeres are inhibited. Later a set of so called teloblastes act as proliferation cells and produce strings of similar cells which correspond to the later segments (G.S.Stent and D.A.Weisblat [14]). The offspring of a teloblast has a programmed number of division steps (proliferation depth). Since in a certain stage the teloblastic proliferations are synchronized, the observed exponent may correspond to the proliferation depth. If after k steps proliferation stops we would have after n generations $\approx (n - a)^k$ new cells (a adjustment). For chickens embryons [13] we got parabolic growth with plausible k-values corresponding to proliferation depth from 9 to 3 (see Fig. 5). The hypothesis concerning the k values is not yet checked by experiments.
Another case of target oriented growth is the ecological succession. Here the inner program comes from the available inventory of species. Forest successions show also parabolic growth and saturation behaviour, and micrococsmic successions, tissues and organs. Especially Schmalhausen in the 30th emphasized the parabolic law. The hyperbolic relation to structure building is not violated because there is no new structure but only a replication of already accumulated information. It may be of great interest to know whether a process is a genuine evolution process with an open pioneer path into the future or a target oriented

process with determined attractor. The growth dynamics gives the answer,

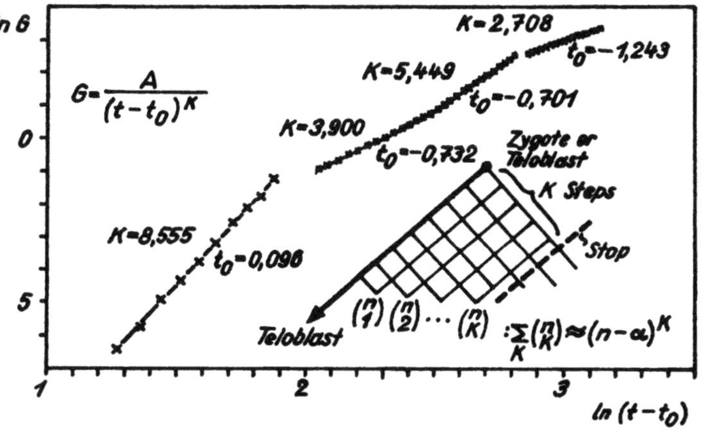

Fig. 5: Parabolic growth pulses in the embryonic development of chickens. Data M.I. Gerskova 13 . Fit: K.F. Albrecht.

The evolon as a combination of power growth and power saturation

By the given examples we demonstrated that the power functions (hyperbola, exponential, parabola) are significant on very different scales. The tower expansion gives another argument for the universality of the power growth according to the equation $\dot{Y}=AY^k$.

It is easy to construct the tower for the corresponding saturation process $\dot{Y} = A\,(B - Y)^l$ which has the same growth type trichotomy as the power growth. The saturation process can be combined with the growth process by multiplication because both processes work independent of each other. Growth works as an internal (intrinsic) property of the system and saturation is in simple cases (in the learning phase) external property of the systems environment. A kind of systems hull (membrane, skin etc.) decouple these processes maximum so that multiplication is justified. Combining these two processes we get the hyperlogistic form of our model:

$$\dot{Y} = Y^k(B - Y)^l$$

If the saturation limit interacts nonlinear with the state Y (again according to our philosophy we use a poweransatz) we get a more general model:

$$\dot{Y} = (Y^k(B - Y^w)^l$$

which we called evolon growth model. We received it historically nearly in the same way as we sketched in this paper [6] [8].

The interpretation of the exponents (hyperbolic = pioneer process, parabolic = iterative process) is completely analog to the case of power growth (the first half of the evolon). The exponents can serve as highly aggregated complexity or cooperativity parameters (Synergetics!)

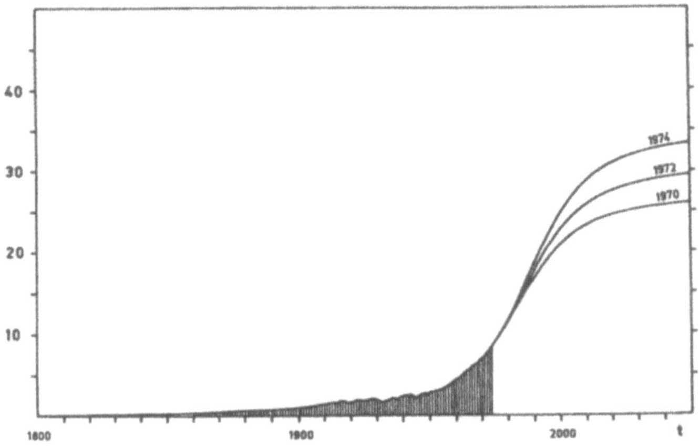

Fig. 6: Evolon growth model of the world primary energy consumption 4 . The bars are real data.

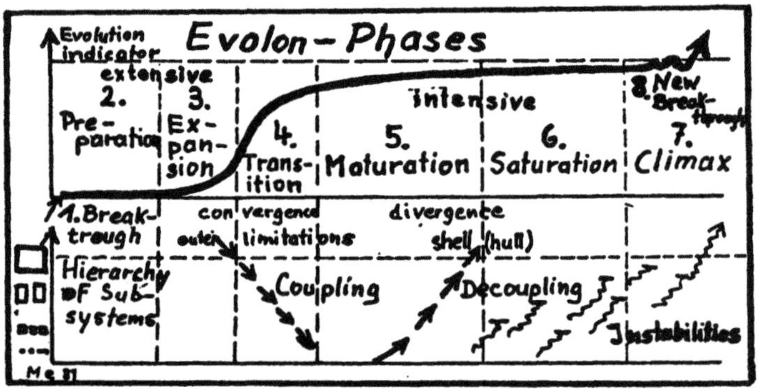

Fig. 7 Processes in an evolon

1. Breakthrough: (Quick and hidden): Above a critical threshold a fluctuation will be amplified. Integration of moduls to a new function, successful matching, new cooperativity, Anagenesis, Innovation
2. Preparation: (latent): stabilization of the type, connecting to resources, for quick use of the identified potential organizing the growth process, overcoming the inhibitions
3. Expansion: (obvious): hyperbolic growth process, constant cooperativity (Founders effect), increasing new diversity under small selection pressure, flow and production oriented, little infrastructure
4. Transition: (distressful): environmental limitations come into effect (density stress) and compel adaptation to the carrying capacity and growth control, internalize the limitation to lower levels
5. Maturation: (soft): refining of existing mechanisms, intensification, coordination, synchronization of subsystems, economy of energy and materials, optimization of rates of corrosion, decomposition, recycling, transport, piping, storage, reliability. In a later stage subsystems become autonomeous moduls for further combination (modularization), decoupling ascends to higher subsystems, hull-formation
6. Saturation: (barock): Nonessential modifications, - diversity, Luxuration, accumulation of reserves and reserve-strategies. Ascending instability germs in subsystems due to decoupling strategies will be damped.
7. Climax: (sterile): steady state, evolution superimposed by fluctuation. With outer creative impulses onset of rigidation, agening?
8. New Instability: Phase 6 or 7 may lead to breakdown or break-through

of the well separarated growth and saturation process. Clearly one can combine several evolon processes (e.g. Evolon + Antievolon) to multi-modal processes.

<u>The connection of the evolon with Volterra systems</u>

We can apply a tower expansion to the whole evolon X(t), if we use already generated signals in form of feedbacks the expansion becomes finite according to the structure design principle [10], [13] and we get a Lotka-Volterra-System which drives an exponential integrator:

$\dot{X}_2 = X_2(K_2X_2 - L_2X_1)$, $\dot{X}_1(K_1X_1 - L_1X_2)$, $\dot{X}_o = X_oK_oX_1$, $X_o(t) = X(t)$

which allows an interesting interpretation. Evolon growth can be decomposed into competing factors (f.e. infrastructure and mass growth) which compet for a common resource(growth potential). The structural design principle allows us only to represent the time dynamics (not the topological structure) in a Volterra system which could generate many other time processes. But this is not the case. Together with the logarithmic and the exponential borderline case the evolon represents the full time-dynamics in an 2-dimensional Lotka-Volterra-system: i.e. allcompeting and symbiotic cooperations of two factors lead to evolon growth behaviour. Together with the general tendency in nature in carrying through the processes in only a few master variables (slaving down the others) the evolon transition process is thus very universal. The regularly observed changes between extensive and intensive phases may express this competition (Fig. 7). Applications are given in [1,4,7]

Literature

[1] Albrecht, K.F. and Mende, W.: Population growth in German towns, IIASA Collaborative Paper CP-84-, to appear

[2] Carr, B.J. and Rees, M.J.: Nature <u>278</u> (1979), 605-612

[3] Iben, I.jr.: Astrophys. J. <u>147</u> (1967), 624 ff

[4] Kriegel, U., Mende, W.: Analysis of world energy consumption by the hyperlogistic evolution model, IIASA Coll. Paper CP 83-22, 1983

[5] Meier, D.L.: Astrophys. J. <u>207</u> (1976), H.2, p. 343-350

[6] Mende, W.: Wachstum und Evolution ökologischer und ökonomischer Systeme, Forschungsbericht IGG der AdW der DDR, IGG 1976

[7] Mende, W.: Untersuchungen zur Phänomenologie und zur Systemtheorie von Wachstums- und Evolutionsprozessen, Forschungsbericht AdW der DDR, IGG, 1982

[8] Mende, W., Peschel, M.: Problems of fuzzy modelling, control and forecasting of time-series and some aspects of evolution, IFAC-Symposium of Control Mechanisms in Bio- and Ecosystems, Plenary Paper, Leipzig, 1977

[9] Mende, W., Peschel, M.: In: Haken, H.: Chaos and Order in Nature, 196-206, Springer-Verlag, Berlin, 1981

[10] Mende, W., Peschel, M.: msr <u>24</u> (1981), H.10, 581-583

[11] Nicolis, G., Prigogine, J.: Self-Organization in Nonequilibrium Systems, Wiley, New York, 1977

[12] Peschel, M., Mende, W.: msr <u>24</u> (1981), H.11, 602-606, <u>25</u> (1982), H.5, 255-257

[13] Peschel, M., Mende, W.: Leben wir in einer Volterra-Welt - Ein ökologischer Zugang zur Angewandten Systemanalyse, Akademie-Verlag Berlin, 1983

[14] Weisblat, D. u.a.: Developmental Biology <u>76</u>, H.1, p. 58-78, (1980)

Band 1

Inverse and Improperly Posed Problems in Differential Equations

Proceedings of the Conference on Mathematical and Numerical Methods, held in Halle/Saale (GDR) from May 29 to June 2, 1979

Edited by Gottfried Anger
296 Seiten, 1979

Band 2

Fundamentals of Computation Theory FCT '79

Proceedings of the Conference on Algebraic, Arithmetic and Categorial Methods in Computation Theory, held in Berlin/Wendisch-Rietz (GDR) September 17—21, 1979

Edited by Lothar Budach
576 Seiten, 1979

Band 3

Decidability and Generalized Quantifiers

By Andreas Baudisch, Detlef Seese, Hans-Peter Tuschik, and Martin Weese
XII, 235 Seiten, 1980

Band 4

Zur Elastizitäts- und Thermoelastizitätstheorie

Potentialtheoretische Methoden in Anwendung auf verschiedene Modelle

Von Lothar Jentsch und Johannes Maul
157 Seiten, 1980

Band 5

Systems Analysis and Simulation 1980

Proceedings of the International Symposium held at Berlin (GDR) September 1—5, 1980

Edited by Achim Sydow
496 Seiten, 1980

Band 6

Nichtlineare diskrete Optimierung

Von Eberhard Girlich und M. M. Kowaljow
218 Seiten, 1981

Band 7

Completeness Problems in the Structural Theory of Automata

By Jürgen Dassow
148 Seiten, 1981

Band 8

Darstellung von Lösungen linearer elliptischer Differentialgleichungen

Von Günther Wildenhain
92 Seiten, 1981

Band 9

Introduction to the Spectral Theory of Operators in Spaces with an Indefinite Metric

By I. S. Iohvidov, M. G. Krein, and H. Langer
120 Seiten, 1982

Band 10

Nonlinear Time-discrete Systems

A General Approach by Nonlinear Superposition

By Michael Gössel
115 Seiten, 1982

Band 11

Preprimal Algebras

By Klaus Denecke
162 Seiten, 1982

Band 12

Ergodic Theory and Related Topics

Proceedings of the Conference held in Vitte/Hiddensee (GDR) October 19—23, 1981

Edited by Horst Michel
232 Seiten, 1982

Band 13

Singularly Perturbed Differential Equations

By H. Goering, A. Felgenhauer, G. Lube, H.-G. Roos, and L. Tobiska
176 Seiten, 1983

Band 14

Leben wir in einer Volterra-Welt?

Ein ökologischer Zugang zur angewandten Systemanalyse

Von Manfed Peschel und Werner Mende
233 Seiten, 1983

Band 15

Markov Decision Problems with Countable State Spaces Optimality Criteria — Algorithms — Clustering

By Hans Michael Dietz and Volker Nollau
174 Seiten, 1983

Band 16

Stochastische Verfahren zur Suche nach dem Minimum einer Funktion

Von Ryszard Zieliński und Peter Neumann
133 Seiten, 1983

Band 17

Invertibility and Asymptotics of Toeplitz Matrices

By Albrecht Böttcher and Bernd Silbermann
200 Seiten, 1983

Band 18

Structural Induction on Partial Algebras

Introduction to Theory and Application of Partial Algebras — Part II

By Horst Reichel
205 Seiten, 1984

Band 19

Algebraic Methods for Toeplitz-like Matrices and Operators

By Georg Heinig and Karla Rost
212 Seiten, 1984

Band 20

Frege Conference 1984

Proceedings of the International Conference held at Schwerin (GDR), September 10—14, 1984

Edited by Gerd Wechsung
408 Seiten, 1984

Band 21

Numerical Analysis of Selected Semilinear Differential Equations

Edited by Christian Großmann, Thomas Riedrich, and Hartmut Schönheinz
203 Seiten, 1984

Band 22

Singular Ordinary Differential Operators and Pseudodifferential Equations

By Johannes Elschner
196 Seiten, 1985

Band 23

Lotka-Volterra-Approach to Cooperation and Competition in Dynamic Systems

Proceedings of the 5th Meeting of UNESCO's Working Group on System Theory held on the Wartburg, Eisenach (GDR), March 5—9, 1984

Edited by Werner Ebeling and Manfred Peschel
272 Seiten, 1985

Bei Fragen zur Produktsicherheit wenden Sie sich bitte an:
If you have any questions regarding product safety,
please contact:

Walter de Gruyter GmbH
Genthiner Straße 13
10785 Berlin
productsafety@degruyterbrill.com